In the Footsteps of Columbus

European Missions to the International Space Station

John O'Sullivan

In the Footsteps of Columbus

European Missions to the International Space Station

John O'Sullivan
County Cork
Ireland

SPRINGER-PRAXIS BOOKS IN SPACE EXPLORATION

Springer Praxis Books
ISBN 978-3-319-27560-4 ISBN 978-3-319-27562-8 (eBook)
DOI 10.1007/978-3-319-27562-8

Library of Congress Control Number: 2016937338

Cover design: Jim Wilkie
Project Editor: David M. Harland

This Springer imprint is published by Springer Nature
The registered company is Springer International Publishing AG Switzerland

To Mary and Lily

Contents

Foreword

While working at ESA/ESTEC in Noordwijk in the 1990s, I witnessed the work performed on the APM, the Attached Pressurized Module, never realizing that several years later I would be selected as a European astronaut to work with the final result of all that effort on the International Space Station. It was a great honour to perform my tasks in outer space for the benefit of the people on our planet. I was privileged to work with very passionate and dedicated people at the ESA centres and European industries and institutes. Together, we could do a lot of 'good work'.

But not only that. Europe is doing incredibly well at achieving a common goal in human space flight. Floating in a state of weightlessness aboard the International Space Station, I often went from one European contribution to another. In the span of several hours I could participate in unloading a docked ATV, do an ESA experiment aboard the Russian module, undertake a photo session in the Cupola, work with cargo in an attached MPLM, install an experiment in the Microgravity Science Glovebox (MSG) in the US Destiny lab and put samples in the Minus Eighty Laboratory Freeze for ISS (MELFI) in the Japanese Kibo module, before ending my day with a press conference in the Columbus module. ESA is clearly present everywhere in the station and that made me proud.

By now, quite a few ESA astronauts have flown to the International Space Station. As an ISS partner, ESA started to send up her astronauts in the Space Shuttle and later with the Soyuz when her astronauts also served on long duration ISS crews. Additionally, there were several short duration Soyuz visiting flights prior to the attachment of the Columbus module. Regardless of the length of their missions, European astronauts performed a wide selection of tasks, ranging from construction, maintenance and repairs, to experiments, education, outreach, operating robotic arms, spacewalks and docking visiting supply craft.

One day during my long duration presence in the International Space Station, working behind a rack in the Columbus module, I heard the noise of metal; yet, it seemed as if it was the creaking of wood caused by the ocean waves on one of Christopher Columbus' wooden ships crossing the oceans centuries ago. I realized I was in an explorer's ship too, but now in the 'ocean of space'.

André Kuipers. (Walter Kallenbach)

This book is primarily about the so called 'second generation' of ESA astronauts. My generation. Those who worked on the ground and in space for the International Space Station. Enjoy this very informative book about a great technological, scientific and human adventure. It is not only full of facts and figures about the different missions, spacecraft and experiments but also includes the great personal stories of those astronauts, their road to space, and living and working on the International Space Station.

André Kuipers
ESA astronaut

Preface

I wrote this book because I wanted to read it. I am a reader and collector of the Outward Odyssey series of books published by the University of Nebraska Press and edited by Colin Burgess. In April 2012 Colin posted on the collectSPACE website that UNP had asked him to explore the possibility of going beyond the 12 books in the series. He was asking the readers for suggestions. My suggestions were a book on the Soyuz/Salyut/Mir missions and a book on the ESA missions to Mir, ISS and on Shuttles. Colin's reply was that the UNP didn't see a market for non-US stories. Demonstrating the depth of the forum users, David J. Shayler posted a reply mentioning Clive Horwood of Praxis. I contacted Clive proposing a book covering the European missions to the ISS... and the rest, as they say, is history.

The history of European human spaceflight is not as straightforward as its American or Russian counterparts. Europe wasn't a competitor in the 'space race'. As a collection of nations with different languages, cultures and goals, the vision for space has been complex. For the first three decades of the space age, Europe was divided by the Iron Curtain. Even today, the European Space Agency does not build or fly a human-rated spacecraft. But despite all these factors there is a rich history of Europeans travelling to space on a variety of spacecraft and performing a variety of missions.[1]

As Europe isn't a single country with a manned space programme, European citizens must 'hitch a ride' to get into space. This has resulted in many different routes to orbit. Before the period covered by this book, astronauts from communist countries and from France had flown on Soviet Soyuz spacecraft to the Salyut and Mir space stations. Later, astronauts from other Western European space agencies and ESA flew to Mir. Western Europeans represented their national space agencies and ESA by flying on NASA Space Shuttle missions. Naturalised US citizens from around the world, including quite a few Europeans, succeeded in joining NASA's astronaut corps by applying to the Johnson

[1] In December 2013, NASA and ESA agreed that the European Service Module (ESM), based on the Automated Transfer Vehicle (ATV), would provide power and propulsion for the first Orion mission, Exploration Mission 1. EM-2 is planned to be a crewed mission.

Space Center in Houston, Texas. And finally, almost as a footnote, there have been several European-born American astronauts.

Within the scope of this book are the European ESA astronauts who have flown to the International Space Station (ISS). Even this story isn't straight forward, because while most flew as members of the ESA astronaut corps, one flew as a French CNES astronaut prior to joining ESA, and others flew as representatives of their national space agencies and not as ESA astronauts.

As a prelude to the 2001–2011 decade of European missions to the ISS, I have written a brief history of European human spaceflight up to 2000. It is written in chronological order and contains flights of the following types:

- Intercosmos missions: Eastern Europeans on Soviet missions to Salyut 6. For simplicity, including Bulgarian Alexandrov's mission to Mir.
- ESA Shuttle missions: ESA astronauts on Space Shuttle missions, including missions that carried the Spacelab module.
- Non-ESA Shuttle missions: CNES/DLR/ASI astronauts on Space Shuttle missions. Some may or may not have flown as ESA astronauts on other missions. They include Payload Specialists not assigned by a space agency; e.g. Dirk Frimout.
- ESA Soyuz missions: ESA astronauts on Soyuz missions.
- Non-ESA Soyuz missions: CNES/ASA/DLR astronauts on Soyuz missions.
- Miscellaneous: Helen Sharman was selected to fly to Mir on a mission funded by private UK companies without the assistance of either the UK government or ESA.

However, I have considered the following technically European human spaceflights to lie *outside* the scope of this book:

- Cosmonauts: For reasons of practicality, I have excluded all cosmonauts from European Russia (west of the Urals), Ukraine, Belorussia, and the Baltic countries. Even though these territories lie in Europe, this would only complicate matters.
- Naturalised and dual citizens: Europeans who gained citizenship of USA/Canada and joined NASA/CSA, or who flew as Payload Specialists on the Shuttle. For example:
    - Lodewijk van den Berg (Netherlands born, US citizen) who flew as Payload Specialist on STS-51B.
    - Michael Foale (UK born, dual UK/US citizen) who has flown to space six times on both Soyuz and Shuttle spacecraft and has lived on Mir and commanded the ISS.
    - Michael Lopez-Alegria (Spanish born, US citizen) who has flown in space four times on both Soyuz and Shuttle spacecraft and has commanded the ISS.
    - Bjarni Trygvasson (Iceland born, Canadian citizen) who flew as a Payload Specialist on STS-85 representing CSA.
    - Piers Sellers (UK born, US citizen) who has flown on three Shuttle missions as Mission Specialist and visited the ISS.

- Nicholas Patrick (UK born, US citizen) who has flown on two Shuttle missions as Mission Specialist and visited the ISS.
- Charles Simonyi (Hungarian born, US citizen) who has flown twice on Soyuz missions to the ISS as a 'space tourist'.

- US citizens born abroad: US citizens by birth, born abroad to US citizen parents:
  - Michael Collins (born in Italy) who flew on Gemini 10 and Apollo 11. He is undeniably the first European born astronaut, but for consistency has to be excluded on the grounds of his American parentage and citizenship.[2]
  - Gregory Johnson (born in the UK) who flew on two Shuttle missions as pilot and visited the ISS.
  - Richard Garriott (born in the UK) who flew on Soyuz TMA-31 to the ISS as a 'space tourist' and second generation spacefarer.

I have used a variety of sources for the material in this book including NASA and ESA mission reports, astronaut biographies and blogs, contemporaneous magazines and reference websites such as the encyclopaedic www.spacefacts.de. As a result, the content of each mission chapter may differ in tone or focus. I have endeavoured to keep a common 'look and feel' to each chapter but, for example, the DELTA mission chapter covers Soyuz training and preparation to good effect, while the Celsius mission chapter covers Shuttle training. Similarly the Cervantes mission chapter describes the very personal experience of Pedro Duque, whereas other chapters may describe the crew space activities more clinically. I hope this variety adds to the enjoyment of the book.

There is a rich story of human spaceflight happening between and around the European missions detailed in this book. Although I have endeavoured to inform the reader of key missions and events in American and Soviet/Russian spaceflight during this time period, these missions lie outside the scope of this book; they are, however, well covered in other Springer-Praxis publications.

When selecting terminology, I've used the term astronaut when describing flights on American spacecraft and cosmonaut for flights on Soviet or Russian spacecraft. The same spacefarer could be described as both over the course of the book, as many Europeans have flown on both American and Russian craft. As yet there have been no European Taikonauts…

There are several reasons why I chose to conclude the coverage of this book with the Promisse mission of André Kuipers:

- A round 10 years elapsed between STS-100 in 2001 and the launch of the Promisse mission on Soyuz TMA-03M in 2011, and a decade is a good period to cover.
- There were ESA launches every year from 2001 to 2011 but none in 2012, making that the first break in 10 years.

[2] A good tip when setting quiz questions!

- The next mission after Promisse by André Kuipers was the Volare mission of Luca Parmitano of the ESA astronaut class of 2009. Thus began a series of missions with the new astronaut school. The last mission for the earlier class was therefore a natural cut off point for this account.
- While I was writing, ESA were launching astronauts to the ISS at such a rate that I had to draw the line *somewhere*, as otherwise I'd never have been able to finish the book!

For mission names, I decided not to use the upper case 'ISS' which ESA insisted on shoehorning into their titles. So, for example, I refer to Odissea rather than OdISSea, because I find the latter to be distracting to the reader.

John O'Sullivan
December 2015

Acknowledgements

I must thank Clive Horwood of Praxis in England and Maury Solomon of Springer in New York for giving me the chance to write this book and for having the patience to wait while I got down to it. I would like to thank David M. Harland in Scotland for his advice to a neophyte author and for editing the manuscript. And I would like to thank Jim Wilkie for understanding my vision and creating the cover. I have endeavoured to provide credits for the images but in some cases the owner couldn't be identified; if anyone with such information contacts the publisher, I shall happily correct a credit in a future edition. And finally, I would like to thank André Kuipers for contributing the Foreword to the book and Michel van Pelt for all his help in arranging this.

About the author

John O'Sullivan BE, Dip Phys Sci, Dip PM, CEng MIEI, PMP, FSP, CMSE® studied Electrical Engineering at University College Cork. He has over 20 years' experience in the automation and control sector delivering solutions to the life-science industry in Ireland. He is a Chartered Engineer with Engineers Ireland and a Project Management Professional with the Project Management Institute. He has always had a fascination with aviation and space, leading him to gain his Private Pilot Licence in 2003 and to study Astronomy and Planetary Science with the Open University. Since 2010 he has been awarded a Certificate in Astronomy and Planetary Science and a Diploma in Physical Science by the OU, as well as a Diploma in Project Management from the Cork Institute of Technology. He was an unsuccessful applicant for the ESA Astronaut Corps in 2008, and lives in East Cork with his wife and daughter.

Acronyms

AAAF	Association Aéronautique et Astronautique de France
ABC	American Broadcasting Company
AF	Assembly Flight
AFB	Air Force Base
AFM	Association Française contre les Myopathies
ANAE	Académie de l'Air et de l'Espace
APU	Auxiliary Power Unit
ARED	Advanced Resistive Exercise Device
ARISS	Amateur Radio on the ISS
ASA	Austrian Space Agency
ASI	Agenzia Spaziale Italiana (Italian Space Agency)
ASTP	Apollo Soyuz Test Programme
ATLAS	Atmospheric Laboratory for Applications and Science
ATV	Automated Transfer Vehicle
BDC	Basic Data Collection
BE	Bachelor of Engineering Degree
BS	Bachelor of Science Degree
B.USOC	Belgian User Support and Operation Centre
CBM	Common Berthing Mechanism
CBS	Columbia Broadcasting System
CD	Compact Disc
CDR	Commander
CDRA	Carbon Dioxide Removal Assembly
CDT	Central Daylight Time
CDTI	Centre for Development of Industrial Technology
CEng	Chartered Engineer
CERN	Conseil Européen pour la Recherche Nucléaire (European Organisation for Nuclear Research)
CET	Central European Time

CETA	Crew and Equipment Translation Aid
CIR	Combustion Integrated Rack
CMG	Control Moment Gyroscope
CNEN	Comitato Nazionale per l'Energia Nucleare
CNES	Centre National d'Etudes Spatiale (Centre for Space Studies)
CNN	Cable News Network
CNR	Consiglio Nazionale delle Ricerche
CNRS	Centre National de la Recherche Scientifique
COGNI	Cognitive Process for 3-D Orientation Perception and Navigation in Weightlessness
COLBERT	Combined Operational Load Bearing External Resistance Treadmill
COL-CC	Columbus Control Centre
CONT	Commission for the Control of Financial Reporting
COSTAR	Corrective Optics for Space Telescope Axial Replacement
COTS	Commercial Orbital Transportation Services
CRISTA-SPAS	Cryogenic Infrared Spectrometers and Telescopes for the Atmosphere - Shuttle Pallet Satellite
CRS	Commercial Resupply Services
CSA	Canadian Space Agency
CST	Central Standard Time
DARA	Deutsche Agentur für Raumfahrtangelegenheiten (German Agency for Space Flight Affairs, now part of DLR)
DCORE/SCORE	Deployer Core Equipment and Satellite Core Equipment
DELTA	Dutch Expedition for Life Science, Technology and Atmospheric Research
D/HSO	Directorate of Human Spaceflight and Operations
DLR	Deutsches Zentrum für Luft- und Raumfahrt (German Centre for Flight and Space Flight) previously DVLR
DM	Descent Module
DSM	Docking and Storage Module
EAC	European Astronaut Centre
EEG	Electroencephalogram
EGNOS	European Geostationary Navigation Overlay Service
ELC	Express Logistics Carrier
EMCS	European Modular Cultivation System
EMET	Investigation of Electromagnetic Emissions by the Electrodynamic Tether
ENEA	Nazionale Energie Alternative
EO	Main Expedition
ESA	European Space Agency
ESEF	European Science Exposure Facility
ESOC	ESA Space Operations Centre
ESTEC	European Space Research and Technology Centre
EPNER	École du Personnel Navigant d'Essais et de Réception
ESRIN	European Space Research Institute

ET	External Tank
ETPS	Empire Test Pilot School
EuTEF	European Technology Exposure Facility
EVA	Extravehicular Activity
FCF	Fluids & Combustion Facility
FGB	Functional Cargo Block
GCTC	Yuri Gagarin Cosmonauts Training Centre
GMT	Greenwich Mean Time
GPC	General Purpose Computer
GPS	Global Positioning System
GSFC	Goddard Space Flight Center
HTV	H-II Transfer Vehicle
IAA	International Academy of Astronautics
IAC	International Astronautical Congress
ICV	Integrated Cardiovascular Experiment
IFSI	l'Istituto di Fisica dello Spazio Interplanetario
IMAX	Image MAXimum
IMDN	Investigation and Measurement of Dynamic Noise in the TSS
ISLE	In-Suit Light Exercise
ISS	International Space Station
ITCS	Internal Thermal Control System
ITRE	Committee for Industry, Research and Energy
IUS	Inertial Upper Stage booster
JAXA	Japanese Aerospace Exploration Agency
JEMRMS	Japanese Experiment Module Remote Manipulator System
JSC	Johnson Space Center
KhSC	Khrunichev State Research and Production Space Centre
KSC	Kennedy Space Center
KTH	Stockholm's Royal Institute of Technology
LEO	Low Earth Orbit
LF	Logistics flight
MBS	Mobile Remote Servicer Base System
MELFI	Minus-Eighty Laboratory Freezer for ISS
MIEI	Member of the Institute of Engineers of Ireland
MISSE	Materials on International Space Station Experiment
MPLM	Multi-Purpose Logistics Module
MS	Master of Science Degree
MSFC	Marshall Space Flight Center
MSG	Microgravity Science Glovebox
MSP	Mission Specialist
NASA	National Aeronautics and Space Administration
NASDA	National Space Development Agency (now part of JAXA)
NATO	North Atlantic Treaty Alliance
NBC	National Broadcasting Company
NBL	Neutral Buoyancy Laboratory

NEEMO	NASA Extreme Environment Mission Operations
NPR	National Public Radio
OBSS	Orbiter Boom Sensor System
ODS	Orbiter Docking System
OESSE	Observations at the Earth's Surface of Electromagnetic Emissions by TSS
OMS	Orbital Manoeuvring System
ORU	Orbital Replacement Unit
OSTEC	Belgian Federal Office for Scientific, Technical and Cultural Affairs
OU	Open University
OV	Orbiter Vehicle
PKE	Plasma Kristall Experiment
PMA	Pressurised Mating Adaptor
PMM	Permanent Multi-purpose Module
PMP	Project Management Professional
PLT	Pilot
POA	Payload Orbital Replacement Unit Accommodation
PPL	Private Pilot Licence
PS	Payload Specialist
PVAA	Photovoltaic Array Assemblies
RCS	Reaction Control System
RETE	Research on Electrodynamic Tether Effects
RKA	Russian Federal Space Agency
ROPE	Research on Orbital Plasma Electrodynamics
RS	Russian Segment
SARJ	Solar Array Rotary Joint
SAW	Solar Array Wing
SCA	Shuttle Carrier Aircraft
SETS	Shuttle Electrodynamic Tether System
SOLO	SOdium LOad in microgravity
SPREE	Shuttle Potential and Return Electron Experiment
SRB	Solid Rocket Booster
SRON	Space Research Organisation Netherlands
SRTM	Shuttle Radar Topography Mission
SSME	Space Shuttle Main Engine
SSRM	Space Shuttle Remote Manipulator
STS	Space Transportation System
TEID	Theoretical and Experimental Investigation of TSS Dynamics
TEMAG	Magnetic Field Experiment for TSS Missions
TMA	Transport Modified Anthropometric
TMST	Theory and Modelling in Support of Tethered Satellite Applications
TNO	Netherlands Organisation for Applied Scientific Research
TOP	Tether Optical Phenomena Experiment
TRRJ	Thermal Radiator Rotary Joint
TSS	Tethered Satellite System

TsUP	RKA Mission Control Centre
TUS	Trailing Umbilical System
ULF	Utilisation and Logistics Flight
UF	Utilisation Flight
UNICEF	United Nations Children's Fund
UNP	University of Nebraska Press
UTC	Universal Time Coordinate
VC	Visiting Crew
VII-FP	Seventh Framework Programme for European Research
WASH	Water, Sanitation and Hygiene
WRS	Water Recovery System
WHC	Waste and Hygiene Compartment

Part 1

Background

1

Before ISS

This introduction briefly covers the history of European human spaceflight prior to the construction of the International Space Station (ISS).

Intercosmos was a Soviet programme to fly 'guest' cosmonauts from the Warsaw Pact and other communist countries to the Salyut 6 space station. There were nine missions, including six Eastern Europeans. Following the Intercosmos missions, there were two flights to the Salyut 7 space station, including France's first space traveller, Jean-Loup Chrétien, and four to the Mir space station which included a second Bulgarian cosmonaut flight after Bulgaria's first mission was unable to dock with Salyut 6, and also a second flight by Chrétien.

The Salyut 6 space station and its successor, Salyut 7, were advances on the previous designs by having two docking ports, one at each end. This facilitated cargo freighters, the routine exchange of crews making long-duration missions and visits by Intercosmos crews. In addition to the propaganda value of the international missions, a visiting crew could leave the new Soyuz spacecraft at the space station and return to Earth in the older one, to ensure that the residents always had a vehicle that was within the limit of its space-worthiness. And of course, the visitors conducted scientific experiments that added to the research programme of the station at no extra cost to the Soviet space programme.

The honour of being the first European in space belongs to Vladimir Remek of Czechoslovakia. Remek was the first person other than American astronauts and Soviet cosmonauts to travel to space. He flew on Soyuz 28 and spent 7 days on board Salyut 6 in March 1978, during which he conducted experiments in pure crystal growth (Morova), semiconductor optics, star brightness measurements (Extinctica), algae culture growth (Chorella) and oxygen concentration in human tissue. Remek, then an air force pilot, was selected as part of the 1976 Intercosmos group that included Czechoslovakian, Polish and East German cosmonaut candidates. There is a joke that Remek returned from space with a red hand, and when asked by the doctors what had happened, he explained that every time he reached for something on the space station, one of his Soviet colleagues would slap his hand and say, "Don't touch anything!"

J. O'Sullivan, *In the Footsteps of Columbus*, Springer Praxis Books,
DOI 10.1007/978-3-319-27562-8_1

The second Intercosmos flight carried Polish air force pilot Mirosław Hermaszewski on board Soyuz 30 in June 1978. He is still the only Polish spacefarer. Again, Hermanaszewski performed a number of experiments on Salyut 6 which included attempting to manufacture cadmium-tellurium-mercury semiconductor material (Serena), researching the most relaxing position to float in space (Relaks), Earth photography (Zierna), cardiovascular studies (Kardiolider), and an experiment to discover why some foods that are delicious on Earth taste like sawdust in weightlessness (Smak).

The last of the 1976 Intercosmos group to fly was Sigmund Jähn, becoming the first East German (and German) in space. His primary experiment was the use of the famous East German Karl Zeiss MKF-6M multi-spectral camera. Other experiments included crystal growth, speech research (Rech) and hearing tests (Audio). Unlike his predecessors, Jähn returned to Earth in a spacecraft other than that in which he was launched. Soyuz 31 docked with the space station on 27 August 1978, but this fresh vehicle was left for the long-duration Soviet crew, the visitors taking the older Soyuz 29 home after 7 days. This involved not only moving the personal seat liners from one vehicle to the other but also balancing weights to position the centre of gravity of the capsule to ensure that it would re-enter the atmosphere in the correct orientation to steer for an accurate landing.

The next Intercosmos group was selected in 1978 with ten candidates from Bulgaria, Romania and Hungary, as well as Cuba and Mongolia. The first to fly in space was Bulgarian air force pilot Georgy Ivanov. He was born with the surname Kakalov but changed it to Ivanov after joining the Intercosmos programme and discovering that Kakalov was an obscenity in Russian.

Ivanov flew on Soyuz 33 with the first Soviet civilian Soyuz commander, Nikolay Rukavishnikov. Despite high winds at the Baikonur launch site, the ascent was uneventful. However, on approach to Salyut 6 the engine which was to fire for 6 seconds to slow their approach actually fired for 3 seconds erratically and then shut down. Mission rules dictated that the docking be aborted and the crew return to Earth as soon as possible. They could have used their backup engine for the docking manoeuvres but the flight controllers decided to err on the side of caution. If the backup engine failed during the second docking attempt, there would be no way home. They ate, and then opened the gifts they were carrying for the station residents and "fortified themselves" with what they found. After a sleepless night, they put their Sokol suits on again and used the backup engine for the de-orbit manoeuvre. It fired for 213 seconds, which was 25 seconds longer than planned and had to be shut down by the crew. As a result of the extended burn, the crew endured a ballistic re-entry which produced a peak deceleration of 10g rather than the normal 4g. The crew landed safely and coincidently near the original target. The cause of the main engine malfunction was later identified as a pressure sensor that prevented propellant from being pumped into the combustion chamber, causing it to burn erratically. To compensate for failing to deliver Ivanov to Salyut 6, in 1988 the Soviets would fly Aleksandr Aleksandrov, his countryman and backup for Soyuz 33, to the new Mir space station.

In May 1980, after being delayed due to the problems of Soyuz 33, Soyuz 36 carried Hungarian pilot Bertalan Farkas to Salyut 6. With the Intercosmos flights becoming routine, there was not much attention paid to this mission outside of Hungary. Once again the international cosmonaut performed a series of experiments. In this case, they included processing gallium arsenide crystals with chromium, making a photographic map of the

Carpathian Basin, and investigating the effect of Interferon before during and after weightlessness. Farkas adjusted to weightlessness more rapidly than his experienced commander Valery Kubasov, who was on his third flight. As before, the visitors exchanged vehicles, returning to Earth in Soyuz 35 and leaving Soyuz 36 in orbit for the long-duration crew.

The next European to fly to space was Romanian aerospace engineer Dimitru Prunariu. He flew to Salyut 6 on board Soyuz 40 in May of 1981. It had been an eventful period in human spaceflight since Soyuz 36. The Intercosmos programme had ventured beyond the Warsaw Pact countries and delivered Vietnamese, Cuban, and Mongolian cosmonauts to the station. Soviet crews flew to the station on the new Soyuz T spacecraft that had solar panels to facilitate longer missions, solid state electronics, new engine systems, and a three-person crew. The Intercosmos missions continued to use the older model. Almost a month before Prunariu's flight, after an absence of nearly 6 years since the last flight by an Apollo spacecraft, NASA resumed its human spaceflight programme with the reusable Space Shuttle Columbia on mission STS-1.

Prunariu flew on Soyuz 40. It was the last flight of the older model of the Soyuz spacecraft, the last mission to dock with Salyut 6, and the last Intercosmos flight. As the long-duration crew had arrived on a new longer-lasting Soyuz T, there was no need for a vehicle swap. The crew experiments included observing Earth's magnetic field.

After Intercosmos and Salyut 6, the Soviet Union continued to partner with foreign countries and the next international cosmonaut was Frenchman Jean-Loup Chrétien of CNES, who became the first Western European and the first Frenchman in space. Chrétien and Patrick Baudry were selected as the first CNES astronaut group in 1980. Neither joined the ESA astronaut corps. Chrétien was assigned to train with Yuri Malyshev as commander, but after disagreements he eventually trained and flew with Vladimir Dzhanibekov. Obviously, Chrétien's position on the crew was lucrative to the Soviets. The flight of Soyuz T-6 to Salyut 7 in June 1982 was an eventful one. The spacecraft turned its engines in the direction of flight in order to slow the approach and then, as the vehicle turned to face its docking port towards the station the computer sensed imminent gimbal lock and so cut the engine, leaving the spacecraft spinning in space.[3] The commander took control and docked manually. Once on board the station, Chrétien performed French experiments which impressed the Soviets compared to the earlier international missions. The Echograph heart monitor, for example, was left on board the station after Chrétien's departure.

The next European to fly was German physicist Ulf Merbold, over a year later in November 1983 on Space Shuttle mission STS-9, setting a number of milestones. Merbold was the first West German in space, the first non-American on an American space mission, and the first ESA astronaut (Chrétien flew for CNES and was never a member of the ESA astronaut corps).[4] Merbold was selected together with Claude Nicollier of Switzerland,

---

[3] Gimbal lock occurs when two of the three axes of a three-axis gyroscope are parallel. When this happens the gyroscope can no longer provide a frame of reference to the spacecraft.

[4] Merbold would also be the first ESA astronaut to fly with the Soviets on Soyuz TM-20. Other Western Europeans flew before him but they either represented national space agencies, such as CNES and DLR, or flew as individuals such as Helen Sharman.

Wubbo Ockels of the Netherlands, and Franco Malerba of Italy as Payload Specialists for the new ESA Spacelab.[5] Spacelab was a reusable laboratory built in Europe under the auspices of ESA. Various components flew on 25 Shuttle missions, including unpressurised pallets. STS-9 was the first mission to carry the first of two Spacelab habitable modules (designated LM1). It lasted 10 days and the six-person crew split into two shifts to ensure maximum productivity. Experiments in physics, astronomy, Earth observation, material science, and space plasma physics were conducted.

As an aside, Dutch born Lodewijk van den Berg flew as Payload Specialist on STS-51B in April 1985. As a naturalised US citizen, he didn't hold Netherlands citizenship at the time of the flight but his story is an interesting one. Van den Berg was a PhD chemical engineer who specialised in crystal growth and worked for defence contractor EG&G. Due to the sensitive nature of his research, he was required to become an American citizen. He designed an experiment for crystal growth in space's weightless environment and NASA agreed to fly it on the Shuttle. NASA decided that, owing to the complexity of the experiment, a qualified Payload Specialist should run the experiment, not a career NASA astronaut. When EG&G couldn't find an eighth candidate to fill the NASA requirement, van den Berg was added to the list in the expectation that he would be eliminated because of his age and eyesight. However, after other candidates were excluded for various medical reasons, van den Berg found himself in the final two and he ultimately flew as the prime candidate for the second manned Spacelab.

In June 1985 France's second astronaut, Patrick Baudry, flew on STS-51G as Payload Specialist representing CNES. Baudry had been Chrétien's backup for Soyuz T-6 and he completed a series of similar biomedical experiments on board the Shuttle. The primary mission was satellite deployment and Baudry was joined by Prince Sultan Salman Abdul Aziz Al-Saud who, as the second son of the Crown Prince of Saudi Arabia, was the first royal astronaut. This was an era before the Challenger disaster, when NASA was keen to fly foreign dignitaries, congressmen, and civilians on the Shuttle.

STS-61A, launched in October 1985, contained no fewer than three European astronauts: West German Ernst Messerchmid (DLR), West German Reinhard Furrer (DLR) and Dutch Wubbo Ockels (ESA). All three were physicists. This was the Spacelab D1 mission that was bought and paid for by the West German government. As such, it was the first NASA manned space mission to be partially controlled from outside the USA; from the DLR facility near Munich. The research programme had experiments into the vestibular (balance) organs in the inner ear, microgravity, materials science, life sciences, communications, and navigation.

After the failure of Georgy Ivanov to reach Salyut 6 during his Bulgarian Intercosmos mission, the Soviets offered Bulgaria a 10 day mission to Mir. Aleksandr Alexsandrov, Ivanov's backup, made this flight on board Soyuz TM-5 in June 1988.[6]

Jean-Loup Chrétien became the first European to return to space when he launched to Mir aboard Soyuz TM-7 in November 1988. Representing CNES, his 30 day Aragatz

[5] Payload Specialists were not required to be members of a NASA astronaut selection group, because they worked specifically with a payload for a single mission.

[6] Not to be confused with Soviet cosmonaut, Aleksandr Pavlovich Aleksandrov, who flew on Soyuz TM-3 in 1987.

mission was cut to 25 days because the launch was delayed in order to accommodate the arrival at Baikonur of French President Mitterrand to view that event. Nevertheless, it was a much longer mission than the 7 day visits by the Intercosmos crews. Chrétien performed a number of ground-breaking French experiments, including making a spacewalk to test the deployment of the ERA carbon-fibre space structure. The unfurling of this experiment was almost a failure until fellow spacewalker Aleksandr Volkov gave it a swift kick! This spacewalk was the first by a European. In fact, Chrétien's return to space and his EVA were both firsts by any non-US/non-Soviet astronaut.

The next European space traveller was not a member of any space agency, yet became the first European woman in space and the first Briton in space. Chemist Helen Sharman was one of 13,000 British applicants to Project Juno, a privately funded mission to Mir. Sharman flew to Mir on board Soyuz TM-12 in May 1991 and could almost be called the first 'space tourist' but for the fact that a Japanese TV company had funded a visit by journalist Toyohiro Akiyama to the station the previous year. It has been suggested that Project Juno didn't raise the necessary funds and the Soviets covered the shortfall in order to enable the mission to proceed. Sharman had few experiments to perform, but assisted with other programmes and enjoyed herself. She was an unsuccessful candidate for the ESA astronaut selections of 1992 and 1998.

In the new economic climate following the collapse of the Soviet Union, the now-Russian space programme continued to offer commercial flights to nations and agencies as a way of keeping the Mir station operating. The next to avail was Austria. Representing ASA, Franz Viehbock became the first "Austronaut" when he flew to Mir on Soyuz TM-13 in October 1991. The unmanned Progress M-9 freighter had already delivered 150 kg of Austrian experimental apparatus. During his 8 day mission Viehbock conducted biomedical and materials processing experiments and made Earth observations.

Ulf Merbold returned to space for a second time on Shuttle mission STS-42 in January 1992, with the International Microgravity Laboratory. As with earlier Spacelab missions, the crew operated two shifts. The IML was a joint effort between NASA, ESA, CSA, CNES, DARA (Germany) and NASDA (Japan). The prime experiments assessed the adaptation of the human nervous system to microgravity. Other experiments included materials processing and crystal growth. At one point, the Shuttle passed within 39 nautical miles (72 km) of Mir, which in 1994 would be Merbold's home for 31 days.

DLR astronaut, Klaus-Dietrich Flade flew on Soyuz TM-14 in March 1992 for a 7 day mission to Mir. He conducted fourteen experiments on board, including biomedical measurements in preparation for the ESA participation in the eagerly awaited American-led Freedom space station. The planned Columbus module eventually became part of the International Space Station. As part of the handover of the Mir resident crews, Flade returned to Earth on Soyuz TM-13 together with Sergei Krikalev and Alexandr Volkov. As a result of the dissolution of the Soviet Union in December 1999, Krikalev had launched on Soyuz TM-12 as a Soviet citizen, served a double-length tour, and returned as a Russian citizen of the Commonwealth of Independent States.

As Flade was preparing to return to Earth, Belgian Payload Specialist Dirk Frimout was launching on STS-45 to spend 8 days on the ATLAS-1 mission. This Atmospheric Laboratory for Applications and Science was part of NASA's 'Mission to Planet Earth' programme. It was originally intended to involve ten missions over the 11 year solar cycle, but was curtailed due to budget limitations. During the mission, Frimout spoke with Belgium's Prince Phillipe when the latter was visiting the Marshall Space Flight Center in Huntsville, Alabama. When Frimout returned to Belgium he was awarded the title of Vicompte/Burggraf (Viscount).

Michel Tognini made France's second visit to Mir in July 1992. After backing up Chrétien for the CNES Argatz mission, Tognini flew Soyuz TM-15 to Mir to perform the 13 day Antares mission. He returned to Earth with the resident (EO-11) crew on Soyuz TM-14.

Also in July 1992, STS-46 carried the joint NASA/ASI Tethered Satellite System (TSS) and the European Retrievable Carrier (EURECA) which was an ESA free flyer. ESA's Claude Nicollier of Switzerland and ASI's Franco Malerba both made their first flights. The deployment of EURECA didn't go as planned, suffering some initial delays and then a shorter than desired period of boosting. However, it was finally manoeuvred into the correct orbit on the 6th day of the mission. The TSS was also troublesome. It was unable to unreel to its intended distance of 20 km because the cable snagged after only 256 metres. This experiment would fly again in 1996, with Nicollier being accompanied by two Italian crewmates.

STS-55 was scheduled to perform the second German Spacelab mission, D2, in February 1993. After delays due to engine problems and the replacement of all three SSMEs, STS-55 flew in April 1993. It carried two German DLR astronauts, Hans Schlegel and Ulrich Walter. Although a German science mission, it also carried experiments supplied by ESA, CNES, NASA and NASDA. As on the D1 mission of 1985, two shifts carried out experiments in fluid physics, materials science, biology, technology, Earth observation, atmospherics, and astronomy.

French flights to Mir sponsored by CNES continued when Jean-Pierre Haigneré was launched on Soyuz TM-17 in July 1993 for the 20 day Altair mission. He continued some of Michel Tognini's experiments as well as experiments delivered by a Progress freighter prior to his arrival. As his visit coincided with a resident crew handover, Jean-Pierre returned to Earth with the retiring crew aboard Soyuz TM-16. He would return to Mir in 1999.

In December 1993 Claude Nicollier returned to space on one of the Space Shuttle's most famous missions. STS-61 was the first servicing mission to the Hubble Space Telescope, and it installed the Corrective Optics for Space Telescope Axial Replacement (COSTAR) that was to correct the spherical aberration of the primary mirror of the telescope launched in 1990 by STS-31. Nicollier controlled the Remote Manipulator System (RMS) arm which grabbed the schoolbus-sized HST and placed it in the Shuttle payload bay. He then used the robot arm to manoeuvre spacewalking astronauts from one work site to another on the telescope. Finally he released the HST into its operational orbit on the 9th day of the mission.

EuroMir was a joint effort by ESA and Russian to give ESA astronauts long-duration experience in advance of the assembly of the ISS in general and the installation on that station of the ESA Columbus module in particular. First up was Ulf Merbold on board Soyuz TM-20 in October 1994, on his third space mission. When the Soyuz failed to dock automatically at the front port of the station, spacecraft commander Alexandr Viktorenko performed a manual docking. Merbold's 31 day EuroMir94 mission got off to a bad start, because a computer failure meant the station could not orientate itself to face its solar panels to the Sun, causing the batteries to run flat and requiring the docked Soyuz to manoeuvre the station to allow solar recharging. To allay fears that the entire Kurs automatic docking system was faulty, on the penultimate day of his visit Merbold and the retiring EO-16 crew boarded Soyuz TM-19, undocked from the rear port, withdrew 190 metres, and redocked using the automatic system. Having proved the problem was isolated to the front port, they departed for good the following day in the old spacecraft.

In November 1994 STS-66 carried the Atmospheric Laboratory for Applications and Sciences for the third time. It had the same experiments as on STS-45 and STS-56. ESA's Jean-François Clervoy flew for the first time as Mission Specialist. A secondary payload was the CRISTA-SPAS (Cryogenic Infrared Spectrometers and Telescopes for the Atmosphere, Shuttle Pallet Satellite). It flew behind the orbiter at a distance of 40–70 km for 8 days. When retrieving the satellite, the orbiter performed a fuel-saving rendezvous technique called R-bar, in preparation for future Shuttle-Mir rendezvous missions.

As preparations for the construction and operation of the ISS in partnership with Russia continued, both NASA and ESA signed agreements with the Russians to employ Mir as a training ground. After Norman Thagard became the first of seven NASA astronauts to take part in long-term increments and STS-71 delivered a Russian crew (EO-19), Thomas Reiter joined the EO-20 crew in September 1995 for the EuroMir95 mission. When lack of funds delayed the construction of Soyuz launch vehicles, it became necessary to extend Reiter's mission from 135 to 179 days, providing a considerable bonus to the European programme. There were three spacewalks during the increment, with Reiter performing the first and third. The first EVA was a scheduled one, with Reiter accompanying Sergei Avdeyev to install the European Science Exposure Facility (ESEF). The second spacewalk, during which Reiter remained in the Soyuz in case a problem required the station to be evacuated, was to reconfigure the docking equipment to receive the Priroda module which was to be launched the following year. The third spacewalk had not been planned prior to launch, and the training was provided by radio. Reiter and Yuri Gidzenko first removed the obsolete Cosmonaut Manoeuvring Unit 'flying backpack' from the airlock and attached it to a nearby external fixture to make space on an overcrowded station, then replaced cassettes on the ESEF. Their attempt to remove a faulty Kurs antenna was not successful. A further mission by Christer Fuglesang (EuroMir97) did not materialise because ESA resources were redirected to the ISS.

The Italian ASI's Tethered Satellite System (TSS) flew for the second time on board STS-75 in February 1996. It was a repeat of the 1992 STS-46 mission and ESA's Claude Nicollier once again accompanied it to space. He was joined by two Italian astronauts, both of whom were on their first flights into space: ESA's Maurizio Cheli and ASI's (later to join ESA) Umberto Guidoni. Although this time the satellite was unreeled to its full 20.5 km extension, the tether snapped and the package was lost.

STS-78 in June 1996 was the longest Shuttle flight to date at 16 days and 21 hours; appropriately, the crew of this Life and Microgravity Spacelab mission researched the effects of long-duration flight in preparation for the ISS missions. CNES astronaut Jean-Jacques Favier was a Payload Specialist.

Neuroscientist Claudie André-Deshays (later to fly as Claudie Haigneré after marrying Jean-Pierre Haigneré) spent 16 days on Mir in August 1996. Her Cassiopée mission conducted cardiovascular and neurosensory experiments, as well as materials processing and station vibration investigations. She had been expected to launch with Gennady Manakov and Pavel Vinogradov, but Manakov failed a medical check and so she flew with the backup crew of Valeri Korzun and Alexandr Kaleri, who remained on Mir as the EO-22 crew.

Reinhold Ewald of the DLR spent an eventful 19 days aboard Mir in February 1997. The automatic docking system failed, so this operation was undertaken manually. Ewald joined the EO-23 crew with studies of the effects of microgravity on human performance, hormonal and cardiovascular functions, and materials processing. On 23 February a faulty oxygen candle in the Kvant module caused a flash fire. After 90 seconds it had blown itself out but there was a lot of smoke. This incident raised further concerns for the safety of the ageing station. As well as the risk of fire and fumes, the fire had blocked access to the Soyuz TM-24 'lifeboat' at the rear that contained Ewald's seat liner. The return to Earth was also out of the ordinary, because the Soyuz Descent Module (DM) was orientated with its hatch facing forward until 160 seconds into the descent due to problems with the thermal blankets, and for some time after the Service Module was jettisoned there was a risk of collision.

Jean-François Clervoy returned to space on STS-84 in May 1997. As well as delivering NASA's Michael Foale to Mir and retrieving his predecessor Jerry Linenger, the Shuttle crew transferred 3,400 kg of supplies to the station, carried out experiments inside the double SpaceHab module that was in the payload bay, and tested the ESA docking system that would later be used by Automated Transfer Vehicles (ATV) when docking with the ISS.

After flights to Salyut 7 and Mir in the 1980s and Buran (Soviet counterpart of the Space Shuttle) pilot-training in the 1990s, Jean-Loup Chrétien became an International Mission Specialist in 1994 as part of NASA's Astronaut Group 15. His final flight was on STS-86, which delivered David Wolf to Mir as replacement for Michael Foale.

CNES continued its collaboration with the Russians by flying the Pégase mission to Mir in January 1998. Owing to a sports injury Jean-Pierre Haigneré's mission went to Léopold Eyharts, who repeated the science programme of André-Deshays.

ESA's Pedro Duque flew on STS-95 in October 1998, which marked John Glenn's return to space at the age of 77. Duque managed ESA experiments in both the SpaceHab module and on the mid-deck of the Shuttle.

As the inaugural ISS modules were being assembled in space in late 1998, the future of Mir was in turmoil. NASA wanted the Russian partners to concentrate exclusively on the ISS, but some people in the Russian programme and private commercial partners hoped to fund Mir as a business for scientific experiments, industrial production, space tourism, and advertising. Soyuz TM-29 flew two fee-paying guests to Mir in February 1999 on what was to be the final official state mission. Commander Viktor Afanasiev launched with Jean-Pierre Haigneré of CNES and Ivan Bella of Slovakia. Haigneré would remain with the EO-27 crew for 188 days and carry out the CNES Perseus programme of experiments. Bella was aboard for nearly 8 days before returning to Earth with the EO-26 crew after completing the Stefanik scientific mission. Haigneré made an EVA to test a hull sealant and to retrieve experiments from the exterior. In July, he spoke to fellow Frenchman Michel Tognini when the latter was in space for the STS-93 mission. On 27 August 1999 the crew undocked and returned to Earth, leaving Mir in autonomous mode. As far as they were concerned, this was the final mission. However, one more crew flew to Mir in April 2000. By that time the enthusiasm to sustain Mir had waned, and EO-28 was the final increment. Mir re-entered the atmosphere on 23 March 2000. Later that year, the first permanent crew of the ISS started an unbroken occupation of the new station.

In July 1999 STS-93 deployed the Chandra X-Ray Observatory and its Inertial Upper Stage (IUS) boosted the telescope into an orbit having an apogee of 133,000 km and perigee of 16,000 km. CNES astronaut Michel Tognini acted as Mission Specialist.

On STS-103 in December 1999 Claude Nicollier completed his fourth spaceflight and his second Hubble Space Telescope servicing mission. He was joined by ESA's Jean-François Clervoy, making his third spaceflight. Both had flown all of their missions on Space Shuttles and as ESA astronauts. In the first ever EVA by an ESA astronaut on a Shuttle, Nicollier joined Michael Foale on the second of the mission's three spacewalks and replaced a computer on the telescope.

Gerhard Thiele flew his only space mission on STS-99 in February 2000 as a representative of ESA. This Shuttle Radar Topography Mission (SRTM) was to employ an extendable 61-metre antenna to obtain high resolution topographical data in order to map Earth's surface. The project had participation from the German space agency DLR, and by that time the DLR's astronauts had been integrated into the ESA astronaut corps.

The next European astronaut to fly in space was ESA's Umberto Guidoni, who became the first European visitor to the ISS. His mission kicks off the coverage in Part 2 of this book, which details visits by European astronauts to that station.

2

Spacecraft

European astronauts have travelled to the ISS in two types of spacecraft: the American Space Shuttle and the Russian Soyuz. They have been resupplied on board the station by payload in the Italian-built Multi-Purpose Logistics Modules (MPLM) carried aboard Shuttles and also by a variety of unmanned vehicles – the Russian Progress, the European ATV, the Japanese HTV, and the SpaceX commercial Dragon spacecraft.[7]

Each type of flight to the ISS uses a different code:

R: Russian Roscosmos flight
A: USA NASA flight
E: European ESA flight
J: Japanese JAXA flight
A/R: Joint USA/Russian flight (financed by USA, built by Russia)
J/A: Joint Japanese/USA flight
UF: Utilisation flight
LF: Logistics flight
ULF: Utilisation/Logistics flight
S: Crew delivery flight on Soyuz
P: Cargo delivery flight on Progress
ATV: Cargo delivery flight on ESA Automated Transfer Vehicle
HTV: Cargo delivery flight on JAXA H-II Transfer Vehicle
SpX: Cargo delivery flight on SpaceX Dragon.

[7] Orbital Science's Cygnus first flew to the ISS in September 2013 and is thus outside the scope of this book.

J. O'Sullivan, *In the Footsteps of Columbus*, Springer Praxis Books,
DOI 10.1007/978-3-319-27562-8_2

SPACE SHUTTLE

The Space Transportation System (STS) was developed by NASA as the next generation of spacecraft after the successful Apollo, ASTP and Skylab programmes. ASTP was a joint mission in 1975 where an Apollo spacecraft docked with a Soviet Soyuz in a foreshadowing of international projects such as Shuttle-Mir and ISS. The Skylab station was made by modifying a third stage of a Saturn V rocket to serve as an orbital laboratory and solar telescope, and a succession of crews were delivered in the last three Apollo capsules to fly. STS was to revolutionise space travel with a reusable 'space plane' that would launch on the back of a booster, operate in low Earth orbit, and glide back to a runway landing. It was to be a multi-purpose spacecraft that would fulfil all of America's space requirements, such as deploying civilian and military satellites, conducting microgravity science experiments, making solar, terrestrial and astronomical observations, and ultimately assembling and servicing the Freedom Space Station.

While it did not achieve the hoped for cost savings associated with reusability and frequent flights, the Shuttle was the workhorse of the American human spaceflight programme for thirty years: 1981 to 2011.

The Shuttle consisted of the Orbiter Vehicle (OV), the External Tank (ET), and the Solid Rocket Boosters (SRB). Together these components were called the 'stack'. It was launched vertically from the Kennedy Space Center in Florida. Between them the twin SRBs contributed 80% of the thrust at liftoff, with the remainder being supplied by the cluster of three SSMEs that drew liquid oxygen and liquid hydrogen from the ET. The SRBs were jettisoned after 126 seconds and the OV/ET continued to accelerate. Approximately 8 minutes after launch, the SSMEs shut down, and shortly after that the ET was jettisoned and left to re-enter the atmosphere. The OV then fired smaller rocket engines to adopt a stable orbit.

Five OVs were built and flown in space. Columbia (OV-102), Challenger (OV-099), Discovery (OV-103), Atlantis (OV-104) and Endeavour (OV-105). Enterprise (OV-101) was a test vehicle that carried out free-flight tests prior to the inaugural spaceflight by Columbia. Enterprise was not rebuilt for spaceflight. Challenger was an engineering test vehicle that was rebuilt for spaceflight. Columbia and Challenger never visited the ISS. Challenger was destroyed during a launch accident in 1986 and Columbia was lost during re-entry in 2003. The other three orbiters delivered European astronauts to the station on eight occasions, starting with Endeavour's STS-100 mission which delivered Umberto Guidoni in April 2001, making him the first European to visit the station.

Endeavour also delivered Philippe Perrin during the STS-111 mission in June 2002 and Roberto Vittori on STS-134 in May 2011. STS-134 was Endeavour's last visit to the ISS and its final flight.

Thomas Reiter became the first long-duration crewmember of the ISS when he was launched on Discovery's STS-121 mission and joined Expedition 13. Both of Christer Fuglesang's missions to the station (STS-116 and STS-128) started and finished on board Discovery. Reiter also returned to Earth on STS-116.

Finally, on the STS-122 mission Atlantis played a major European role when it delivered the ESA Columbus module to the ISS along with two ESA astronauts, Hans Schlegel and Léopold Eyharts; the latter joined the Expedition 16 station crew.

In total, 12 Shuttle missions delivered Italian-built Multi-Purpose Logistics Modules (MPLM) to the station between 2001 and 2011.

Space Shuttle Discovery, STS-128. (NASA)

SOYUZ

The Soyuz manned spacecraft has been operational in many guises since 1967 and is still the mainstay of human spaceflight to the ISS. Developed by the Soviet Union after the Vostok/Voskhod spacecraft, the Soyuz could be said to be the equivalent of NASA's Apollo spacecraft because it also carried three people to orbit. Although unmanned variants made circumlunar flights, it was never used to fly a crew into lunar orbit. However it has been used for Earth orbital operations, the ASTP docking, and ferrying cosmonauts to and from the Salyut, Mir, and ISS space stations.

The Soyuz spacecraft is launched on the eponymous rocket from the Baikonur Cosmodrome in Kazakhstan and consists of three parts:

- The Orbital Module that provides accommodation for the crew during their mission.
- The Descent Module that carries the crew into orbit and returns them to Earth.
- The Service Module that contains the instruments and engines and has solar panels attached.

PROGRESS

Progress is an unmanned version of Soyuz and shares its architecture and design. It is a cargo freighter and has been used to deliver supplies to the Salyut, Mir, and ISS space stations. The Descent Module of the Soyuz was replaced with an unpressurised propellant and refuelling compartment. It can deliver up to 2,230 kg of cargo to the ISS, to which it docks automatically.

The Progress engines can be used to boost the ISS orbit. The entire craft burns up on re-entering the atmosphere, taking away trash from the station.

ATV

The Automated Transfer Vehicle is ESA's unmanned cargo spacecraft, five of which were launched to the ISS between 2008 and 2014. The design was based on the MPLM module, fitted with a propulsion system. It docked automatically with the Russian section of the ISS and could deliver up to 7,667 kg of cargo.

The ATV's engines could be used to boost the ISS orbit. It would burn up on re-entry, taking away trash from the station.

H-II

The H-II Transfer Vehicle is JAXA's unmanned cargo spacecraft. The design consists of four parts:

- The Pressurised Logistics Carrier (PLC) that contains the Common Berthing Mechanism to mate with the ISS and enable station crew to gain access.

The Soyuz TMA-7 'crew taxi'. (Wikipedia)

A Progress freighter approaching the ISS. (NASA)

ATV-2 'Johannes Kepler'. (ESA)

- The Unpressurised Logistics Carrier (UPLC) that contains the Exposed Pallet which can be transferred to the exterior of the ISS by robotic arm.
- The Avionics Module.
- The Propulsion Module.

An HTV grappled by Canadarm2. (NASA)

A Dragon capsule grappled by Canadarm2. (NASA)

The HTV has a payload of 6,000 kg, including 5,200 kg carried in the PLC. An HTV doesn't dock automatically, it approaches the ISS and is grappled by the station's robot arm and berthed by the ISS crew. It burns up on re-entry, taking away trash from the station. Five vehicles were launched between 2009 and 2015, with plans for more at the time of writing.

DRAGON

The Dragon spacecraft was developed by SpaceX as part of NASA's Commercial Resupply Services (CRS) programme where commercial companies design, build, and operate vehicles under contract.[8] This was a departure from the previous model where the government, via NASA, own the spacecraft.

The design consists of two parts:

- The blunt-cone pressurised ballistic capsule that can return to Earth and is re-usable.
- The unpressurised cargo-carrier trunk section that is equipped with two solar arrays.

The Dragon approaches the ISS and is grappled by the station's robot arm and berthed by the ISS crew. It can deliver up to 3,310 kg of cargo. In a valuable service, it can also return cargo to Earth. A total of eight were launched to the ISS between 2010 and 2015; all successful apart from the last one which was lost when the launch vehicle failed. At the time of writing it is intended to resume Dragon flights in 2016.

[8] In 2008 SpaceX and Orbital Sciences were awarded contracts for 12 and 8 resupply missions to the ISS respectively.

Part 2

Missions

3

STS-100

Mission

ESA Mission Name:	None (NASA Mission STS-100)
Astronaut:	Umberto Guidoni
Mission Duration:	11 days, 21 hours, 31 minutes
Mission Sponsors:	ESA
ISS Milestones:	ISS 6A, 10th crewed mission to the ISS

Launch

Launch Date/Time:	19 April 2001, 18:40 UTC
Launch Site:	Pad 39-A, Kennedy Space Center
Launch Vehicle:	Space Shuttle Endeavour (OV-105)
Launch Mission:	STS-100
Launch Vehicle Crew:	Kent Vernon Rominger (NASA), CDR
	Jeffrey Shears 'Bones' Ashby (NASA), PLT
	Chris Austin Hadfield (CSA), MSP1
	John Lynch Phillips (NASA), MSP2
	Scott Edward Parazynski (NASA), MSP3
	Umberto Guidoni (ESA), MSP4
	Yuri Valentinovich Lonchakov (RKA), MSP5

Docking

STS-100

Docking Date/Time:	21 April 2001, 13:59 UTC
Undocking Date/Time:	29 April 2001, 17:34 UTC
Docking Port:	PMA-2, Destiny Forward

J. O'Sullivan, *In the Footsteps of Columbus*, Springer Praxis Books,
DOI 10.1007/978-3-319-27562-8_3

Landing

Landing Date/Time:	1 May 2001, 16:11 UTC
Landing Site:	Runway 22, Edwards Airforce Base
Landing Vehicle:	Space Shuttle Endeavour (OV-105)
Landing Mission:	STS-100
Landing Vehicle Crew:	Kent Vernon Rominger (NASA), CDR Jeffrey Shears 'Bones' Ashby (NASA), PLT Chris Austin Hadfield (CSA), MSP1 John Lynch Phillips (NASA), MSP2 Scott Edward Parazynski (NASA), MSP3 Umberto Guidoni (ESA), MSP4 Yuri Valentinovich Lonchakov (RKA), MSP5

ISS Expedition

ISS Expedition:	Expedition 2
ISS Crew:	Yuri Vladimirovich Usachyov (RKA), ISS-CDR James Shelton Voss (NASA), ISS-Flight Engineer 1 Susan Jane Helms (NASA), ISS-Flight Engineer 2

3.1 THE ISS STORY SO FAR

Prior to the arrival of STS-100 there had been 11 flights to the ISS, including two unmanned Assembly Flights, two expedition deliveries, Shuttle and Soyuz spacecraft launches. Also there were three supply flights by unmanned Progress freighters.

In November 1998, flight ISS 1AR delivered the Zarya module into orbit on a Russian Proton-K rocket. Zarya, also known as the Functional Cargo Block (FGB), was built by the Khrunichev State Research and Production Space Centre (KhSC) but was funded and owned by the USA.

On flight ISS 2A STS-88 in December the Space Shuttle Endeavour delivered Unity, the first American node, that was to be docked with Zarya. This short pressurised cylinder had six Common Berthing Mechanisms (CBM), one on each end and four at 90° intervals around its side. Attached to Unity for the launch were two Pressurised Mating Adaptors (PMA). PMA-1 was to permanently link Zarya to Unity, and PMA-2 was to be used for future Shuttle dockings, being relocated as necessary.

In May 1999, on flight ISS 2A.1, STS-96 temporarily stowed sections of the US Orbital Transfer Device crane and the Russian Strela crane on the outside of the ISS, as well as delivering cargo from the pressurised SpaceHab module that resided in the cargo bay. Then in May 2000, after the Russian Mir space station had been de-orbited, STS-101 on flight 2A.2a visited the ISS Zarya/Unity complex with cargo and supplies to keep the ISS operational. Specifically four new batteries were installed as well as smoke detectors, cooling fans, power distribution and communications equipment.

In July 2000, the first Russian funded segment arrived at the station. The long awaited Zvezda module was launched on flight ISS 1R using an unmanned Proton rocket and

The configuration of the ISS during STS-100 consisted of Zarya, Unity, Zvezda, the Z1 truss segment and the P6 truss/solar array segment. (NASA)

docked autonomously with the Zarya module. ESA contributed the Data Management System that would control the entire complex until those functions could be transferred to the NASA Destiny laboratory module. The ISS was now ready for its first crew.

In September 2000, flight ISS 2A.2b saw STS-106 deliver cargo to the station and prepare the ISS for Expedition 1. This included unloading cargo from the Progress vehicle that had been docked since August. In October 2000, STS-92, ISS 3A, delivered the Z1 truss segment and PMA-3. The Z1 truss segment was attached to the zenith (top) CBM of Unity to temporarily accommodate various facilities during the early phase of the assembly process, and PMA-3 was attached to the nadir (bottom, Earth-facing) CBM of Unity. As Space Shuttle Discovery undocked, it left the ISS unmanned for the last time.

In October 2000, Soyuz TM-31 as flight ISS 2R delivered Expedition 1 to the ISS and began over 15 years (and counting) of continuous occupation.

In December 2000, STS-97, ISS 4A, delivered the P6 truss. Although ultimately destined for the port side of the ISS, the P6 truss was installed on the Z1 truss segment. It contained two Photovoltaic Array Assemblies (PVAA), each of which unfurled a pair of Solar Array Wings (SAW).

The Destiny module arrived on STS-98 as part of flight ISS 5A and was attached directly to Unity after PMA-2 had been removed from Unity and temporarily stowed on the front of Z1. Once Destiny was in position, PMA-2 was installed on its far end. The addition of Destiny increased the habitable volume of the station to 36.9 cubic metres, making it the most voluminous spacecraft in history.

Space Shuttle Discovery flew STS-102, ISS 5A.1, in March 2001. This was a combined logistics and crew transfer mission that ferried Expedition 2 to the station and returned Expedition 1 to Earth, and carried the first Multi-Purpose Logistics Module (MPLM), named Leonardo, in its payload bay. This was mated to the nadir CBM of Unity after PMA-3 had been moved to the port CBM. Once the cargo had been transferred into the station, the MPLM was returned to the Shuttle for return to Earth. Three MPLMs had been built by Alenia Spazio on behalf of ASI; Leonardo and Raffaello would be flown many times but Donatello would remain on the ground.

3.2 UMBERTO GUIDONI

Early Career

Umberto Guidoni was born in Rome, Italy, on 18 August 1954. As a teenager with a telescope and a fascination for space he followed his passion and studied astrophysics. He graduated with a doctorate in astrophysics from the University of Rome, La Sapienza, in

Umberto Guidoni. (NASA)

1978, then conducted his postdoctoral research in plasma physics at the Comitato Nazionale per l'Energia Nucleare (CNEN) in Frascati from 1979 to 1980. In 1983 he transferred to the Nazionale Energie Alternative (ENEA) to research nuclear fusion and later photovoltaic cells. His brief foray in the field of energy didn't last long and in 1984 he was again researching space at the l'Istituto di Fisica dello Spazio Interplanetario (IFSI), a laboratory of the Consiglio Nazionale delle Ricerche (CNR), which was also in Frascati. It was there that he took a step closer to becoming an astronaut, since his field of research was the Research on Electrodynamic Tether Effects (RETE).

In 1989, as project scientists on RETE, Guidoni and his colleague Franco Malerba were selected to join the ASI astronaut office and to train as NASA Payload Specialists to take RETE into space on the Shuttle. Malerba ultimately became the first Italian citizen into space when he flew on STS-46 and ran the Tethered Satellite System (TSS-1) experiment. Unfortunately for Malerba the experiment wasn't a success because the tethered satellite jammed after unreeling to only 256 metres of its planned 20 km. Guidoni was aboard STS-75 for the TSS-1R mission.

Previous Mission

STS-75

The prime objective of the STS-75 mission in February 1996 was to repeat STS-46's TSS experiment. Umberto Guidoni, on his first spaceflight, joined ESA's Claude Nicollier, who was flying in space for the third time; one of his previous missions being STS-46/TSS-1.

The specific mission objectives of TSS-1R were:

- Characterise the current-voltage response of the TSS/OV combination.
- Characterise the satellite's high-voltage sheath structure and current collection process.
- Demonstrate electric power generation.
- Verify tether control laws and the basic dynamics of a tether.
- Demonstrate the effect of neutral gas on the plasma sheath and current collection.
- Characterise the TSS radio frequency and plasma wave emissions.
- Characterise the TSS dynamic-electrodynamic coupling.

Investigations included:

- Deployer Core Equipment and Satellite Core Equipment (DCORE/SCORE).
- Research on Orbital Plasma Electrodynamics (ROPE).
- Research on Electrodynamic Tether Effects (RETE).
- Magnetic Field Experiment for TSS Missions (TEMAG).
- Shuttle Electrodynamic Tether System (SETS).
- Shuttle Potential and Return Electron Experiment (SPREE).
- Tether Optical Phenomena Experiment (TOP).
- Investigation of Electromagnetic Emissions by the Electrodynamic Tether (EMET).
- Observations at the Earth's Surface of Electromagnetic Emissions by TSS (OESSE).
- Investigation and Measurement of Dynamic Noise in the TSS (IMDN).
- Theoretical and Experimental Investigation of TSS Dynamics (TEID).
- Theory and Modelling in Support of Tethered Satellite Applications (TMST).

The STS-75 crew with Umberto Guidoni second from the left. (www.spacefacts.de)

Deployment of the satellite was delayed until Flight Day 3 due to computer problems. After the tether had unreeled to 19.7 km of a planned 20.5 km, it snapped and the satellite was lost. There was insufficient propellant to perform a rendezvous. The satellite was observed optically for the remaining mission duration. The subsequent NASA/ASI investigation decided that a breach in the insulation had allowed a current arc to pass from the copper wire to ground, burning away the tether.

In May 1996, despite previously having flown as a Payload Specialist, Guidoni joined the NASA Astronaut Group 16, known as the Sardines because there were so many of them, as an International Mission Specialist. In August 1998 he finally joined the ESA astronaut corps. In September 2001 he was assigned to the European Space Research and Technology Centre (ESTEC) at Noordwijk in the Netherlands to work on payloads destined for the ESA Columbus ISS module.

3.3 THE STS-100 MISSION

STS-100 Mission Patches

Unlike the other official ESA missions to the ISS, Umberto Guidoni's STS-100 mission did not have an ESA mission name or an associated mission patch. He wore the NASA mission patch designed for that Shuttle mission. This was in the shape of a NASA EVA

The STS-100 crew with Umberto Guidoni in the centre. (NASA)

helmet, with reflections of Endeavour in Earth's shadow and the ISS catching the dawn sunlight. The patch contained both '100' for the Space Shuttle mission number and '6A' for the ISS Assembly Flight designation. The MPLM Raffaello was depicted in the payload bay, together with a pallet carrying the Space Station Remote Manipulator System (SSRMS, named Canadarm2) and a communications antenna for the station. The flags of the four participating nations (USA, Russia, Italy and Canada) were present at the base of the helmet and the names of the astronauts were listed around the border.

The ASI patch included the name of Umberto Guidoni and celebrated the delivery of MPLM Raffaello to the ISS. It showed the ISS in its planned final configuration, and the logo of ASI was prominent.

STS-100 Mission Objectives

As Assembly Flight 6A, the prime objectives of STS-100 were as follows:

- Deliver the first component of the Space Station Remote Manipulator System (SSRMS) named Canadarm2. The Mobile Remote Servicer Base System (MBS) was to be delivered on STS-111 to enable Canadarm2 to travel the length of the truss system (once that was assembled). When the Special Purpose Dextrous Manipulator (SPDM), called Dextre, arrived on STS-123 in 2008 it would provide

STS-100 Mission Patch. (NASA)

Canadarm2 with a very able tool to perform maintenance and assembly tasks. Dextre would have lights, cameras, and a variety of tools, and it would be able to steady itself with one 'hand' while tightening a bolt.

- Deliver the Space Station Ultra-High Frequency communication antenna.
- Carry MPLM Raffaello, mate it with the station, transfer cargo, then return the module to Earth.

Timeline

Flight Day 1 – Thursday, 19 April 2001

Umberto Guidoni's second mission to space was on Space Shuttle Endeavour launched from Pad 39-A at the Kennedy Space Center. He was the only European member of the crew, and he was seated on the orbiter's mid-deck. The others were commander Kent Rominger (NASA), pilot Jeffrey Ashby (NASA), Mission Specialists Canadian Chris Hadfield (CSA), John Phillips (NASA), Scott Parazynski (NASA), and Russian Yuri Lonchakov (RKA).

Raffaello Guidoni Patch. (www.spacefacts.de)

The ISS was occupied by the Expedition 2 commander Yuri Usachyov (RKA) and flight engineers James Voss (NASA) and Susan Helms (NASA).

After achieving preliminary orbit, the crew configured systems for on-orbit operations and opened the payload bay doors.

Flight Day 2 – Friday, 20 April 2001

Each morning in space, Shuttle astronauts were woken by music selected by their family and friends. On the first morning of STS-100 they woke to *Then the Morning Comes* by Smashmouth, chosen for Phillips.

Hadfield and Parazynski checked the tools and hardware that was to be used during Endeavour's approach to the ISS, while Rominger and Ashby installed a centreline camera in the Orbiter Docking System. The engines were fired to achieve the correct orbit for rendezvous. Hadfield and Parazynski checked the spacesuits they were to wear during their two EVAs. Guidoni and Ashby prepared for the logistics of transferring cargo from the MPLM to the ISS and also verified operation of the robot arm which would grab and move Raffaello. Lonchakov filled two large containers of water for transfer to the station.

The water was a byproduct of the fuel cells which provided the electrical power to run the Orbiter and would be a valuable resource for the station.

Aboard the ISS, Usachev, Helms and Voss prepared for Endeavour's arrival and packed up items that were to be returned to Earth.

Flight controllers reported that the Russian segment's carbon dioxide removal system had degraded functionality due to a clogged filter screen. Although this was not an immediate problem for the station crew or the visitors, the residents were asked to prepare to enact repairs if requested.

Flight Day 3 – Saturday, 21 April 2001

Saturday saw the crew woken by *Danger Zone* by Kenny Loggins, which was the *Top Gun* soundtrack hit. This was in honour of Rominger. Rominger controlled the docking with the ISS as both spacecraft were 243 miles over the southern Pacific Ocean. He flew Endeavour to a position 180 metres directly below the station, then followed an arc to a point about 90 metres in front of it. Then the Shuttle eased slowly towards PMA-2, on the front of Destiny, with the docking occurring at 8:59 a.m. Central Time.

The hatches were not opened immediately, due to the difference in atmospheric pressure between the Shuttle and the ISS. Endeavour's cabin pressure had been lowered to 10.2 psi on the previous day, in preparation for a spacewalk that Hadfield and Parazynski were to make on Sunday. The atmosphere in the ISS was the usual 14.7 psi. The PMA served as an airlock to transfer items between the vehicles without opening both hatches simultaneously. Guidoni and his colleagues retrieved a battery drill left there by the Expedition crew, and put in place new items from Earth: Four water containers, computer equipment, fresh food, and IMAX film stock. Spacewalk planning was conducted on both sides of PMA-2, and spacewalk coordinator Phillips helped Hadfield and Parazynski to verify the suits.

Meanwhile, Expedition 2's Usachev and Helms tested the Vozdukh carbon dioxide removal system after its poor performance and confirmed that it was operating within normal parameters.

Flight Day 4 – Sunday, 22 April 2001

After being woken by Canadian Stan Rogers' *Take It From Day to Day*, the Shuttle crew prepared for the first spacewalk. The song was played for Hadfield who would be making the first spacewalk by a Canadian.

During the 7 hour, 10 minute spacewalk, Guidoni and Ashby used the Shuttle's robotic arm to lift Canadarm2 out of the payload bay and transfer it to its new position on the station's Destiny module. Phillips supported the spacewalkers as they connected the power and communications cables between the arm and the station. Then they unfolded it from its launch configuration and tightened the bolts to keep its booms extended. They had to use their pistol grip tools in manual mode because they weren't confident of the torque in automatic mode. Before transferring Canadarm2, they installed a new UHF antenna on the Destiny module. Within the ISS, Helms and Voss successfully exercised Canadarm2.

Flight Day 5 – Monday, 23 April 2001

To look forward to the opening of the hatches, the Shuttle crew was woken to the sound of *Both Sides Now* sung by Judy Collins, chosen for Jeff Ashby. The hatches were opened at 4:25 a.m. and Rominger led the Endeavour crew in the station. Cargo carried in the Shuttle was transferred to the ISS through this hatch, with Guidoni supervising as 'loadmaster'. However, the majority of the cargo destined for the ISS was in the MPLM Raffaello and would be offloaded the next day.

Expedition 2's Helms controlled the 'walk off' of Canadarm2 as it extended, grabbed the electrical grapple fixture on Destiny and then, on receiving power and commands through that new connection, disconnected itself from the pallet on which it had been delivered. This manoeuvrability was to permit the robot arm to wander across the outside of the station in order to help the crew perform maintenance and construction tasks.

At this time, Ashby, assisted by Guidoni, transferred the Italian Raffaello module from the Shuttle payload bay to a CBM on the ISS Unity node.

Meanwhile, Parazynski and Hadfield checked out their tools and spacesuits in preparation for their second spacewalk. The hatches were closed at 2:26 p.m. after 10 hours of joint operations to allow the Shuttle's cabin pressure to be lowered again for the next day's spacewalk.

As a final task, Rominger and Ashby fired Endeavour's thrusters to raise the orbit of the ISS from 237.8 to 240.3 statute miles. This, and two subsequent planned reboosts, raised the orbit of the station ready for the arrival of Soyuz TM-32.

Umberto Guidoni was the first European on the ISS. (NASA)

Flight Day 6 – Tuesday, 24 April 2001

In honour of Parazynski's planned spacewalk, the crew woke to *What A Wonderful World* sung by Louis Armstrong. With the hatches closed, the Shuttle and station crews worked both independently and together.

Aboard the ISS, the Expedition 2 crew opened the hatches between Raffaello and Unity and spent the first part of the day transferring supplies and equipment to the station.

At 7:34 a.m., Hadfield and Parazynski began their second spacewalk, completing the connections between Canadarm2 and the Destiny module. They also removed an antenna and transferred a spare Direct Current Switching Unit from the Shuttle to an equipment storage rack on the outside of Destiny. During the removal of the antenna, Hadfield lost hold of an electrical connector which floated off and became stuck behind a thermal blanket on a docking port. Ground control determined that it wouldn't cause a problem in the future, so it was abandoned. The spacewalk was completed at 3:15 p.m., and it was decided not to pursue the option of a contingency third spacewalk.

Voss and Helms controlled Canadarm2 as it lifted its first payload, namely the Spacelab pallet on which it had been transported in the Shuttle's payload bay.

After the spacewalk was completed, the Shuttle's cabin pressures was raised and the hatches were opened again, with Endeavour's crew entering the ISS at 5:15 p.m.

Flight Day 7 – Wednesday, 25 April 2001

The wakeup call was *Con te Partiro* sung by Andréa Bocelli, which was played in honour of Italian Umberto Guidoni.

The ISS had three Command and Control Computers, and overnight Mission Control had lost the connection with C&C-1. After unsuccessful attempts overnight to overcome the problem remotely, in the morning the crew was requested to transfer data. When this also failed, the computer was power-cycled and some functionality was successfully tested. Owing to these computer issues the following precautions were taken: Communications were routed through Endeavour, handoff of Canadarm2's transfer pallet to the Shuttle's arm was postponed, and a second reboost of the station using Shuttle thrusters was postponed. Mission Control informed the crew that their engineers would continue to reset the computers overnight.

The crew performed the tasks that didn't require computer control; namely offloading the cargo from Raffaello to the station, supervised by Guidoni.

Flight Day 8 – Thursday, 26 April 2001

This morning's wakeup call was *Behind the Fog*, a Russian folk song chosen for cosmonaut Yuri Lonchakov.

Overnight Mission Control had recycled power on all three Command and Control Computers but only one came back online. This meant postponing the handoff of the pallet and the unberthing of the Raffaello module. Shortly after the wakeup call, Helms announced that communications between the ground and US systems on the station were re-established. The remaining cargo was offloaded from Raffaello while computer troubleshooting continued both on the ground and in space. The astronauts also started to load

Raffaello with cargo that was to be returned to Earth. In all, nearly 4,000 pounds was transferred into the ISS and 1,600 pounds was offloaded.

Due to the lost time, and to help solve the computer problem, Mission Control were attempting to extend the STS-100 mission by two days. The extension relied on the Russians agreeing to postpone the launch of Soyuz TM-32, which was being prepared at the pad at Baikonur.

A complex computer resynchronisation was planned, involving two fault protection computers in the Unity node. Mission Control intended to do this while the crew slept.

Flight Day 9 – Friday, 27 April 2001

The morning's wakeup call wasn't chosen for a crewmember in orbit, but rather the STS-100 Ascent and Entry Flight Director Leroy Cain. The crew awoke to *Buckaroo* by Don Cain of Dubuque, Iowa, who was Leroy's father.

NASA and Rosaviakosmos reached an agreement to delay the launch of Soyuz TM-32, and so an extension of one day was granted to STS-100.

The flight controllers in Houston had successfully synchronised all three computers on the station overnight and although only one C&C on Destiny was online, the backup in Unity was operational, so it was decided to proceed with unberthing Raffaello. Guidoni and Parazynski employed Canadarm2 to grapple the MPLM and transfer it back to the Shuttle payload bay.

It was determined that C&C-1 had a failed hard drive. The crew installed a backup, and the faulty laptop was returned to Earth for inspection.

It was decided to perform the handoff of Canadarm2's pallet the following day. It had been planned to rehearse the berthing of an airlock module, but most of this activity was cancelled and only the tasks related to Shuttle robotic arm camera views would now be performed.

Flight Day 10 – Saturday, 28 April 2001

The crew were woken by *Dangerous*, sung by The Arrogant Worms, chosen for Chris Hadfield by his wife.

While waiting for approval to move the pallet to the Shuttle's payload bay, the crew performed a truncated dress rehearsal for the delivery of the Quest Joint Airlock by STS-104, on which the Space Vision System would be integral to the berthing procedure. This alignment system employed station cameras and the Shuttle's robotic arm.

Soyuz TM-32 was launched on this day, carrying a visiting crew whose main objective would be to swap 'lifeboats', because the Soyuz had an on-orbit lifetime of approximately six months. The visiting crew planned to leave their craft and return in Soyuz TM-31, which was at the end of its safe life. The new mission was notable for carrying the first 'space tourist' to the ISS – the American businessman Denis Tito.

With Expedition 2's Helms in control of Canadarm2 and Hadfield operating the STS-100 arm, the 'handshake in space' was a success. Canadarm2 transferred its own launch pallet to the Shuttle's arm, which stowed it in the payload bay for return to Earth. This was the first robotic-to-robotic transfer in space.

Flight Day 11 – Sunday, 29 April 2001

The last morning docked to the ISS saw the crew awakened with *Miles from Nowhere* by Cat Stevens, played for Ashby.

After interventions from the ground and the crews in space, all three of the Command and Control Computers (one primary and two backups) were online and operational, although one of the backups had a failed hard drive.

After undocking and release at 12:34 p.m. Central Time, Endeavour backed away from the ISS as they flew 240 miles above the Pacific Ocean, east of Australia. At 138 metres separation, pilot Ashby initiated a three-quarter circle flyaround of the ISS and Lonchakov activated the IMAX camera to film the ISS. At 1:28 p.m., Ashby performed the separation burn and Endeavour departed the station.

Flight Day 12 – Monday, 30 April 2001

The crew prepared for their return to Earth, with commander and pilot verifying flight control surfaces and thrusters and with Guidoni and the other Mission Specialists stowing equipment. On being woken by the soundtrack to the movie *Gladiator*, they also enjoyed some time off after what had been a busy 11 days in space.

Due to the chance of bad weather at the Shuttle Landing Facility in Florida at the projected time of arrival, Edwards Air Force Base in California was alerted by Mission Control to prepare to receive the Shuttle in case that should become necessary.

Flight Day 13 – Tuesday, 1 May 2001

After waking to *Truth* by Spandau Ballet, chosen by the family of Rominger, Endeavour and its crew landed at Edwards Air Force Base at 11:11 a.m. Central Time; it was the 48th time that a Shuttle had set down there.

3.4 POSTSCRIPT

Subsequent Missions

STS-100 was Umberto Guidoni's final spaceflight.

Umberto Guidoni Today

From 2004 to 2009 Guidoni was an elected Member of the European Parliament (MEP). Until 2009 he was a member of the parliamentary group of the European Left (Nordic Green Left), a member of the Committee for Industry, Research and Energy (ITRE), the Temporary Committee on Climate Change, the Commission Environmental, Health and Food Safety (ENVI), and the Commission for the Control of Financial Reporting (CONT). He was a speaker at the Seventh Framework Programme for European Research (VII FP) and presented a report on the European Research Area.

In 2007 he presented a radio show on RAI2 called *From Sputnik to the Shuttle* for four weeks, and then in 2009 he presented a series on the Apollo missions.

Umberto Guidoni at the National Award for Science Communication, Rome, December 2014. (www.associazioneitalianadellibro.it)

Guidoni has written the following books:

Il giro del mondo in ottanta minuti (*Around the world in eighty minutes*), Di Renzo Editore, 1998, 2002.
Libro dell'anno 2001 (*Book 2001*), AA.VV., Treccani 2002.
Un passo fuori (*Step out*), Laterza, 2006.
Idee per diventare astronauta (*Ideas to become an astronaut*), Zanichelli, 2006.
Martino su Marte (*Martino on Mars*), Editoriale Scienza, 2007.
Dallo Sputnik allo Shuttle (*From Sputnik to the Shuttle*), Sellerio, 2009.
Astrolibro dell'universo (*Astrolibro universe*), Editoriale Scienza, 2010.
Dalla Terra alla Luna (*From the Earth to the Moon*), Di Renzo Editore, 2011.
Così extra, così terrestre (*So extra, so Earth*), Editoriale Scienza, 2013.
70 Ore nel Futuro (*70 Hours in the Future*), AA. VV., Rêverie, 2013.
Viaggiando oltre il cielo (*Travelling over the sky*), Rizzoli BUR, 2014.

He has also received the following honours:

- In April 1996, after his first space mission on STS-75 he was made Commendatore Ordine al Merito della Repubblica Italiana (Commander of the Order of Merit of the Italian Republic) by Oscar Luigi Scalfaro, the Italian president at that time.
- In May 2001, after his second space mission on STS-100 he was made Grande Ufficiale Ordine al Merito della Repubblica Italiana (Grand Officer of the Order of Merit of the Italian Republic) by Carlo Azeglio Ciampi, who succeeded Scalfaro as Italian president in 1999.
- Also in 1996 and 2001 he received Space Flight Medals from NASA and in 2002 he received the Exceptional Service Medal.
- In 1996 the asteroid 10605 Guidoni was named after him.

4

Andromède

Mission

ESA Mission Name:	Andromède
Astronaut:	Claudie Haigneré, née André-Deshays
Mission Duration:	9 days, 20 hours, 0 minutes
Mission Sponsors:	CNES/ESA
ISS Milestones:	ISS 3S, 14th crewed mission to the ISS

Launch

Launch Date/Time:	21 October 2001, 08:59 UTC
Launch Site:	Pad 1, Baikonur Cosmodrome, Kazakhstan
Launch Vehicle:	Soyuz TM
Launch Mission:	Soyuz TM-33
Launch Vehicle Crew:	Viktor Mikhailovich Afanasiev (RKA), CDR
	Claudie Haigneré (CNES, ESA), Flight Engineer
	Konstantin Mirovich Kozeyev (RKA), Flight Engineer

Docking

Soyuz TM-33	
Docking Date/Time:	23 October 2001, 10:44 UTC
Docking Port:	Zarya Nadir
Soyuz TM-32	
Undocking Date/Time:	31 October 2001, 01:38 UTC
Docking Port:	Pirs

J. O'Sullivan, *In the Footsteps of Columbus*, Springer Praxis Books,
DOI 10.1007/978-3-319-27562-8_4

Landing

Landing Date/Time:	31 October 2001, 04:59 UTC
Landing Site:	Near Arkalyk, Kazakhstan
Landing Vehicle:	Soyuz TM
Landing Mission:	Soyuz TM-32
Landing Vehicle Crew:	Viktor Mikhailovich Afanasiev (RKA), CDR Claudie Haigneré (CNES, ESA), Flight Engineer Konstantin Mirovich Kozeyev (RKA), Flight Engineer

ISS Expeditions

ISS Expedition:	Expedition 3
ISS Crew:	Frank Lee Culbertson, Jr. (NASA), ISS-CDR Vladimir Nikolayevich Dezhurov (RKA), ISS-Flight Engineer 1 Mikhail Vladislavovich Tyurin (RKA), ISS-Flight Engineer 2

4.1 THE ISS STORY SO FAR

After Umberto Guidoni returned to Earth on board STS-100, the next flight to the ISS made history. After failing to be launched to Mir as part of a deal with MirCorp, American aerospace and investment management multi-millionaire Denis Tito flew to the ISS on Soyuz TM-32. This flight, costing Tito $20 million, was the first of many organised by Space Adventures Ltd. ISS crew transfers were now being carried out by Space Shuttle, but the ISS required a Soyuz 'lifeboat' to be available at all times. The operational lifetime of a Soyuz was six months, so they needed to be replaced at regular intervals. As with Mir, the Russians would send a passenger with professional cosmonauts flying as commander and flight engineer. This enabled the cash-strapped Russian space programme to take advantage of the space tourism dollars.

In July 2001, STS-104 delivered another vital component to the ISS on Assembly Flight 7A, when the Quest Joint Airlock was attached to the starboard side of the Unity node, opposite PMA-3. Prior to this, American spacewalks could be made only when a Shuttle was docked, as US spacesuits were too bulky to exit the Russian transfer chamber on the Zvezda module. Quest was 'joint' in the sense that it could facilitate both US and Russian spacewalks, and without a Shuttle being present. Two EVAs were conducted from Atlantis as part of the STS-104 mission, but the third EVA became the first to be made from Quest.

The peculiarly labelled Assembly Flight 7A.1 in August 2001 saw Discovery, STS-105 deliver the Expedition 3 crew to the ISS, unload and load the MPLM Leonardo, and bring the Expedition 2 crew home. During their week at the station the crew also performed EVAs to prepare for future missions, including installing the Early Ammonia Servicer with additional coolant for the station and installing heaters and handrails to Destiny in preparation for the later arrival of the S0 truss segment.

The configuration of the ISS during STS-105 (previous mission) showing Destiny and the Quest Joint Airlock. (NASA)

In September 2001, the Russian Pirs airlock was launched on board an unmanned Soyuz rocket and delivered by a modified Progress vehicle. When docked to the nadir port of Zvezda, Pirs could serve as an airlock for spacewalks and act as a docking port for Soyuz/Progress craft.

4.2 CLAUDIE HAIGNERÉ

Early Career

Born in 1957 in Le Creusot, France, Claudie Haigneré (née André-Deshays) studied medicine at the Faculté de Médecine (Paris-Cochin) and at the Faculté des Sciences (Paris-VII). She obtained certificates (CES Certificats d'Etudes Spécialisées) in biology and sports medicine in 1981, in aviation and space medicine in 1982, and in rheumatology in 1984. She was awarded a diploma (DEA Diplôme d'Etudes Approfondies) in biomechanics and physiology of movement in 1986, and finally received her PhD in neuroscience in 1992. From 1984 to 1992 she worked in the Rheumatology Clinic and the Rehabilitation Department and Sport Traumatology at Cochin Hospital in Paris. In parallel, from 1985 to 1990 she worked in the Neurosensory Physiology Laboratory at the Centre National de la Recherche Scientifique (CNRS) in Paris. Her special research interests were how the human motor and cognitive systems adapted to the state of weightlessness.

In 1985 CNES recruited its second group of astronaut candidates comprising of Claudie André-Deshays, Jean-François Clervoy, Jean-Jacques Favier, Jean-Pierre Haigneré (her future husband), Frédéric Patat, Michel Tognini and Michel Viso.

From 1990 to 1992 Claudie was responsible for French and international space physiology and medicine programmes in the CNES Life Sciences Division in Paris. As backup cosmonaut to Jean-Pierre Haigneré for the Franco-Russian Altair mission in July 1993, she monitored the biomedical experiments from the Mission Control Centre in Kaliningrad, near Moscow. From September 1993, Claudie was responsible for coordinating the scientific programme of the Franco-Russian Cassiopée mission and for the French experiments of the ESA EuroMir94 mission.

In December 1994 she was assigned to the Cassiopée mission as Research Cosmonaut and started training in Star City near Moscow on 1 January 1995. The 16 day mission took place from 17 August to 2 September 1996.

In 1997 Claudie worked in Moscow as the French representative of the Franco-Russian company Starsem. In May 1998 she was selected as backup for Jean-Pierre Haigneré for the Franco-Russian Perseus mission to Mir in February 1999. She trained for EVA and qualified as Cosmonaut Engineer for both the Soyuz spacecraft and the Mir space station. During that mission, she was crew interface coordinator at the Mission Control Centre near Moscow.

In July 1999 she became the first woman to qualify as a Soyuz Return Commander, enabling her to command a three-person Soyuz capsule during an emergency return from space.

On 1 November 1999 Claudie joined the European astronaut corps along with Philippe Perrin and Michel Tognini, when the remaining CNES astronaut team was amalgamated into the ESA corps. She participated in ESA development projects for the European Microgravity Facilities for Columbus and supported the medical activities in the Agency's Directorate of Manned Spaceflight and Microgravity. In January 2001, she started training in Star City near Moscow for her assignment for the Andromède mission.

Previous Mission

Soyuz TM-24 Cassiopée

Claudie André-Deshays (before she married fellow CNES astronaut Jean-Pierre Haigneré) undertook the Cassiopée mission to Mir in August 1996.

The primary objectives of the flight were to deliver Valeri Korzun and Alexandr Kaleri as part of Mir Expedition 22 and replace the Soyuz 'lifeboat'. She returned to Earth with Expedition 21's Yuri Onufriyenko and Yuri Usachev ("the two Yuris" in the Russian press) on Soyuz TM-23.

The Soyuz TM-24 crew with Claudie André-Deshays on the left. (www.spacefacts.de)

The Soyuz TM-33 crew with Claudie Haigneré on the right. (www.spacefacts.de)

Whilst aboard Mir for two weeks, André-Deshays carried out experiments in cardiovascular and neurosensory science. She recorded vibrations while the station hosted a larger than usual crew, and also performed a materials processing experiment.

4.3 THE ANDROMÈDE MISSION

Andromède Mission Patches

Claudie Haigneré chose artist Virginie Enl'art to design the mission patch. The name Andromède was derived from Greek mythology and was aptly the daughter of Cassiopeia, the name used for Claudie's previous mission to Mir. In that myth, Cassiopeia claimed that her daughter was more beautiful than Poseidon's sea nymphs, the Nereids. Poseidon chained Andromède to a rock as a sacrifice to the sea monster Cetus, but she was saved by Perseus.

The female figure on the patch bore a resemblance to a space traveller, but she was a cave dweller from the Tagus valley and represented 4,000 years of human progress. The chains that bound her were reminiscent of DNA. The symbols around the figure, whilst primitive, gave a hint of technology with images of satellite dishes and the space station. The gold crescent represented the Moon. The yellow sphere represented the Sun. The light and dark blues represented the transition from the sky within the atmosphere to the black

Andromède Mission Patch. (www.spacefacts.de)

of space, while the silver represented technology. The text listed the name of the mission and the organisations involved; namely the French space agency CNES, the Russian space agency Rosaviakosmos, and the rocket company RSC Energia.

The Soyuz TM-33 mission patch depicted the replacement of the Soyuz 'lifeboat'. To achieve this the crew delivered Soyuz TM-33 for the Expedition 3 crew to use in the event of evacuating the ISS, and themselves returned home in Soyuz TM-32, which had been delivered 6 months earlier. The flag colours represented Russia and France with red, white and blue in different orientations.

Andromède Mission Objectives

The Andromède mission had two objectives: To exchange the Soyuz 'lifeboat' at the ISS and to carry out a scientific and technical research programme organised by the French space agency CNES.

Soyuz TM-33 Mission Patch. (www.spacefacts.de)

Science

The following experiments were to be performed as part of the Andromède mission:

- IMMEDIAS
- Cardioscience
- COGNI
- EAC
- Aquarius
- GCF
- Spica-S
- F/PKE
- Mirsupio.

IMMEDIAS

Objective: Observation and recording of cloud formations, regions of natural and man-made atmospheric pollution, and areas on Earth with bad environmental conditions.

Tasks: Conducting visual observations and making a photographic record of specified areas in Europe, Asia, Africa, plus regions of current interest (forest fires, dust clouds in deserts, active volcanoes, smog over towns, etc.).

Equipment Used:

- Nikon F5 still camera and Nikon Coolpix 990 digital still camera with wide-angle lens WCE 63.
- EGE-2 unit for storing digital data from the PCMCIA card of the Nikon Coolpix 990, as well as a special atlas programme used for preliminary search of areas specified for survey and capable of displaying an Earth map.

Expected Results: Obtaining digital images and photographs on film of cloud formations, areas of natural and man-made atmospheric pollution, desertification areas, and other areas of Earth with bad environmental conditions.

Cardioscience

Objective: Research into changes occurring in the human cardiovascular system in the initial period of adaptation to weightlessness.

Tasks: Taking measurements to investigate vegetative regulation of arterial pressure and heartbeat rate.

Equipment Used: The PORTAPRES device for measuring finger blood pressure and the EGE-2 unit.

COGNI

This experiment was formally the Cognitive Process for 3-D Orientation Perception and Navigation in Weightlessness.

Objective: A study of a cosmonaut's perception of their orientation and movement in three-dimensional space under zero gravity.

Tasks:

- Gaining new knowledge about the problems of a cosmonaut determining his/her location whilst moving about the space station during the 3-D Navigation study.
- An evaluation of the effects that gravity has on a cosmonaut's perception and memorization of his/her orientation and location during the Visual Orientation study.

Equipment Used:

- Viewing hood/screen frame, lens tube with a mask, bracket, trackball, earphones, microphone, a small keyboard with a support, and three restraining straps.
- EGE-2 unit, power supply, cables, PCMCIA card and hard drive.

Expected Results: Obtaining inputs to permit a study of the individual characteristics of a cosmonaut's psychomotor functions in the spaceflight environment.

EAC

Objective: The application of new computer technologies to improve the psychological condition of a cosmonaut during a long-duration space mission.

Tasks: Activation of specific brain areas responsible for visual associations of the cosmonaut related to their home and family on Earth in order to further increase their capacity for work. Also the analysis of the cosmonaut's in-orbit condition by tests using special procedures.

Equipment Used: The EGE-2 unit with a hard drive containing a picture album and a questionnaire.

Aquarius

Objective: To study growth and development of biological objects under zero gravity. Specifically:

- A study of development of xenopus larvae (Xenopus laevis; a kind of frog) of different ages under exposure to spaceflight on the Soyuz spacecraft and the Russian segment of the ISS.
- A study of the effects of microgravity on otolith crystallisation at different phases during the development of Pleurodeles waltl larvae (a kind of newt or salamander).
- A study of cytoskeleton, secretion, chromosome migration using bakery yeast (Saccharomyces cerevisiae) as a model.

Tasks:

- Taking measurements of vestibular-ocular reflex of xenopus larvae.
- Conducting an analysis of the morphology and functioning of inner ear receptor cells, as well as the morphology of xenopus larvae otoliths.
- A study of the metabolism and enzymes of Pleurodeles waltl larvae which developed in space.
- A study of molecular mechanisms that determine the cell division polarization.

Equipment Used:

- Thermostatically controlled transportation container AQUA-1 to deliver biological samples to the ISS, store embryos in zero gravity, and then their fixation at certain stages of development.
- Transportation container AQUA-2 to support the return of the experiment results to Earth.
- DSR-PDIP camcorder.

Expected Results: Twelve dishes with live embryos, seven dishes with embryos held fixed at different phases in their development and one dish with bakery yeast, plus temperature/pressure data and video recording of embryo behaviour at the start and at the end of their stay on board the ISS.

GCF

The GCF experiment was a continuation of the studies that began on STS-95 when three experiments were performed using APCF crystallisation units to test a crystallisation technique that used counter-diffusion in zero gravity. The task now was to use the already developed crystallisation technique on new equipment, namely the Granada Crystallisation Facility unit.

Objective: Protein crystallisation in zero gravity to study spatio-temporal oversaturation patterns in the course of protein crystal nuclei formation and growth using the technique of counter-diffusion in gel.

Tasks: Developmental testing of the crystallisation process in the GCF unit with the counter-diffusion technique using X-ray capillaries as a protein chamber.

Equipment Used: Thermal insulation transport container GCF-1, plus the GCF-2 kit for returning the experiment results to Earth.

Spica-S

The Spica-S experiment was a continuation of the EXEC and SPICA experiments conducted on Mir. In order to improve the quality of the results, a new set of scientific equipment was developed to study the spectra of particles (electrons, protons, heavy ions). The results were of interest to Russian and French researchers.

Objective: A study of the effects of space radiation on last-generation (commercial available) off-the-shelf electronic components.

Tasks:

- Measuring the dynamic behaviour of electronic components.
- Measuring the radiation environment inside the Russian segment of the ISS during the course of the experiment.
- Obtaining the statistical data to improve evaluation models of the risks to which the components are subjected when exposed to space radiation.
- A feasibility study for the use of last-generation off-the-shelf components in spaceflight.

Equipment Used: SPICA-S equipment.

F/PKE

Objective: A study of the growth of plasma-dust structures in zero gravity. A study of particle cloud behaviour and internal flow structure in plasma-dust crystals.

Task: To conduct two series of measurements that were needed to study the physics of the particles contained in low-pressure plasma, and to study quasi-stable plasma structures formed out of these particles.

Equipment Used: Experimental unit Plasma Crystal-3, plus two TEAC video tape recorders that were included in the on-board Telescience equipment.

Expected Results: Video of the plasma crystal formation process, plus a PCMCIA card of digital data on the experiment parameters (gas pressure, high-frequency radiated power, the size of dust particles from which the crystal is formed, etc.).

Mirsupio

Objective: Developmental testing of auxiliary equipment to assure a cosmonaut's comfortable stay in space during long-duration missions.

Task: The use of a multipurpose bag in various situations and recording a qualitative assessment of its ease of use.

Equipment Used: Multipurpose Mirsupio bag.

Expected Results: Evaluation of how comfortable the bag was in use

Timeline

Training January to August 2001

Claudie Haigneré documented her training at the Gagarin Cosmonaut Training Centre at Star City in her diary on the ESA website.

It started with 200 hours of theory and classroom training on the Soyuz systems. "I have already been through four training periods at Star City, so this time the work was much easier. I was able to concentrate mainly on technical systems and on orbital dynamics – getting into orbit, manoeuvring, approach, docking, de-orbiting and descent." Next was approximately 150 hours of training on the Russian segment of the ISS. Her experience on Mir systems, on which some ISS modules are based, helped, allowing her to focus on the newer Russian components of the station. Once she passed her theoretical exams there was practical training and simulation. Star City has both Soyuz and Russian ISS module simulators.

From June onward, the Andromède Soyuz TM-33 crew trained together and they spent a week at CNES in Toulouse training on the scientific experiments.

Back at Star City, she trained in the Soyuz simulator practising flight manoeuvres and learning the best response to a variety of 'non-nominal' situations. This consisted of 15 sessions of four hours each. After that there were another 15 sessions on docking and performing a manual re-entry.

"Right from the beginning, we've also had straightforward physical training, two or three times a week," wrote Claudie. "There are also medical examinations every three months. The first was when we started training, the second to confirm the crew choice, and the final one will be a few weeks before launch."

Sixty days prior to the scheduled launch, Claudie was examined by the medical team for the Basic Data Collection (BDC) to establish a baseline. Cosmonauts undergo this twice; firstly at 60 days and then at 30 days before launch. This baseline defines reference points against which the physiological changes resulting from the mission can be measured.

In August, Roberto Vittori and Frank de Winne arrived at Star City to begin training as prime and backup respectively for the Marco Polo mission on Soyuz TM-34.

Training September 2001

September saw a change of location for Claudie as she began her training on the American modules of the ISS at NASA's Johnson Space Center in Houston, Texas. She wrote in her blog, "Theory is never enough. For real confidence, you have to see and touch. Even though we will work almost exclusively in the Russian section, we will have the run of the whole station and we have to know our way around. It was also good to get to know the ground controllers and the CapComs, the people we'll be talking to from orbit. And to meet astronauts from America, Europe, Canada and Japan: it brings home just how international the project is."

On meeting other ESA trainees: "Scattered between Russia, Germany, the Netherlands and the USA, we don't often have a chance to get together. Even though we're all in the European Corps of Astronauts, our lives are very different. It was good to meet again in a relaxed way, talk to people's wives and husbands. And when you see how big their children have grown, you realise how quickly time passes."

As September drew to a close: "We're now in the last stages of preparation. It's quite intense. So that we will be ready for the last week in September, we have an exam practically every day. For the Soyuz, there are tests on manual piloting for approach, docking and descent for the whole crew. There are exams on every part of the flight and aboard the station. An exam on a typical day's work. Medical exams, too. After a series of tests in September, the GMK (the Russian medical commission) will give us our final flight clearance. We've just been trying out our spacesuits for size and checking for leaks. Everything is ready."

Training October 2001

Claudie and the other crewmembers had to pass a series of flight qualification exams on the operation of the Soyuz spacecraft.

1) Manual approach: Simulation of the last 5 km of the approach to docking. If the automatic Kurs system breaks down, Claudie, as the flight engineer, needs to set up and operate the backup laser system and call out the speed and distance while the commander carries out the approach under manual control of the thrusters.
2) Manual docking: Claudie watches the approach and keeps time as the commander docks.
3) Manual re-entry: In the event that the automatic system fails, the commander or engineer has to adjust the roll rate of the capsule manually to control the trajectory and deceleration force. They are expected to land within 10 km of the target and experience no more than 4g of deceleration.
4) Full mission simulation: Claudie wrote, "That involves just about everything. We had to go through every step of the mission. Orbital insertion; the first flight checks; orbital correction manoeuvres; approach; docking; undocking; descent. It took about ten hours, in our spacesuits, watched over by a team of experts who took turns devising more and more complex breakdowns and problems to throw at us. We came out of the simulator haggard and exhausted. But we still had a full debriefing session ahead of us where those same experts made us explain and justify every action we had taken."

These exams were conducted in a centrifuge. All the tests were passed successfully. After being passed medically fit, the crew were able to take a few days off and relax with their families.

Monday, 15 October 2001

On 9 October the prime and backup crews flew (on separate aircraft) from Chkalovski military airport to the Baikonur Cosmodrome in Kazakhstan.

After a welcome from Russian military and RKK Energia dignitaries the two crews were driven on separate buses to the Cosmonaut Hotel, a blue bus for the prime crew and a yellow one for the backup crew.

As Claudie wrote, "I have been coming to Baikonur for ten years now, and these are still the same buses. Somehow, these familiar rituals are deeply comforting. I appreciate the constancy. I know this place. That road leads to the launch pads. This road goes to the Cosmonaut Hotel. There are the camels, dromedaries really, a feature of the Kazakh steppe. There's the usual militia truck driving ahead of us. Over there are the communication relay towers that will track our lift off. All of this is familiar to me, and I can feel that almost imperceptibly I am slipping into another universe."

On 10 October the crew checked out the Soyuz TM-33 spacecraft that would take them to the ISS. This involved entering the capsule wearing the actual spacesuits and checking whether they could fit comfortably and work in that environment. Firstly the seats were raised to the landing position, 20 cm higher than launch position to allow for landing shock absorbers. They checked that there was enough clearance between their knees and the control panels.

Claudie wrote, "The space inside the fully-loaded capsule is so cramped that we almost feel there has been some mistake. There's no way all three of us will fit in here. But we do, somehow. We have to pay very careful attention to all the projecting pieces of equipment, since we don't want to damage anything or to tear a hole in our spacesuits. It's a very necessary test but it isn't easy because we have to combine extreme care with the physical abilities of a contortionist."

With the seats placed in the launch position, they checked that they could reach all of the controls and that their visibility was acceptable, then practised finding a comfortable position in the cramped cockpit.

They returned to Star City for three days to pack and prepare, and then returned to Baikonur on 16 October for the real thing.

Flight Day minus 1 – Saturday, 20 October 2001

At a press conference held at Baikonur on 20 October, Claudie Haigneré gave a gift on behalf of ESA to each of her crewmembers, a hand-painted matrioschka doll (made by the Moscow-based company Serena Technology Ltd.) which depicted Soyuz 'taxi' Commander Victor Afanasiev, Flight Engineer Konstantin Kozeyev, and herself.

Flight Day 1 – Sunday, 21 October 2001

Soyuz TM-33 launched on schedule at 10:59 CET. As was traditional the crew took toys with them in the capsule. These served as good luck charms and indicated the onset of weightlessness by beginning to float freely. Claudie Haigneré had taken a teddy bear as well as books and photographs, including one of her daughter. At 11:08 Claudie reported to the Mission Control Centre, "Thank you to all, we are now in orbit, everything went very well."

Shortly thereafter the solar arrays were deployed. The Andromède mission to the ISS was under way.

Flight Day 2 – Monday, 22 October 2001

ESA was represented at the Mission Control Centre (TsUP) in Moscow by astronaut Reinhold Ewald. As he explained, "Actually the programme that the crew performs in the Soyuz capsule after the first couple of orbits is not too crowded. It takes two days of orbital mechanics to chase the ISS before the docking. Most of this time, the crew spends checking the systems of the space vehicle and resting. The capsule mostly flies in an orientation towards the Sun to recharge the batteries. You can feel the spin as a slight centrifugal force constantly tugging at you in the otherwise forceless environment. The pace of work picks up when in the 34th orbit, after two days, they initiate the needed corrections of the orbit to approach and finally dock with the space station.

"Radio contact and some television transmission of data with the TsUP is only possible when the Soyuz is passing over Russian territory. For each orbit there are only 10–20 minutes to check how the flight is progressing. But seven or more of the 16 daily orbits around Earth are without radio contact. The crew takes a rest or tries to sleep in the cramped environment.

"After the two days in the Soyuz, you are longing for a more spacious environment and most of all a hot tea and a good meal. In the Soyuz, you only have cold canned food and tepid water to drink. The moments before the docking are very tense, although the crew has rehearsed this moment many times on the ground. Here Claudie, as a flight engineer, will serve a pivotal role in helping Victor Afanasiev to control the automatic approach, and in the event of any deviation, to switch to manual mode for the last minutes. This, again, will have taken them hours of training to quickly interact with each other in this critical situation."

Flight Day 3 – Tuesday, 23 October 2001

Soyuz TM-33 docked with the nadir port of the Zarya module at 12:44 CEST on Tuesday 23 October 2001. Ninety minutes later, Claudie was the first of the newcomers to enter the station, becoming the first European woman to board the ISS. French Prime Minister Lionel Jospin was present at the TsUP in Moscow to witness the historic moment and he sent his congratulations.

The Soyuz TM-33 crew transferred cargo to the station and relocated their seat liners to the Soyuz TM-32 spacecraft which had arrived in May and would take them home. Claudie set up her sleeping quarters in the Quest Joint Airlock. According to Ewald, "Before taking a well-earned night's rest, Claudie will set up the Spica experiment. This experiment is meant to test electronic equipment under space conditions. Aquarius, the biological experiment will also be serviced. Her colleague Konstantin Kozeyev, flight engineer number 2 of the Soyuz spacecraft, will prepare tomorrow's run of the Plasma Kristall experiment."

Flight Day 4 – Wednesday, 24 October 2001

Claudie started work on the COGNI experiment, designed to better understand how the brain uses gravity in the process of perceiving and representing three-dimensional space. All three Andromède crewmembers were involved in this experiment, in which they looked at computer-generated images and then said what they had seen. The results were to be compared to measurements obtained on the ground prior to the flight. A problem with a keyboard hampered the operation of this experiment.

The Aquarius and Cardioscience experiments were also conducted. Aquarius explored the earliest development of amphibians and yeast in weightlessness. The Cardioscience experiment investigated the cardiovascular deconditioning phenomenon under such conditions.

COGNI and Aquarius were performed in conjunction with school children and university students as part of an educational project.

Flight Day 5 – Thursday, 25 October 2001

For part of the mission Claudie worked with students, and on this day she spoke with French Minister for Research, Roger-Gérard Schwartzenberg. She told him that she was particularly keen to encourage younger children to explore new worlds such as that of weightlessness in space.

Claudie investigated the fault with the COGNI keyboard and tested the Mirsupio bag. Previously carried aboard the Franco-Russian Perseus flight in 1999, the Mirsupio bag was produced in response to problems encountered when working in microgravity. Elastic pockets made it possible to store and retrieve small objects without using a zip or any other fastening system. A transparent flap allowed an astronaut to see what was inside each pocket.

Claudie was interviewed by French TV and explained that her experiences on Mir in 1996 were helping adjust to weightlessness this time.

Flight Day 6 – Friday, 26 October 2001

With the keyboard problem rectified, the COGNI experiment was able to get properly underway. The Plasma Kristall Experiment (PKE) was yielding impressive images showing patterns of macroscopic particles in plasma.

On this day, the International Space Station was awarded the 2001 Prince of Asturias Award for International Cooperation. The combined Andromède and Expedition 3 crew spoke with His Royal Highness Don Felipe of Spain from an award ceremony in Oviedo, the capital of the Principality of Asturias in northern Spain, which was attended by the ISS partner space agencies and Spain's own astronaut Pedro Duque.

Flight Day 7 – Saturday, 27 October 2001

Claudie participated in a Q&A with students at the Cité de L'Espace in Toulouse. As it was a rest day, she carried out some light duties. This included data backups, reconfiguration of the LSO experiment, an Earth observation experiment that looked for sprites in the higher atmosphere (flashes of light from electrical discharges between thunderstorms and the ionosphere; in effect, upward lightning bolts), and monitoring the PKE. She also spoke with fellow French ESA astronaut Michel Tognini serving as the CapCom at Mission Control in Houston.

Flight Day 8 – Sunday, 28 October 2001

On Sunday morning Claudie hosted a guided tour through the ISS, using a recording she had made the previous day with her crewmate Victor Afanasiev. The video showed impressive scenes, starting with a view of the Soyuz attached to the station with Canadarm2 in the background, through the modules of the Russian segment and the Unity node into the Destiny module and finally into the Quest airlock on the side of Unity, where she had set up her living and sleeping facilities.

She executed the penultimate Aquarius experiment run and performed both parts of the COGNI experiment, then answered questions on the Plasma Kristall Experiment in a space-to-ground radio session.

Flight Day 9 – Monday, 29 October 2001

In wrapping up Claudie's experimental work, the Cardioscience run had to be curtailed due to low battery power. The complete data set for the PKE was saved for analysis by scientists following her return to Earth.

Flight Day 10 – Tuesday, 30 October 2001

Alain Labarthe, Head of the Andromède project at CNES, spoke to Claudie offering thanks and best wishes for the return trip.

Reinhold Ewald in the TsUP offered his praise, "On behalf of EAC and the European astronauts, I could add that Claudie's performance as flight engineer of the Soyuz spacecraft and her role as scientist on board Space Station has set milestones for future undertakings. Her exemplary status as a European woman astronaut will encourage young women to choose a career in science, and perhaps also in space flight. In this she has truly represented the ideals laid down in the European Astronaut Charter."

Flight Day 11 – Wednesday, 31 October 2001

The Andromède crew boarded Soyuz TM-32 on Tuesday 30 October, taking with them the GCF in which protein crystals had been grown and the Aquarius containers with their frog and salamander larvae. The hatches were closed at 23:37 CET, and the undocking was at 02:39 CET on 31 October. The capsule landed 180 km from Dheskasgan in northern Kazakhstan at 05:59:26 CET.

Claudie Haigneré in 2014. (Wikipedia)

4.4 POSTSCRIPT

Subsequent Missions

Andromède was Claudie Haigneré's final spaceflight.

Claudie Haigneré Today

Claudie has been awarded the following honours:

- Honorary Member of the Société Francaise de Médecine Aéronautique et Spatiale Corresponding Member of the International Academy of Astronautics (IAA).

- Honorary Member of the Association Aéronautique et Astronautique de France (AAAF), Member of the Académie de l'Air et de l'Espace (ANAE).
- Member of the French Académie des Technologies.
- Member of the French Académie des Sports.
- Member of the French Académie des Sciences de l'outre mer.
- Member of the Belgian Académie des Sciences et Techniques.
- Docteur Honoris Causa of the École polytechnique fédérale de Lausanne, Switzerland.
- Docteur Honoris Causa de la Faculté de Mons (Belgique).
- Professeur Honoris Causa of the University of Beihang in Peking, China.
- Patron of the Cité de l'Espace in Toulouse.
- Patron of the Institut de Myologie de la Pitié-Salpétrière of the Association Française contre les Myopathies (AFM).
- Grand Officier de l'Ordre de la Légion d'Honneur and Chevalier de l'Ordre National du Mérite.
- Russia's Order of Friendship.
- Russia's Medal for Personal Valour.
- Commander of the Order of Merit of the Republic of Germany.
- France's Médaille de l'Aéronautique.
- The Grand Siècle Laurent Perrier Prize (1996).
- The Henry Deutsch de la Meurthe Prize from the Académie des Sports (1998).
- The Louise Weiss Prize (2006).

On entering French politics, from June 2002 to April 2004 Claudie served as Minister for Research and New Technologies, and from April 2004 to June 2005 she served as Minister for European Affairs and Secretary General for Franco-German Cooperation. From November 2005 to September 2009 she was Senior Adviser to ESA's Director General in European space policy. And then from October 2009 to February 2015 she was the CEO of Universcience, a French science museum that combined the Cité des Sciences and the Palais de la Découverte in Paris. In addition, Claudie is part of the innovation jury (Concours Mondial de l'innovation), and is now chairing the scientific council of the Chair du Collège des Bernardins (2014–2016) in Paris entitled Digital humanism. She has held directorships of France Telecom and Sanofi-Aventis, and has served on the board of Airbus, Loreal and Fondation de France.

In February 2015 she returned to ESA as an adviser to the Director General.

5

Marco Polo

Mission

ESA Mission Name:	Marco Polo
Astronaut:	Roberto Vittori
Mission Duration:	9 days, 21 hours, 25 minutes
Mission Sponsors:	ASI/ESA
ISS Milestones:	ISS 4S, 17th crewed mission to the ISS

Launch

Launch Date/Time:	25 April 2002, 06:26 UTC
Launch Site:	Pad 1, Baikonur Cosmodrome, Kazakhstan
Launch Vehicle:	Soyuz TM
Launch Mission:	Soyuz TM-34
Launch Vehicle Crew:	Yuri Pavlovich Gidzenko (RKA), CDR Roberto Vittori (ASI/ESA), Flight Engineer Mark Richard Shuttleworth (NASA), Spaceflight Participant

Docking

Soyuz TM-34	
Docking Date/Time:	27 April 2002, 07:55 UTC
Docking Port:	Zarya Nadir
Soyuz TM-33	
Undocking Date/Time:	5 May 2002, 00:31 UTC
Docking Port:	Pirs

J. O'Sullivan, *In the Footsteps of Columbus*, Springer Praxis Books,
DOI 10.1007/978-3-319-27562-8_5

Landing

Landing Date/Time:	5 May 2002, 03:51 UTC
Landing Site:	Near Arkalyk, Kazakhstan
Landing Vehicle:	Soyuz TM
Landing Mission:	Soyuz TM-33
Landing Vehicle Crew:	Yuri Pavlovich Gidzenko (RKA), CDR
	Roberto Vittori (ASI/ESA), Flight Engineer
	Mark Richard Shuttleworth (NASA), Spaceflight Participant

ISS Expeditions

ISS Expedition:	Expedition 4
ISS Crew:	Yuri Ivanovich Onufriyenko (RKA), ISS-CDR
	Carl Erwin Walz (NASA), ISS-Flight Engineer 1
	Daniel Wheeler Bursch (NASA), ISS-Flight Engineer 2

5.1 THE ISS STORY SO FAR

Since Soyuz TMA-33 there had been two Space Shuttle missions to the ISS with different objectives.

First, in December 2001 STS-108, Utilisation Flight UF1, delivered cargo using MPLM Raffaello, delivered the Expedition 4 crew, and returned the Expedition 3 crew to Earth. A single EVA from the Shuttle's airlock installed insulation blankets on the rotating mechanisms of the solar arrays and also executed 'get ahead' tasks in preparation for future missions. The STS-108 crew also delivered flags from the three sites of the 9/11 terrorist attacks.

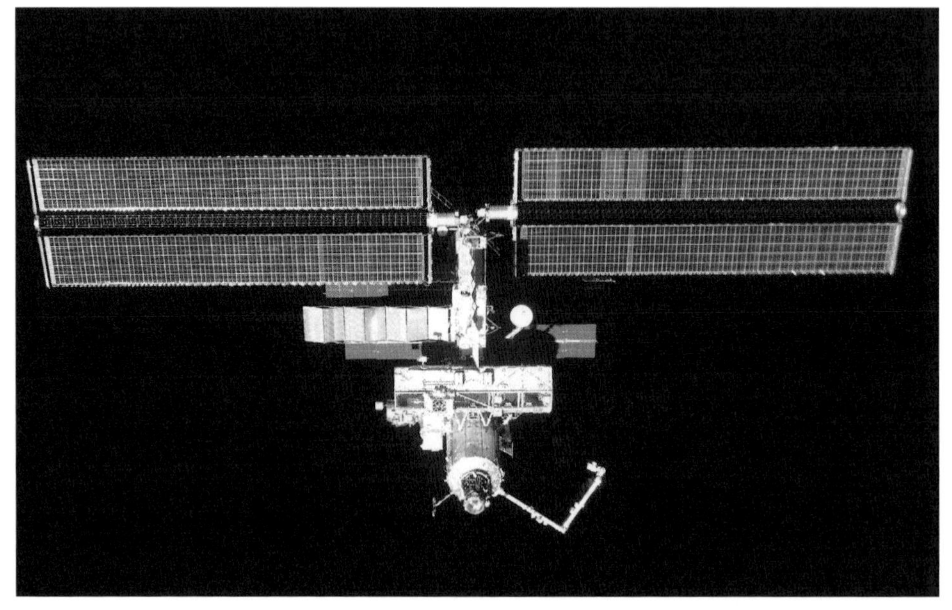

The configuration of the ISS during STS-110 (previous mission) with the S0 truss on the Destiny module. (NASA)

Then in April 2002, STS-110, Assembly Flight 8A, delivered the S0 truss and mounted it onto the Destiny module as the central section of what, in time, would become the backbone of the solar array truss system. In four spacewalks by two pairs of astronauts, the S0 and the Mobile Transporter (MT) were installed. The MT would enable Canadarm2 to operate from up to 10 different work sites. This mission saw Jerry Ross become the first person to launch into space seven times, and also exceed the American spacewalk record after spending 58 hours and 18 minutes on nine EVAs accumulated over four missions.

5.2 ROBERTO VITTORI

Early Career

Roberto Vittori was born in Viterbo, Italy in 1964. He attended Italy's Aeronautics Academy, from which he graduated in 1989, and then he joined the Italian Air Force where he flew the Tornado GR1 with the 155th Squadron, 50th Wing, Piacenza, Italy from 1991 to 1994. In 1992 he was promoted to Squadron Commander.

Roberto Vittori. (ESA)

In 1995 he attended the US Navy Test Pilot School at Patuxent River. In 1997 he attended the US Air Force Flight Safety School and from 1997 to 1998 he served as Flight Safety Officer at the Italian Experimental Flight Section (Reparto Sperimentale di Volo). He was also a teacher of aerodynamics for the Italian Air Force's Accident Investigation Course, and project pilot for the Eurofighter. From 1996 to 1998 he was the Italian representative in the Beyond Visual Range Air-to-Air Missile research and development programme. It was while flying there that he heard in 1998 that ASI were looking for astronaut candidates.

Vittori applied and his application was successful. In one year he joined ASI, the ESA astronaut corps, and NASA's Astronaut Group 17; the latter being known as the Penguins. He was relocated to NASA's Johnson Space Center in Houston, Texas and entered the 1998 astronaut class in training for the Space Shuttle and the International Space Station. After completing his Mission Specialist training he performed technical duties in the Space Shuttle Operations Systems Branch, and Robotics, Future Vehicle and International Space Station Branch. In August 2001 he took up training as flight engineer at the Gagarin Cosmonaut Training Centre in Star City, Russia, in preparation for his first spaceflight on Soyuz TM-34 to the ISS for the Marco Polo mission. Returning to Houston, he supported the New Generation Space Vehicles Branch. After Space Shuttle Columbia was lost in 2003, Roberto served in the Tiger Teams accident investigation.

In October 2004 Roberto resumed training at Star City for his second mission to the International Space Station on board Soyuz TMA-6 for the Eneide mission.

During his career Vittori:

- Graduated from the Italian Air Force Academy in 1989 with a degree in Aeronautical Science.
- Completed basic training with the US Air Force at Reese Air Force Base in Texas in 1990.
- Graduated from the US Navy Test Pilot School in 1995.
- Completed the Italian Air Force's Accident Prevention course (Guidonia AFB, Italy) and Accident Investigation course (Kirtland AFB, New Mexico, US) between 1996 and 1997.
- Graduated from the NATO Defence College Senior Course 108 in 2006.
- Completed a master's degree in physics in December 2007.

Previous Missions

N/A

5.3 THE MARCO POLO MISSION

Marco Polo Mission Patches

The very simple design showed the Soyuz approaching the station from the viewpoint of the ISS. The mission title 'Marco Polo' was chosen to honour the 13th century Italian merchant traveller, and was featured in large text in the middle of the patch. The logos of the four participating agencies were also displayed, namely Agenezia Spaziale Italiana,

The Soyuz TM-34 crew with Roberto Vittori on the right. (www.spacefacts.de)

Italian Space Agency (ASI); Rosaviakosmos, Russian Aviation and Space Agency; European Space Agency (ESA); and Energia, which is the Russian S. P. Korolev Rocket and Space Corporation.

The Soyuz TM-34 mission patch was dominated by the ISS, which was depicted in the originally planned final configuration with the cancelled Russian Science Power Platform. The primary mission objective was to replace the Soyuz 'lifeboat' at the ISS so both the new spacecraft Soyuz TM-34 and the return vehicle Soyuz TM-33 were listed, as were the names of the crew heading for the station.

Marco Polo Mission Objectives

The Marco Polo mission had two objectives. To exchange the Soyuz 'lifeboat' at the ISS and to carry out a scientific and technical research programme organised by ASI and ESA.

Science

The following experiments were to be carried out as part of the Marco Polo mission:

- CHIRO
- ALTEINO
- VEST
- BMI.

Marco Polo Mission Patch. (www.spacefacts.de)

CHIRO

Objective: To study the effects of microgravity on the mechanism of muscular contraction and the working capacity of a cosmonaut's hand in flight.

Tasks: The measurement of isometric strength of the hand, as well as finger grip strength during performance of strictly regulated exercises with hand and finger dynamometers in order to assess possible deterioration of a cosmonaut's working capacity.

Equipment Used:

- Hand dynamometer HGD, finger dynamometer PFD, coupler unit SCU, data and power cables.
- EGE laptop computer, DSR PDIP camcorder, PCMCIA memory cards and Mini-DV cassettes to record the data from the experiment and then be returned to Earth for analysis.

Expected Results: The data needed to improve standard procedures for evaluating the working capacity of ISS crews in space, and to facilitate the development of measures to prevent adverse effects on crew health and working capacity.

Soyuz TM-34 Mission Patch. (www.spacefacts.de)

ALTEINO

Objective: To study the mechanisms and effects of heavy charged particles in cosmic radiation and the phosphenes (flashes of light) that they create on visual analysers and on the central nervous system of a cosmonaut, affecting his capacities as an operator in the space environment.

Tasks: Radiology studies in the field of space particle physics, plus electrophysiology studies of the level of total effects of spaceflight on the functional state of a cosmonaut's central nervous system.

Equipment Used:

- AST spectrometer to measure angle, charge, and energy properties of the heavy ions (nuclei) in cosmic radiation.
- Halley Package (a portable electroencephalograph) for pick-up, amplification, and analogue-to-digital conversion of electrophysiological signals (up to 16 channels), recording and storing data on PCMCIA cards. It comprised the Halley device, a Dictaphone, an event indicator, a mask and a belt.
- ALTEINO-PM Kit of a headpiece for an electroencephalograph, plus batteries and electrodes.
- PCMCIA memory cards and MC-60 mini-cassettes for the Dictaphone.

Expected Results: Obtaining data on the radiation safety of cosmonauts during long-duration and long-range missions, as well as the dynamics of their central nervous system functional state and variations in the level of their capacity as operators under these conditions. More precisely determining the level of radiation protection in the working compartments of the Russian segment of the ISS.

VEST

Objective: To evaluate a new integrated system of garments made of different types of materials for use in the space environment.

Tasks:

- Wearing a VEST garment by Roberto Vittori specially designed for the mission in the Russian segment of the ISS. He was provided with three types of the VEST personal garments for use during different types of activity: one to be worn throughout his working day, a second during exercise (fitness) cycles utilising standard exercise hardware (the VB-3 stationary bicycle and dynamometers), and a third while sleeping.
- Obtaining the cosmonaut's opinion about his psychological and physiological well-being, the wearability of the garment, its aesthetic qualities, efficiency of thermal stability, and physical hygiene on board the station.

Equipment Used:

- VEST Kit for delivery and stowage on board the Russian segment of the ISS, plus stationary bicycle VB-3, dynamometers, a DSR-PD1P camcorder and video-cassettes, and a Nikon F5 camera and 35-mm film.
- VEST Kit for return to Earth.

Expected Results: Verifying the functionality of the new integrated garment system VEST, including its ergonomic qualities in spaceflight to facilitate a reduction in both the mass and volume of clothing intended to be used on long-duration space missions to the ISS.

BMI

Objective: To demonstrate the serviceability of the Arterial Pressure Meter in the space environment.

Tasks: Taking measurements of a cosmonaut's arterial pressure and heart rate.

Equipment Used:

- Arterial Pressure Meter Kit consisting of an arterial pressure measuring unit, a cuff, a cuff case, an attachment belt, an adhesive plaster, and a return bag.
- Camcorder DSP-PD1P and video-cassette.
- Batteries.

Expected Results: Demonstration of the ability of the portable device to measure a cosmonaut's arterial pressure and heart rate during miscellaneous activities in the space environment. If successful, it would be employed by future medical experiments and/or standard medical examinations.

Timeline

Training March April 2002

During his training in preparation for the launch of Soyuz TM-34, Roberto Vittori wrote a diary for the ESA website. In March 2002, at Star City, near Moscow, he wrote, "Right now, most of our training is concentrated on the Soyuz spacecraft itself. We spend many hours a week in the simulator, learning to deal with any kind of emergency. And of course we have to train as a crew, not just three individuals."

About his 'space tourist' crewmate, South African software millionaire Mark Shuttleworth, he said, "Mark is a pleasure to work with. He's a great guy: young (just 29) enthusiastic and he has been going through some training with Yuri and I. That's the difference between professional astronauts and the rest. Mark has certainly done the necessary mission training, but he doesn't have the same experience as Yuri and myself. This will be my first space flight, but I have been preparing for it for the last four years. In fact, considering when I started in aviation, you could say I have been preparing for fifteen years."

After the final exams on Monday, 8 April 2002, Vittori was in Star City under quarantine awaiting the scheduled launch date of 25 April. When asked about quarantine, Vittori said, "We are not really sealed off. It's simply common sense. We don't want anyone to catch a cold or a dose of influenza in the last few days. So we stay out of the way of most of the rest of the base personnel."

Flight Day 1 – Thursday, 25 April 2002

After suiting up and checking suit functionality, the crew participated in a press conference. Then they were driven by bus to the launch pad. At 06:00 CEST they entered the Soyuz TM-34 capsule and then lifted off on schedule at 08:26 for a 2 day pursuit of the ISS. This was the final flight of the TM model of the spacecraft because the improved TMA would soon be introduced with a 'glass cockpit' and the ability to accommodate heavier and taller astronauts.

Vittori's brief report on the first orbit was, "I'm feeling fine." He helped the commander to verify the functionality of the life support systems of the spacecraft and its docking systems. Once these tests were completed, the crew removed their uncomfortable spacesuits and opened the hatch to the orbital module.

The initial orbit was an elliptical one with an apogee of 216 km and a perigee of 189 km, but they soon made a main engine burn to set up the orbit from which they would initiate the rendezvous with the ISS.

The Marco Polo mission involved a substantial ground element. ESA astronauts Paolo Nespoli and Reinhold Ewald shared the duties of Crew Interface, handling communications with Vittori in space. The mission managers were Fabio Bracciaferri and Simonetta Di Pippo, and the science coordinator was Maria Kristina Falvella working at the Russian Mission Control Centre (TsUP) near Moscow on behalf of ASI.

Flight Day 2 – Friday, 26 April 2002

A thankfully uneventful day was spent correcting the orbit to catch the ISS and reporting their status to the TsUP. The orbit correction on the 17th revolution went as planned. Before sleeping, the crew spoke with their families. Vittori confirmed to his wife that he was feeling good. She told him their two sons had enjoyed watching the launch.

Flight Day 3 – Saturday, 27 April 2002

At 09:55 CEST, Soyuz TM-34 made a fully automatic docking on the nadir port of the Zarya module. After opening the hatches, they were greeted by the Expedition 4 crew and welcomed aboard the ISS.

Later that morning, the combined crew were addressed by Gianfranco Facco-Bonnetti, the Italian Ambassador to Russia, and by Jörg Feustel-Büechl, the Director of the ESA Human Spaceflight and Microgravity programme.

Vittori was congratulated as the first Italian to reach the ISS as a cosmonaut in a Soyuz spacecraft, and the cooperation between ESA, ASI, and the Russians was celebrated. This would be the start of a long and fruitful collaboration between ASI and all of the partners in the ISS because Italy fabricated some of the station modules and Italian astronauts and cosmonauts played a vital role in the assembly and operation of the station and the execution of its science programme.

Prior to going to sleep, Vittori set up the ALTEINO radiation spectrometer.

Roberto Vittori moves through the Zvezda module. (NASA)

Flight Day 4 – Sunday, 28 April 2002

As the previous weeks had seen the arrival and departure of STS-110 and the transfer on April 20 of Soyuz TM-33 from the nadir port of Zarya to the Pirs module to clear the way for Soyuz TM-34, the Expedition 4 crew took a day of rest. But for Vittori it was the first full day of his ambitious science programme.

He continued the ALTEINO experiment, using himself as a subject. He measured the effects of radiation on the electrical activity of the brain utilising the Halley device, a cap with skin electrodes, and a registration unit. He began the CHIRO experiment, which measured the force of handgrips in order to record the changes of muscular and neurological activity in weightlessness. As he carried out his various tasks, he evaluated the VEST garments. And whenever the opportunity presented itself he took photographs of Italy passing below.

Flight Day 5 – Monday, 29 April 2002

The ALTEINO experiment continued, with two sessions using the electroencephalograph (EEG), a device which registers the electrical activity of the brain. Vittori also started a second 48 hour run for the radiation spectrometer.

Gidzenko operated the Plasma Kristall Experiment, a Russian-German-French cooperation which investigated transient order phenomena of dust particles in a gas environment that could be observed only in weightlessness.

Flight Day 6 – Tuesday, 30 April 2002

This day saw another televised linkup with more VIPs for the combined crew. This time the link with Brussels enabled the astronauts to field questions posed by Romano Prodi, the European Commission President, and Antonio Rodotà, the Director General of ESA.

When asked how the mission was going so far, Yuri Gidzenko replied, "We have a lot of things to do. The experiments are going well. Everything is going according to schedule."

When Rodotà asked Vittori about living and working on the ISS, he replied, "I am impressed with the technology on the Space Station. I know this is the result of years of work by engineers from all over the world and also from Europe. ESA's contribution is huge – it is very evident here today. I am very, very excited about working here. Every day there is something new to learn, it's an unbelievable experience."

Later Vittori spoke with school children in Rome hosted by ESA astronaut Umberto Guidoni, his predecessor who was the first European to visit the ISS in April 2001. He gave a televised tour of the station and the two docked Soyuz spacecraft.

He also worked on the VEST and CHIRO experiments.

Flight Day 7 – Wednesday, 1 May 2002

The day was spent mainly preparing the Soyuz TM-33 spacecraft for return to Earth. The ALTEINO experiment was considered a success, so planned work was cancelled in order to allow Vittori more time to pack.

Fifty kilograms of payload, including 15 kg of ASI/ESA experiment material were stowed in the cramped Soyuz capsule.

Flight Day 8 – Thursday, 2 May 2002

This was another day of VIP calls. Mark Shuttleworth gave former South African President Nelson Mandela, known as 'Madiba', a description of his work on board the station. One orbit later, Vittori chatted by television with Pier Ferdinando Casini, President of the Camera dei Deputati (the Italian Chamber of Deputies), and with Sergio Vetrella, President of the ASI, thanking them for their good wishes and showing them around the ISS.

When a technical issue arose with the data storage of the CHIRO computer, the ESA/ASI team at the TsUP set to work on a solution.

Flight Day 9 – Friday, 3 May 2002

This was another day spent packing the Soyuz TM-33 capsule. Some experiments, such as VEST, were complete and ready for packing. After further investigation of its computer the CHIRO experiment was packed. ALTEINO, which had been the first experiment to be started, was the last to be shut down and packed.

In the afternoon, the cosmonauts received a newly written improvised procedure from the TsUP to guarantee that the CHIRO data would be available for return to Earth.

Flight Day 10 – Saturday, 4 May 2002

Vittori and his two crewmates undocked Soyuz TM-33 from the Pirs module at 02:28 CEST.

Flight Day 11 – Sunday, 5 May 2002

The Marco Polo crew safely descended to Earth, ending a 10 day mission with a textbook touchdown near Arkalyk on the steppes of Kazakhstan at 10:55 a.m. local time.

5.4 POSTSCRIPT

Subsequent Missions

See the Eneide and DAMA missions.

Roberto Vittori Today

Roberto Vittori has received the following honours and awards:

- Academic award at the Undergraduate Pilot Training, Reese Air Force Base, Texas, USA.
- Honour student at the Test Pilot School, Patuxent River, Maryland, USA.

- Honour student at the Flight Safety School, Kirtland Air Force Base, New Mexico, USA.
- Italian Air Force Long Service Medal (1997).
- Gold Medal to the Aeronautical Value awarded by the President of the Italian Republic (2002).
- Special recognition as Commendatore della Repubblica awarded by the President of the Italian Republic (2005).

Roberto Vittori (on the right) at the 55th Anniversary of the Frecce Triclore, the Italian Air Force aerobatic display team. Umberto Guidoni is on the left and Franco Malerba is in the centre. (www.uspa24.com)

Between February 2006 and August 2008, Vittori was detached to the Italian Air Force under an agreement with ESA, during which he served on the Board of ASI's Technical Scientific Committee.

Vittori is currently serving as Space Attaché at the Embassy of Italy and heads the ASI Office in Washington DC.

6

STS-111

Mission

ESA Mission Name:	None (NASA Mission STS-111)
Astronaut:	Philippe Perrin
Mission Duration:	13 days, 20 hours, 35 minutes
Mission Sponsors:	CNES/NASA
ISS Milestones:	ISS UF2, 18th crewed mission to the ISS

Launch

Launch Date/Time:	5 June 2002, 21:22 UTC
Launch Site:	Pad 39-A, Kennedy Space Center
Launch Vehicle:	Space Shuttle Endeavour (OV-105)
Launch Mission:	STS-111
Launch Vehicle Crew:	Kenneth Dale 'Taco' Cockrell (NASA), CDR
	Paul Scott 'Paco' Lockhart (NASA), PLT
	Franklin Ramon Chang-Diaz (NASA), MSP1
	Philippe Perrin (CNES), MSP2
	Valeri Grigoriyevich Korzun (RKA), MSP3, ISS-CDR
	Peggy Annette Whitson (NASA), MSP4, ISS-Flight Engineer
	Sergei Yevgeniyevich Treshchyov (RKA), MSP5, ISS-Flight Engineer

Docking

STS-111

Docking Date/Time:	7 June 2002, 16:25 UTC
Undocking Date/Time:	15 June 2002, 14:32 UTC
Docking Port:	PMA-2, Destiny Forward

J. O'Sullivan, *In the Footsteps of Columbus*, Springer Praxis Books,
DOI 10.1007/978-3-319-27562-8_6

Landing

Landing Date/Time:	19 June 2002, 17:58 UTC
Landing Site:	Runway 22, Edwards Airforce Base
Landing Vehicle:	Space Shuttle Endeavour (OV-105)
Landing Mission:	STS-111
Landing Vehicle Crew:	Kenneth Dale 'Taco' Cockrell (NASA), CDR
	Paul Scott 'Paco' Lockhart (NASA), PLT
	Franklin Ramon Chang-Diaz (NASA), MSP1
	Philippe Perrin (CNES), MSP2
	Yuri Ivanovich Onufriyenko (RKA), MSP3, ISS-CDR
	Carl Erwin Walz (NASA), MSP4, ISS-Flight Engineer 1
	Daniel Wheeler Bursch (NASA), MSP5, ISS-Flight Engineer 2

ISS Expeditions

ISS Expedition:	Expedition 4
ISS Crew:	Yuri Ivanovich Onufriyenko (RKA), ISS-CDR
	Carl Erwin Walz (NASA), ISS-Flight Engineer 1
	Daniel Wheeler Bursch (NASA), ISS-Flight Engineer 2
ISS Expedition:	Expedition 5
ISS Crew:	Valeri Grigoriyevich Korzun (RKA), ISS-CDR
	Peggy Annette Whitson (NASA), ISS-Flight Engineer
	Sergei Yevgeniyevich Treshchyov (RKA), ISS-Flight Engineer

6.1 THE ISS STORY SO FAR

There were no missions between the departure of Roberto Vittori on Soyuz TM-33 and the arrival of Philippe Perrin on STS-111.

6.2 PHILIPPE PERRIN

Early Career

Philippe Perrin was born in Meknes, Morocco in 1963 but grew up in Avignon in the Provence region of France. He graduated from the Ingénieur Polytechnicien Programme (a degree in engineering) at the Ecole Polytechnique in Paris in 1985. While at the school, he completed his national military service in the French Navy and spent 6 months at sea in the Indian Ocean.

After joining the French Air Force in 1985 he was assigned to the 33rd Reconnaissance Wing at Strasbourg AFB (1987–1991). He served in Africa and Saudi Arabia flying the Mirage F1CR. He was temporarily detached to the French Space Agency (CNES) in 1992 and sent to Star City near Moscow to receive two months of cosmonaut training. He earned

The configuration of the ISS during STS-111 with the Mobile Base System installed on the S0 truss. (NASA)

his Test Pilot Licence in 1993 from the Ecole du Personnel Navigant d'Essais et de Réception (EPNER), the French Test Pilot School at Istres Air Force Base. In 1993 he reported to the 2nd Air Defense Wing of Dijon AFB as the Senior Operations Officer (Operation Southern Watch). In 1995, he returned to the Bretigny Test Centre, as Chief Pilot Deputy and was in charge of the development of the Mirage 2000–5.

In 1996 Perrin joined NASA Astronaut Group 16, known as the Sardines, and began training as a Mission Specialist. He was initially assigned technical duties in the Spacecraft Systems/Operations Branch of the Astronaut Office and worked on man–machine interface issues in various programmes, including the Shuttle upgrade, the X-38 (to develop the technology for a prototype emergency Crew Return Vehicle for the ISS) and the ESA ATV (Automated Transfer Vehicle). In 1999 he joined the ESA astronaut corps and worked on

Philippe Perrin. (Wikipedia)

engineering support at the ATV Control Centre in Toulouse. After a single spaceflight in 2002 he left the astronaut corps in 2004 and became an Experimental Test Pilot with Airbus in Toulouse.

Previous Missions

N/A

6.3 THE STS-111 MISSION

STS-111 Mission Patches

The STS-111 patch was dominated by the orbiter launching to the ISS. Its left wing bore 'UF2' for the second ISS Utilisation Flight. The 'MBS' on the right wing was for the Mobile Base System supplied by Canada. The vehicle was rising on the three pillars created by the '111' of the mission number and passed through the circular astronaut logo that formed the orbit of the ISS – which was coloured red, white and blue in different sequences to represent the colours of the flags of USA, Russia and Costa Rica; the latter honouring Chang-Diaz. Italy was visible on Earth to indicate the home of the MPLM being carried. Ten stars represented the ten astronauts and cosmonauts on-orbit during the flight, and the large star at the top of the patch represented Mission Control at the Johnson Space Center, Texas. The names of the Shuttle crew were listed in the upper border, and the retiring Expedition 4 crew and the new Expedition 5 crew were at the bottom.

STS-111 Mission Objectives

As Utilisation Flight UF2, the prime objectives of STS-111 were:

- Deliver the Mobile Base System (MBS) for the Remote Servicer. The Space Station Remote Manipulator System (SSRMS), known as Canadarm2, had been delivered by STS-100. MBS was to enable Canadarm2 to travel the length of the truss system once that was completed. In 2008 the STS-123 mission would deliver the Special Purpose Dextrous Manipulator (SPDM), called Dextre.
- Carry MPLM Leonardo, mate it with the station, transfer cargo, and then return the module to Earth.
- Deliver the Expedition 5 crew to the station and return the Expedition 4 crew to Earth.

The STS-111 crew with Philippe Perrin on the left. (NASA)

STS-111 Mission Patch. (NASA)

Timeline

Flight Day 1 – Wednesday, 5 June 2002

Endeavour launched at 4:23 p.m. CDT on mission STS-111 carrying French astronaut Philippe Perrin, who was flying not under the auspices of ESA but through a bilateral CNES agreement with NASA.

Within 9 minutes Endeavour was in obit. After settling in, checking the systems and stowing their suits and other equipment, the crew prepared for a planned 12 day mission that would be stretched to 15 days.

Flight Day 2 – Thursday, 6 June 2002

The crew arose at 6:23 a.m. Central Time by the mission's first wake up call, *Gettin' Jiggy Wit It* by Will Smith, which was chosen for Valeri Korzun.

Perrin had a busy day. Firstly, he and Ken Cockrell activated the Shuttle robotic arm and used its cameras to survey the items in the payload bay – the Leonardo MPLM, the MBS, a replacement wrist roll joint for Canadarm2, and debris shields for the Zvezda module. Then Perrin and Franklin Chang-Díaz checked out their spacesuits in preparation for scheduled spacewalks. Finally, these two men set up the centreline camera that would be used by the commander to observe the docking mechanisms in the final phase of the approach to the ISS.

Meanwhile, Cockrell and Paul Lockhart fired Endeavour's thrusters three times during the day to adjust the speed at which the vehicle was closing in on the station.

Costa Rican-born Chang-Díaz took part in a live linkup with Abel Pacheco, the President of Costa Rica, and John Danilovich, the American Ambassador to that nation, as well as reporters from Spanish language TV stations Univision and Telemundo. This was Chang-Díaz's seventh and final spaceflight, tying the record held by NASA's Jerry Ross. At the time of writing no other astronaut has beaten their joint record; in fact no currently active astronaut has flown six times.

Flight Day 3 – Friday, 7 June 2002

As Endeavour approached within 900 miles of the ISS at 4:30 Central Time, Mission Specialist Peggy Whitson was the recipient of the wakeup call *American Woman* by Lenny Kravitz.

During the final approach and docking, it was all hands on deck. Perrin monitored the rendezvous navigation displays using a laptop computer. Chang-Diaz oversaw operations of the Shuttle's docking system. Whitson used the handheld laser rangefinder to provide supplemental range and closing rate information to Cockrell, who was flying the craft from the rear console of the flight deck. Endeavour docked with the forward port of the Destiny module at 11:25 Central Time and the hatches opened at 2:08 p.m.

The spacewalks on the STS-100 mission had been carried out from the Shuttle, requiring different atmospheric pressures to be maintained on each side of the hatch and postponing joint operations until after the first spacewalk. Since then, STS-104 had delivered the Quest Joint Airlock. So for STS-111 Perrin and Chang-Diaz transferred their spacesuits and tools to the ISS and then set up and tested the communications between their suits and the Quest systems.

The ISS crew exchange was officially achieved when the Expedition 5 crew relocated their Soyuz seat liners and Sokol re-entry suits to the station. The Expedition 4 crew simultaneously stowed theirs on Endeavour and joined the Shuttle crew.

A problem with the Flash Evaporator System Primary B controller occurred later. This device was to spray excess supply water into the inside of a trash-can-shaped vessel which was wrapped by Freon coils. The heat in the coils would cause the water to flash into vapour and be vented overboard, hence disposing of excess heat and excess supply water. As there was a Primary A controller, this failure did not directly affect the mission.

Flight Day 4 – Saturday, 8 June 2002

The first full day of joint operations started at 4:23 Central Time with *I Have a Dream* by ABBA for Treschev.

Shuttle Commander Cockrell used Endeavour's robot arm to transfer the Leonardo MPLM to the Unity node. With the berthing achieved at 9:28 a.m., the hatch was opened at 4:30 p.m. and the crew set about the task of transferring of more than 5,600 pounds of cargo.

Perrin and Chang-Díaz, assisted by Lockhart, checked out their EVA space suits and tools, and reviewed procedures for Sunday's spacewalk.

A problem occurred when one of the four Control Moment Gyroscopes (CMG) on the ISS failed. After assessing the situation, Mission Control decided that a bearing had failed and reconfigured the station to rely on the three remaining gyros.

Flight Day 5 – Sunday, 9 June 2002

Perrin and Chang-Diaz exited the Quest airlock at 10:27 a.m. to start what would be a busy 7 hour 12 minute spacewalk. Lockhart supported spacewalk operations while Cockrell monitored their activities using the cameras on the Shuttle robotic arm. They first installed a Power and Data Grapple Fixture to the P6 unit. This would later be used to transfer the P6 to its permanent position on the port end of the main truss structure.

Whitson and Korzun transported Chang-Díaz on the end of Canadarm2 from the payload bay over to PMA-1, which connected the Unity node to the Zarya module, where he stowed six micrometeoroid debris shields. These shields were to be installed on Zvezda by Whitson and Korzun in August, as part of an ISS spacewalk.

In a task added the previous night, Chang-Díaz then conducted a visual and photographic inspection of the failed CMG on the station's Z1 truss. The other three gyroscopes were functioning normally and the station could actually maintain its attitude with only two such units operating.

The spacewalkers then removed the thermal blankets from the MBS in the payload bay. Whitson and Walz used Canadarm2 to position it close to the Mobile Transporter on the station's truss. It was left there overnight to ensure it would be thermally conditioned in advance of Monday's operation to attach it to the station.

Perrin and Chang-Díaz re-entered the Quest airlock at 5:41 p.m. Central Time, ending a busy and successful day.

Flight Day 6 – Monday, 10 June 2002

Whitson and Walz resumed yesterday's operation by attaching the MBS onto the Mobile Transporter, already in place on the S0 truss. Mission Control then remotely closed the latches to secure the MBS to the transporter. During a later mission, Canadarm2 would 'walk' itself from its original position on the Destiny module to the MBS in order to be able to assist the assembly of the truss structure.

Both crews continued to unload the Leonardo MPLM. Although the Expedition 5 crew became the official residents of the station the previous Friday, there was now a ceremony to mark the change of command.

Starting at 4:53 p.m. Central Time the Shuttle executed a 1 hour reboost manoeuvre to increase the station's altitude by over a mile.

Flight Day 7 – Tuesday, 11 June 2002

Perrin and Chang-Díaz embarked upon their second spacewalk from the Quest airlock at 10:20 a.m. Central Time. Their first task was to connect video, data, and power cables between the MBS and its transporter. Then Mission Control remotely commanded the connection of the umbilical attachments between the MBS and the rails of the S0 truss along which the transporter ran. Next the spacewalkers installed the Payload ORU Accommodation (POA), identical to the end effector of Canadarm2. They concluded by bolting the MBS to the transporter. The spacewalkers re-entered the airlock at 3:20 p.m. and repressurised it.

After Mission Control had verified that all the connections on the MBS were operating correctly, Canadarm2, which had been supplying electrical power to the MBS, was disconnected. The arm was then repositioned in readiness for the third and final spacewalk of the mission, during which its wrist roll joint was to be replaced.

In addition to handover conferences between the two Expedition crews, the Shuttle crew loaded unneeded items into the MPLM for return to Earth, with the latter task proceeding ahead of schedule.

Flight Day 8 – Wednesday, 12 June 2002

The joint operations continued with the stowage of unneeded cargo from the station into the MPLM and onto the mid-deck of the Shuttle. There was a televised news conference with American, French and Canadian broadcasters.

In the second of three reboosts, Endeavour's thrusters were fired to raise the orbit of the station by another mile.

Flight Day 9 – Thursday, 13 June 2002

On the Road Again by Willie Nelson was selected as the wakeup call by Walz's family to celebrate his imminent return to Earth after spending six months on the ISS.

Perrin and Chang-Diaz's third spacewalk of this mission was undertaken to repair a failed primary communication channel which was believed to be the result of a short circuit within the wrist roll joint of Canadarm2. The problem, which arose in March, caused the arm's brakes to engage, but a software modification and the use of the backup channel enabled the arm to be used in a fully operational mode in the interim. They first removed the arm's latching end effector (LEE) and attached it temporarily to a handrail on the Destiny module. They released six bolts connecting the wrist roll joint to the wrist yaw joint and the seventh bolt connecting power, data, and video umbilicals. Perrin took the extracted unit to the Shuttle payload bay and returned with the new joint. Once the new wrist joint was in place, the LEE was reinstalled. Bursch and Korzun verified that Canadarm2 was fully operational. The two spacewalkers then re-entered the Quest airlock to conclude the final spacewalk of the mission, which had lasted 7 hours and 17 minutes.

Philippe Perrin during the second spacewalk of the STS-111 mission. (NASA)

Expedition 5's Whitson and Treschev spent the day completing the transfer of cargo to the MPLM and the Shuttle.

Late in the day, an attempt by Mission Control to provide power to Canadarm2 from the MBS was not successful. This was diagnosed as a software issue that shouldn't cause any delay in Canadarm2's programme of operations.

Flight Day 10 – Friday, 14 June 2002

In order to celebrate US Flag Day (the Stars and Stripes flag had been adopted on 14 June 1777) the Shuttle crew was awakened to the sound of the *Star Spangled Banner*.

Endeavour's thrusters were fired for the last of three reboosts which raised the orbit of the station by a combined total of approximately six miles.

Perrin used the Shuttle's robot arm to return the MPLM to the payload bay at 3:11 p.m. Central Time. Leonardo had carried a total of 8,062 pounds of supplies and equipment to the station, including a new science rack to house microgravity experiments and a glovebox that would allow the station to conduct experiments which required physical isolation. The MPLM was returning with 4,667 pounds of equipment and supplies that were no longer needed aboard the station. More than 1,000 pounds of equipment was also returned in the Shuttle mid-deck.

Flight Day 11 – Saturday, 15 June 2002

It was departure day for the crew of Endeavour and their Expedition 4 passengers, and the crew were awakened at 3:30 Central Time by *Hello to All the Children of the World*, which had been chosen for Bursch by his son's schoolmates.

After final farewells, the hatches were closed at 7:23 a.m. Central Time and Endeavour undocked from the ISS at 9:32 a.m. As it departed, Whitson rang the ship's bell aboard the ISS and announced, "Expedition Four departing, Endeavour departing." Bursch replied, "Smooth sailing, Peggy." After a flyaround of the station, Lockhart made a final separation burn and Endeavour left the ISS behind at 11:15 a.m.

Flight Day 12 – Sunday, 16 June 2002

On what should have been the last full day of the mission, Endeavour's crew were awakened at 3:23 a.m. to *Where My Heart Will Take Me*, the theme from *Star Trek: Enterprise*, performed by Russell Watson.

While Cockrell, Lockhart, and Chang-Diaz tested the spacecraft's flight controls and thrusters in preparation for returning to Earth, Perrin and the Expedition 4 crew packed up gear and installed the seats on the mid-deck for the returning station crew.

Onufriyenko, Bursch, and Walz spoke to Fox News, WOIO-TV of Cleveland (Walz's hometown), and WICZ-TV of Vestal, NY (Bursch's hometown).

Flight Day 13 – Monday, 17 June 2002

The wakeup call at 3:23 a.m. was the University of Texas Marching Band playing *The Eyes of Texas* in honour of Cockrell and Lockhart, both alumni.

Neither of the day's two landing opportunities at the Kennedy Space Center was taken because of clouds and rain there. As Endeavour had sufficient consumables to extend the mission to Thursday, it was decided not to activate the backup landing site at Edwards AFB in California. The preference was always to land at the Kennedy Space Center, where an orbiter was prepared for its next mission. If the orbiter were to land at Edwards, then it would have to be flown to Florida on the back of a Boeing 747 Shuttle Carrier Aircraft (SCA), which cost money and imposed a delay in the preparations for the next mission.

Flight Day 14 – Tuesday, 18 June 2002

When the mission ran beyond its planned duration, it is presumed they ran out of wake up calls. This morning's backup wakeup call was at 2:30 a.m., and was *Sojourner* by Matt Gast, Mission Control's own lead timeliner/scheduler of crew activities.

Showers in Florida and gusting winds in California resulted in another postponement of the return to Earth.

Flight Day 15 – Wednesday, 19 June 2002

Showing that Houston had a sense of humour, the crew woke at 1:23 a.m. to *I Got You Babe* by Sonny and Cher, from the soundtrack of the movie *Groundhog Day*.

As weather still prevented a landing in Florida, Endeavour glided down to a landing at Edwards at 12:58 p.m. CDT.

6.4 POSTSCRIPT

Subsequent Missions

STS-111 was Philippe Perrin's only space mission.

Philippe Perrin Today

After his flight, Perrin returned to France to work as a support astronaut on the Automated Transfer Vehicle programme. Having flown in space for CNES, not for ESA, he resigned from the European astronaut corps in 2004 to become a test pilot for Airbus.

Perrin has been award the following honours:

- Awarded his pilot's wings 'first of his class' in 1996.
- Recipient of two French Air Force awards for Flight Safety in 1989.
- Recipient of the French Overseas Medal (Gulf War in 1991).
- Recipient of two French National Defence Medals.
- Chevalier, Légion d'Honneur.
- Officier, Légion d'Honneur.

Philippe Perrin as an Airbus test pilot. (laurencemassat2012.wordpress.com)

Perrin is the only French astronaut to have received the Légion d'Honneur prior to flying in space because he became chevalier whilst as a military pilot. After his space mission he was elevated to the officier class.

In 2012 Perrin was the suppléant (or alternative) to National Assembly candidate Laurence Massat Guiraud-Chaumeil, meaning that he would have replaced her in the assembly had she been elected and subsequently elevated to government.

7

Odissea

Mission

ESA Mission Name:	Odissea
Astronaut:	Frank De Winne
Mission Duration:	10 days, 20 hours, 53 minutes
Mission Sponsors:	ESA/OSTC
ISS Milestones:	ISS 5S, 20th crewed mission to the ISS

Launch

Launch Date/Time:	30 October 2002, 03:11 UTC
Launch Site:	Pad 1, Baikonur Cosmodrome, Kazakhstan
Launch Vehicle:	Soyuz TMA
Launch Mission:	Soyuz TMA-1
Launch Vehicle Crew:	Sergei Viktorovich Zalyotin (RKA), CDR
	Frank De Winne (ESA), Flight Engineer
	Yuri Valentinovich Lonchakov (RKA), Flight Engineer

Docking

Soyuz TMA-1	
Docking Date/Time:	1 November 2002, 05:01 UTC
Docking Port:	Pirs
Soyuz TM-34	
Undocking Date/Time:	9 November 2002, 20:44 UTC
Docking Port:	Zarya Nadir

Landing

Landing Date/Time:	10 November 2002, 00:04 UTC
Landing Site:	80 km NE of Arkalyk

J. O'Sullivan, *In the Footsteps of Columbus*, Springer Praxis Books,
DOI 10.1007/978-3-319-27562-8_7

Landing Vehicle:	Soyuz TM
Landing Mission:	Soyuz TM-34
Landing Vehicle Crew:	Sergei Viktorovich Zalyotin (RKA), CDR
	Frank De Winne (ESA), Flight Engineer
	Yuri Valentinovich Lonchakov (RKA), Flight Engineer

ISS Expeditions

ISS Expedition:	Expedition 5
ISS Crew:	Valeri Grigoriyevich Korzun (RKA), ISS-CDR
	Peggy Whitson (NASA), ISS-Flight Engineer 1
	Sergei Yevgeniyevich Treshchyov (RKA), ISS-Flight Engineer 2

7.1 THE ISS STORY SO FAR

After STS-111 there was one mission to the ISS prior to Frank De Winne's arrival on Soyuz TMA-1. In October 2002 STS-112, Assembly Flight 9A, delivered the S1 truss. During three spacewalks, this was attached to the starboard end of the central S0 truss. Also delivered was the first of two Crew and Equipment Translation Aids (CETA), a

The ISS with the S1 truss segment and radiator on the left, photographed by STS-113 (next mission). (NASA)

human-powered tool cart that was mounted on the MBS rail. The S1 truss included a radiator panel that was unfurled during the first EVA. The Thermal Radiator Rotary Joint (TRRJ) was also installed to provide the mechanical energy needed to rotate the radiator.

7.2 FRANK DE WINNE

Early Career

Frank De Winne was born in Ghent, Belgium on 25 April 1961. He graduated from the Royal School of Cadets in 1979 and obtained a master's degree in engineering from the Royal Military Academy in 1984. He trained at the elementary flying school of the Belgian Air Component at Goetsenhoven, then flew Mirage 5s. In 1991 he completed the Staff Course at the Defence College in Brussels, earning the highest distinction. In 1992 he attended the Empire Test Pilot School in Boscombe Down in England, where he won the McKenna Trophy as best student.

In December 1992, De Winne was assigned as an Air Force test pilot and from January 1994 until April 1995 he was responsible for flight safety of the 1st Fighter Wing, operating from Beauvechain. Between April 1995 and July 1996 he was attached as senior test pilot to the European Participating Air Forces at Edwards AFB in California, working on the mid-life update of the F-16 fighter aircraft, focusing upon testing its radar. From 1996 to August 1998 he was senior test pilot in the Belgian Air Force, responsible for all test programmes and for pilot-vehicle interfaces for future aircraft/software updates.

On 12 February 1997 De Winne lost both the engine and instruments whilst flying his F-16 over Leeuwarden. Rather than bail out and risk the aircraft crashing into a populated area, he stayed with it and landed successfully at a nearby airbase. In response, Lockheed Martin gave De Winne its Joe Bill Dryden Semper Viper Award, making him the first non-American to be so honoured.

From August 1998, De Winne commanded 349 Squadron operating from Kleine Brogel, and flew seventeen combat missions over the Balkans during NATO Operation Allied Force.

Frank De Winne was selected as an ESA astronaut candidate in 1998, along with Léopold Eyharts, André Kuipers, Paolo Nespoli, Hans Schlegel, and Roberto Vittori. He provided technical support for the X-38 project from within the Directorate of Manned Spaceflight and Microgravity, located at the European Space Research and Technology Centre in the Netherlands. In August 2001 he went to Star City near Moscow to start training as a Soyuz flight engineer and an ISS crewman.

Frank De Winne's education was:

- Graduated from the Royal School of Cadets, Lier, Belgium.
- AIA Prize for the best thesis on his master's degree in telecommunications and civil engineering from the Royal Military Academy, Brussels, Belgium, 1984.
- Distinction in Staff Course at the Defence College in Brussels, 1991.
- Graduated from the Empire Test Pilot School in Boscombe Down, England with the McKenna Trophy, 1992.

Previous Missions

N/A

7.3 THE ODISSEA MISSION

Odissea Mission Patches

The Odissea mission name was the first of many ESA missions to incorporate 'ISS'. In this case, the mission was in honour of Odysseus, the Greek king of Ithaca who took ten years to return home after the Trojan war.

The Odissea crew with Frank De Winne on the right. (www.spacefacts.de)

At the centre of the patch was a circle for Earth, with both the ISS and the Soyuz orbiting it. The black pillar represented Europe positioned between the USA and Russia. The black, yellow and red represented Belgium and the red, white and blue represented the other nations. The mission objectives of Exploration, Science and Technology were in the outer ring. Along the bottom were the logos of the four participating agencies, namely Rosaviakosmos, Russian Aviation and Space Agency; Energia, the Russian S. P. Korolev Rocket and Space Corporation; European Space Agency (ESA); and the Belgian Federal Office for Scientific, Technical and Cultural Affairs (OSTC) that funded De Winne's mission and the related experimental programme.

Odissea Mission Patch. (ESA)

Soyuz TMA-1 Mission Patch. (www.spacefacts.de)

The Soyuz TMA-1 patch evolved from an earlier design for the planned inclusion of Lance Bass as a 'space tourist' on the flight. After Bass' withdrawal, the blue and white Soyuz spacecraft rising on a white curve was retained.

Perhaps due to the change in crew, this version, which was designed by Dima Shcherbinin, a TsPK employee, arrived too late to be worn by the crew in their official photographs or on their space suits at launch. The Odissea patch was worn in flight.

Strangely the Soyuz was depicted rising on a curve that bore the name 'Soyuz TM-34', whereas in fact that spacecraft was used for their descent, with Soyuz TMA-1 being left for the Expedition crew. The silhouette of the ISS was featured in the background. The Rosaviakosmos, Russian Aviation and Space Agency, and European Space Agency (ESA) logos were incorporated at the top and bottom of the patch respectively.

Odissea Mission Objectives

The Odissea mission had two objectives. To exchange the Soyuz 'lifeboat' at the ISS and to carry out a scientific and technical research programme that was organised by ESA and the Belgian Federal Office for Scientific, Technical and Cultural Affairs.

The following experiments were to be carried out as part of the Odissea mission:

- VITAMIN D
- RHOSIGNAL
- RAMIROS
- MESSAGE
- GCF-B
- PROMISS
- CARDIOCOG
- NEUROCOG
- SYMPATHO
- VIRUS
- SLEEP
- DCCO
- ZEOGRID
- NANOSLAB
- COSMIC
- LSO-B
- EDUCATION.

VITAMIN D

Objectives: Characterisation of the effect of microgravity on the mechanism of action of Vitamin D in mammalian osteoblasts.

Tasks:

- Study of microgravity-induced alterations in 1,25(OH)2 vitaminD3 regulated gene expression. The mouse osteoblastic cell line MC3T3-E1 was to be used and the gene expression was to be determined using real-time PCR.

- Investigate the mechanisms that underlies microgravity-related alterations in gene expression.

Equipment Used: The Aquarius-B hardware including:

- Container for Transport/Ascent (CTA-B) – an AQUARIUS-B incubator with a set of five plunger boxes containing osteoblastic cells, combined in a single container.
- Container for Transport/Return (CTR-B) – a passive container to return to Earth a set of B-containers with cultivated osteoblastic cells.
- Electronics Control Unit (ECU).

Expected Results: An ECU memory card and the B-Container/VITAMIN D with five plunger boxes containing cultivated mouse osteoblastic cells.

RHOSIGNAL

Objectives: Investigate functional alterations of the small GTPases of the Rho family (Rho GTPases), central molecular switches of the intracellular signalling pathways, in human fibroblasts subjected to microgravity.

Tasks: A study of functional alterations in fibroblasts subjected to microgravity conditions by using immunomorphological procedures.

Equipment Used: Identical to the Vitamin D experiment.

Expected Results: An ECU memory card and the B-Container/RHOSIGNAL with five plunger boxes containing cultivated human osteoblastic cells.

RAMIROS

Objectives: Analysis of the biological effects of heavy particle (HZE) radiation on primary mammalian tissue in space in order to understand how single cells and their environment deal with HZE, in order to make a contribution to radiation safety guidelines for human space activities.

Task: Detection and study of occurrence of different DNA lesions (hallmarks of cancer and ageing) in spaceflight.

Equipment Used: Identical to the Vitamin D experiment.

Expected Results: An ECU memory card and the B-Container/RAMIROS with five plunger boxes containing cultivated human fibroblasts.

MESSAGE

Objective: An investigation of the impact of space environmental conditions on the microbial motility and genetic processes (gene expression, mutation, the repair and rearrangement of DNA, gene transfer, pathogenicity enhancement, etc.).

Task: Growing the biological strain Ralstonia metallidurans CH34.

Equipment Used:

- MSS Container 1 – a thermal insulating container consisting of three numbered experimental bags (MSS Sample 1, 2, 3) and an ice pack. Each numbered bag contained two Falcon tubes: one tube containing two capillaries filled with solid bacterial culture, the other one containing two microvials filled with liquid bacterial culture.
- MSS Container 2 with three numbered experimental bags (MSS Sample 4, 5, 6); identical to MSS Container 1.
- MSS Container A – a passive container that had eight Petri dishes with solid bacterial culture and temperature data logger.
- MSS Container B – a passive container in which to return MSS Samples 1–6 and the ice pack.
- Freezer Kriogem-3M to provide specified temperature conditions.

Expected Results: Incubated bacterial cultures in kits MSS-1, MSS-2 and MSS-3 (twelve capillaries, twelve microvials and eight Petri dishes).

GCF-B

Objective: Growth of crystals of biological macromolecules using the counter-diffusion crystallisation technique in space in order to compare the quality of the crystals grown in space with those grown on Earth.

Task: To verify the results of the first flight of the Granada Crystallisation Facility (GCF) experiment performed in October 2001 by the Andromède mission.

Equipment Used: Transport container GCF-1 and Crystallisation facility GCF-2, with 23 Granada Crystallisation Boxes (GCB) accommodating capillaries (six items per box) with protein solutions.

Expected Results: Grown protein crystals located in crystallisation facility GCF-2.

PROMISS

Objective: To study protein growth processes in microgravity conditions by using a counter-diffusion crystallisation technique.

Tasks:

- Measurement of the parameters of the growing protein crystals.
- Measurement of the composition changes (depletion zone) of liquid around the growing protein crystals.
- Use a holographic microscope to make an optical investigation of crystal growth mechanisms in microgravity conditions.
- Detailed analysis and a quantitative interpretation of the relationship between the quality of the obtained crystals and the environment in which they were produced.

Equipment Used:

- Microgravity Science Glovebox (MSG) in the Destiny module.
- PROMISS Experiment box and Wheel Kit, including a wheel with six reactors with samples of protein solutions.
- PROMISS Kit 1, including camera clamp, microdrives and harnesses.
- PROMISS Kit 2, including mini DV videotapes and PCMCIA cards.

Expected Results: Six grown protein crystals in the PROMISS Wheel, as well as seven videotapes and the PCMCIA card of PROMISS Kit 2.

CARDIOCOG

Objective: To study alterations of the human cardiovascular system in weightlessness revealed at the level of peripheral arteries, and vegetative regulation of arterial pressure and heart beat rate.

Tasks:

- Investigate the effects of microgravity on the cardiovascular system and interactions between the cardiovascular and respiratory systems.
- Investigate alterations of cognitive system and stress reactions caused by conditions in space.

Equipment Used:

- Cardioscience Kit, Sphygmomanometer Tensoplus for measuring arterial pressure, cosmonaut hearing protection, EGE-20 with control unit, EGE-30 with power supply, and CARDIOCOG consumables.
- SONY DCR-PC120 camcorder with accessories, Mini DV videotape, and a returnable hard drive.

Expected Results: Information about the cardiovascular system (electrocardiogram, arterial pressure, respiration rate, etc.) recorded on a hard disc, PCMCIA card (backup copy) and flight log sheets, and mini DV videotapes showing the experiments performed.

NEUROCOG

Objective: To study integration of visual, vestibular, and proprioceptive information for perception of spatial orientation.

Task: To assess the effect of spaceflight conditions on perception and the process of memorizing the orientation and location when performing the 'Virtual Turns' and 'Orientation' investigations.

Equipment Used:

- COGNI Kit, Galley Kit, Galley-RM Kit, Galley-INF Kit with PCMCIA cards, EGE-20 with control unit, EGE-30 with power supply, and NEUROCOG Kit with science hardware.
- SONY DCR-PC120 camcorder with accessories, Mini DV videotape, and a returnable hard drive

Expected Results: Records of electroencephalograms and electrooculograms on PCMCIA cards and information on perception rates and accuracy recorded on a hard disc and on PCMCIA card (backup copy), plus mini DV videotape recordings.

SYMPATHO

Objective: Verification of hypothesis about alterations of sympathoadrenal activity in spaceflight.

Task: To study activity of sympathetic system by means of the laboratory analysis of venous blood samples taken in flight.

Equipment Used:

- Plasma-3 Kit, Accessories Kit Plasma-3 to take samples, Centrifuge Plasma-3 to separate blood into plasma and serum, Freezer Kriogem-3 to freeze samples, and Consumables Kit Plasma-3.
- Return Container KB-3 to return frozen samples to ground.

Expected Results: Blood samples (two per test subject).

VIRUS

Objective: To determine the frequency of induced reactivation of latent viruses, latent virus shedding, and clinical disease after exposure to the physical, physiological, and psychological stressors that are associated with the space environment.

Task: Laboratory analysis of saliva samples taken prior to, during, and after the flight to investigate alterations in the immune system, and the potential for reactivation and dissemination (shedding) of latent viruses.

Consumables: A Saliva Kit with a set of saliva samplers.

Expected Results: Saliva samples returned in the Saliva Kits.

SLEEP

Objective: To further understand the effects of spaceflight on sleep and the development of effective countermeasures for both short-duration and long-duration missions.

Tasks: To continuously monitor the subject's motor activity and to record the light levels in the crew area.

Equipment Used: A Sleep Kit with an Actilight watch for automatic recording of the subject's motor activity and light exposure, and a Sleep Logbook to record subjective memories of each sleep period.

Expected Results: Information recorded by the Actilight watch and the Sleep Logbook.

DCCO

Objective: To measure the isothermal diffusions in binary and ternary mixtures that are representative of actual crude oils.

Tasks: To study mass transfer phenomena and make a quantitative assessment of diffusion coefficients.

Equipment Used:

- Microgravity Science Glovebox (MSG) in the Destiny module.
- DCCO Experiment Container including interferometer and nine cells with liquid samples.
- DCCO Kit including Harness (Power & Data), Clamp, Camera Mount.
- DCCO Experiment Container.
- Videotapes and PCMCIA cards.

Expected Results: Processed samples (in nine cells) in DCCO experiment container, plus videotapes and the PCMCIA card.

ZEOGRID

Objective: To investigate the induced organisation of nanoscopic zeolite slabs.

Tasks: A study of the induced organisation of nanoscopic zeolite slabs, as well as a determination of the ZEOGRID structure.

Equipment Used: The ZEOGRID Assembly, including a working chamber for 21 cells plus samples.

Expected Results: Processed samples (in 21 cells) in the ZEOGRID Assembly.

NANOSLAB

Objective: To study the aggregation mechanism and kinetics of ZSM-5 and Silicalite-1 nanoslabs into ZSM-5/Silicalite-1 hybrid phases under microgravity conditions.

Task: To obtain kinetic profiles of the formation of ZSM-5/Silicalite-1 phases in microgravity.

Equipment Used:

- Microgravity Science Glovebox (MSG) in the Destiny module.
- NANOSLAB Sample container including 10 assemblies, each containing three cells with the experimental liquid samples, plus the NANOSLAB Electronics Box.

Expected Results: Processed samples (in 30 cells) in NANOSLAB Assembly.

COSMIC

Objective & Task: An investigation of the microstructure formation of Ti-Al-B compressed powder samples during self-propagating high-temperature combustion synthesis.

Equipment Used:

- Microgravity Science Glovebox (MSG) in the Destiny module.
- Sample Container including 10 reactors with material samples, an electronics box, mounting plate, MSG Videotape recorder, camera mount, videotapes and PCMCIA cards.

Expected Results: Processed samples (in 6 reactors) in the sample container, plus videotapes and the PCMCIA cards.

LSO-B

Objective: To study optical radiation in the atmosphere and ionosphere of Earth related to the thunder activity and seismic processes (optical glows in the upper atmosphere and ionosphere that accompany storm phenomenon are called 'sprites' and 'elves').

Tasks:

- Measure the spatial and spectral distribution of 'sprites' and 'elves' and correlate this data with lightning radiation intensity.
- Test new procedures for nadir space measurements.
- Statistically analyse the frequency of sprite occurrences and their origin based on observations lasting tens of hours.

Equipment Used: The LSO hardware including micro-cameras on a rotational bracket, an electronics unit and cable set, an EGE-10 control unit, an EGE-30 power supply, and a removable hard drive.

Expected Results: Science information recorded on a hard drive that is returned to Earth.

EDUCATION

Objective: To demonstrate some of the physical phenomena of orbital flight for educational purposes (VIDEO). To support the community activity of amateur ('ham') radio using the facilities of the ISS (ARISS).

Tasks:

- To video De Winne demonstrating the behaviour of liquid in weightlessness (e.g. air bubbles in a water drop and water drops on a solid surface) and the behaviour of items in zero-gravity (e.g. a glider and solid object, and 'drops' of solid); VIDEO.
- To make real-time radio transmissions from the ISS to enable students from the selected Belgian schools to ask questions of their astronaut; ARISS.

Equipment Used:

- SONY DCR-PC120 camcorder with accessories, plus a mini-DV (PDVM-40ME) videotape.
- Radio-amateur communication hardware and any items available for demonstration of physical phenomena.

Expected Results: Videotape mini-DV (PDVM-40ME).

Timeline

During his training for the launch of Soyuz TMA-1, Frank De Winne contributed to a diary hosted by the ESA website. There follows a summary of his training and travels from July through to launch at the end of October.

Training Thursday, 4 July 2002

After a holiday at home in Belgium, De Winne returned to Star City on Monday, 1 July 2002. He had some Russian language lessons, but most of the week was devoted to re-entry procedures training in the Soyuz TMA simulator. Although De Winne and his Soyuz TMA-1 crewmates were intending to return to Earth in Soyuz TM-34 after leaving the newer spacecraft at the station as a 'lifeboat' for the residents, they needed to train for re-entry in the TMA model in case of an emergency that prevented them reaching the ISS.

It was a week for celebrations as the international contingent marked first Canada Day on 1 July, led by Chris Hadfield, who was the Canadian Director of Operations for NASA at the Yuri Gagarin Cosmonaut Training Centre (GCTC), and then US Independence Day on 4 July.

Also this week, De Winne practised manual docking procedures in the simulator and underwent a cardiogram.

Training Friday, 12 July 2002

This week De Winne was measured for his flight suit; the one piece overall-type suit the crew were to wear on board the ISS. Made by the same team that made the cosmonaut's suits, he was very proud of the Belgian flag colours on the breast. It was sent up ahead of him on an unmanned Progress freighter.

He also conducted his first training session with the new software for the Soyuz TMA spacecraft, which had a glass cockpit with digital displays.

Training Wednesday, 24 July 2002

This week saw a start to planning the daily itinerary on board the station for the Odissea mission. Also De Winne and his crewmates practised helicopter rescue in the Hydro Laboratory that is normally used to train for spacewalks. It can accommodate full-sized ISS module replicas, but this time it was used to practise being winched to safety by helicopter after an ocean splashdown.

And finally this week the crew rounded off the first in a series of medical tests. In order to interpret the data collected during the mission, the same tests were baselined several months beforehand. These test results would assist the doctors and scientists in understanding the effects on the human body of flying in space. For example, one of the tests looked at the influence of weightlessness on the number of virus cells in the blood, which in turn could affect the immune system.

Training Monday, 29 July 2002

De Winne was in the Netherlands at the Academisch Medisch Centrum (Academical Medical Centre) of the University of Amsterdam for the AORTA experiment that was to investigate the problems that astronauts have in remaining in an upright position following their return to Earth.

Training Saturday, 3 August 2002

At the end of July, De Winne, Soyuz TMA-1 commander Sergei Zalyotin and backup commander Yuri Lonchakov spent four days in the Belgian User Support and Operation Centre (B.USOC) in Ukkel near Brussels, training on the medical experiments that were to be conducted on the station.

De Winne wrote, "Some of the training sessions are very interesting. I am able to practise using the programmes and the computers that I will be using during my flight. We also did some baseline data collection. Scientists are making a number of measurements to compile a database with our data. Later on, the data will be compared with the experiments which I carry out on board the International Space Station."

There were two main groups of experiments. CARDIOCOG consisted of four experiments which would investigate how the cardiovascular and cardiorespiratory functions changed in space and what happened after returning to Earth. NEUROCOG investigated changes of the brain functions in space.

"During our stay in Belgium I showed Sergei together with some friends the central Market Square of Brussels, and, of course, the Manneken Pis on a very beautiful summer evening. Sergei also visited Bruges, which he liked very much. Another place we visited was the Russian House in Brussels. This is a small piece of Russia in Belgium which I did not even know existed; now I know it is there I will certainly visit it again in the future."

Training Monday, 5 August 2002

Back in Moscow, the crew were given samples of the food and drinks that would be available on the station. De Winne was pleasantly surprised with the large choice, and that Russian space food tasted very good. The ISS had the facility to warm cans of food and only the sauces and ketchup were still served in tubes. Among the choices available were meat in tomato sauce, chicken with eggs, mashed potatoes with onions, vegetable ratatouille, tea with sugar, and fruit juice.

Training Sunday, 18 August 2002

On 9 August, De Winne visited RKK Energia in Korolyev, near Moscow, to be shown the design and assembly of a Soyuz rocket system. On the 14th he visited the Zvezda factory for an initial fit-test of the Sokol pressure suit he was to wear in the Soyuz capsule. He was to return in September to test the suit in vacuum conditions. This was also the day that his scientific hardware was accepted by Energia and handed over for delivery to the ISS on an unmanned Progress before the launch of Soyuz TMA-1.

Sunday was a rare day off from training, and De Winne accompanied Dutch ESA astronaut André Kuipers to a Russian military air show at Star City.

Training Sunday, 25 August 2002

Amongst other things, this week saw the crew train on emergency and evacuation procedures on the ISS. In the event of a leak, a depressurisation or a fire there were various procedures, some of which involved using the Soyuz to evacuate the station.

De Winne and Zalyotin also learned how to take a blood sample from each other, because that is what they were to do as part of a number of experiments on the station.

Training Monday, 26 August to 8 September 2002

In Houston for two weeks, De Winne was able to see how NASA approached training for the ISS. He wrote, "At the Johnson Space Center everything seems to be bigger than in Russia. The halls in which we train are bigger and there is a full mock-up of the Space Station. In Russia there is only a model of the Russian ISS segment. Here in the United States though, the mock-ups of the Russian elements are somewhat less accurate." He received training similar to the Russian training of the previous week on emergency procedures, but this time for the American segment of the station.

De Winne was trained on the computer systems which would be his communications lifeline with ground control. He also familiarised himself with the medical supplies and equipment on board. His main training was with the Microgravity Science Glovebox (MSG). He spent more time with the US simulator at JSC than he had with the one at ESTEC in the Netherlands, and ended by carrying out a fully integrated simulation with all of the ground control centres, just as he would once in space.

On Saturday, 31 August, De Winne, Zalyotin, and Lonchakov took time off to visit Galveston Bay, take a ferry ride, watch dolphins swim, and enjoy a typical American meal with his Russian colleagues.

Lance Bass, the singer and member of the NSYNC pop group, was in JSC training to accompany De Winne and Zalyotin as a 'space tourist' but it was announced on 3 September that Bass would not fly aboard Soyuz TMA-1, therefore Lonchakov took his place on the crew.

Training Monday, 9 September 2002

Back in Moscow, De Winne had his Baseline Data Collection (BDC) medical examination to obtain data that would be compared with the results of the experiments that he would conduct in space and from examinations after his return to Earth. He was also medically approved to launch.

Training Monday, 16 September 2002

It was a busy few days in Star City, where De Winne participated in a website chat for ESA and was part of the pre-flight press conference to present the crew. He also tested his space suit and rehearsed his previous training tasks.

Training Wednesday, 18 September 2002

De Winne wrote, "Today I'm back again in the Zvezda factory. This time my space suit or *skafandr*, as they used to say here in Russia, is being tested in vacuum in a so-called baro-chamber. For me this is a bit awkward. I have to stay in my chair for two hours without moving, while my suit is inflated. There are several reasons for this test. First of all, one wants to be sure that the space suit is totally OK; that all the minor changes that had to be done have been made correctly. It is also a kind of psychological training – I feel what it is like to sit in vacuum with my suit inflated. Next week my *skafandr* will be sent to Baikonur in preparation for my launch."

Training Friday, 2 October 2002

At this stage of training and preparation, the crew were having vestibular training three times a week on a revolving-chair to help them to acclimatise to microgravity and avoid space sickness, which is a problem for many astronauts during their first few days in space.

On Tuesday, 24 September, De Winne practised using models of the Aquarius, ZEOGRID, GCF, and Message experiments. He had received briefings in Brussels, but this marked his first hands-on experience. On Wednesday, De Winne and Zalyotin received their certificates allowing them to take blood samples. The next day, they had an examination in the centrifuge which contained part of the Soyuz trainer in order to simulate descent and landing manually under g-force.

De Winne wrote, "I had my exam in the centrifuge yesterday. During the coming days I will have several other exams; among others accomplishing a manual rendezvous with the Space Station in the event of a computer failure and an exam about the flight programme, or *polyota* as it is called here. I don't have to know by heart all of its details, but I have to have a general picture of what I have to do during the mission and when I have to do it."

Training Friday, 11 October 2002

This week the crew performed simulations of every aspect of the flight, including Soyuz TMA entry, the verification of all systems, the first days in space, and rendezvous and docking with the ISS. There were also simulations of undocking and the landing. All the training and examinations were conducted under observation by an official commission.

On Thursday they received a visit from Yuri Koptev, Head of Rosaviakosmos, the Russian space agency, who gave them a reception and wished them well.

Training Wednesday, 16 October 2002

On Tuesday, 15 October, an unmanned Russian Soyuz U rocket failed at the Plesetsk Cosmodrome in the north of Russia. It was due to launch the Foton M-1 mission with an ESA suite of 44 experiments, but one of the five engines failed to start and the rocket crashed 15 seconds after a sluggish lift off.

Although the manned Soyuz TMA-1 spacecraft would use the Soyuz FG rocket, an investigation was ordered and De Winne's flight was delayed. Nevertheless, on Wednesday, 16 October, the crew travelled to Baikonur Cosmodrome in Kazakhstan.

Training Thursday, 17 October 2002

On this day the crew examined their actual spacecraft for the first time, and once again donned their Sokol suits and demonstrated that they could enter the capsule safely.

Flight Day 1 – Wednesday, 30 October 2002

At 03:11 UTC on 30 October Soyuz TMA-1 was successfully launched from the historic Pad 1 at the Baikonur Cosmodrome.

De Winne reflected in his blog, "What a ride on the rocket! Although we got up early, I'm feeling quite fit. A lot of work though on the first day, so not much time to write."

Flight Day 2 – Thursday, 31 October 2002

Continuing his blog, De Winne wrote, "We really got some sleep tonight. And I took my first pictures of this beautiful planet. I had my first meal in space, chicken omelette – really tasty. Sergei and Yuri are a great crew."

Flight Day 3 – Friday, 1 November 2002

At 05:01 UTC Soyuz TMA-1 docked with the Pirs module of the ISS. After being welcomed by the Expedition 5 residents, the newcomers held a conference call with friends and family. They received congratulations from two Belgian officials, Minister of the Economy and Scientific Research Charles Picqué and Government Commissioner of Scientific Issues Yvan Ylieff. Also on the call to give their congratulations were ESA

Director of Human Spaceflight Jörg Feustel-Büechl and ESA Director of Strategy and External Relations Jean-Pol Poncelet.

Flight Day 4 – Saturday, 2 November 2002

After some initial problems, the NEUROCOG experiment operated correctly. De Winne wrote, "We have found a quiet place where we can work without disturbing the main crew because they are also very busy packing and preparing for the next Shuttle arrival. As well as the NEUROCOG experiment, we're busy with blood sampling for Sympatho, Message, DCCO and Promiss monitoring. After three days in space, I could finally wash my hair and have an almost normal morning toilet (washed with a wet towel and washed my hair, but it's already messed up again due to the paste for the electrodes of the NEUROCOG experiment). Still, it felt very good."

Initial problems getting the Promiss experiment operational caused a concern, but with the help of Expedition 5 flight engineer Peggy Whitson and ground support it was made operational.

Flight Day 5 – Sunday, 3 November 2002

The whole visiting crew jointly performed the CARDIOCOG experiment on Sunday. Also De Winne participated in events in his home town of Sint-Truiden in Limberg, Belgium including a session with a school.

Flight Day 6 – Monday, 4 November 2002

On this morning De Winne carried out a second set of the NEUROCOG experimental runs. Again there were some small problems, but ground control helped and the experiment was successfully completed. He made a 'morning movie' of getting up, washing, and having breakfast. According to De Winne this was, "Really good fun. I hope that it can be used later to show how life is going on board the Station."

Flight Day 7 – Tuesday, 5 November 2002

De Winne did some work on the NANOSLAB experiment in the Destiny module. He also participated in the ARISS amateur radio communication exercise by talking to schoolchildren. He missed a chance to photograph (or even see) Belgium. As he explained in his blog, "The weather was fine, but during the passage over Belgium I was sitting by the radio, and unfortunately there is no window there. But between talking to the scientists and engineers of tomorrow and looking out through the window, the choice was quickly made."

He halted the DCCO experiment, and in the evening participated in conferences with Belgian TV stations.

Frank De Winne in the Zvezda module. (NASA)

Flight Day 8 – Wednesday, 6 November 2002

The visitors performed the CARDIOCOG experiment for the final time. This was undertaken for four universities using apparatus from CNES that had already been used during the Andromède mission of Claudie Haigneré.

That evening De Winne wrote in his blog, "Unfortunately one of the experiments (NANOSLAB) in the Microgravity Science Glovebox isn't running as planned. Ground control gave help trying to fix it, but so far no success. But this is also space! One hundred percent success is not yet for today in space. We are still learning a lot as we go on. And I'm sure that we will also learn a lot from this!"

Flight Day 11 – Saturday, 9 November 2002

At 20:44:05 UTC, De Winne and his crewmates undocked Soyuz TM-34 from the nadir port of Zarya. This 'taxi' had been in place for 6 months. They left behind Soyuz TMA-1 as a fresh 'lifeboat' for the ISS crew. No one knew at the time that this would be required. When the loss of the Columbia orbiter on 1 February 2003 at the end of the STS-107 mission grounded the remaining Shuttle fleet for the foreseeable future, the Expedition 6 crew of Ken Bowersox, Nikolai Budarin and Don Pettit, who arrived aboard Endeavour in late

November, found themselves in space with no prospect of a ride home on STS-114, so at the end of their mission in May 2003 they departed in Soyuz TMA-1 and landed in Kazakhstan.

Flight Day 12 – Sunday, 10 November 2002
Frank De Winne came back to Earth on Sunday, 10 November with a flawless night landing near the town of Arkalyk on the steppes of Kazakhstan at 06:04 local time (00:04 GMT).

7.4 POSTSCRIPT

Subsequent Missions

See the Oasiss mission.

Frank De Winne Today

De Winne currently serves as Head of the European Astronaut Centre of the European Space Agency in Cologne, Germany. He chaired the technical committee of the second EU-ESA Space Exploration Conference in Brussels in 2010. He also holds the following posts:

- Goodwill Ambassador for UNICEF Belgium.
- Chairman of the Forum Space and Education.
- Member of the SESAR Scientific Committee.

De Winne has been awarded the following honours:

- Officier in de Orde van Oranje-Nassau by the Dutch Queen for showing leadership during operation Allied Force (NATO operations in the former Yugoslavia).
- Medal of Friendship by the Russian Federation.
- Officer of the Order of Léopold.
- Officer of the Order of the Crown.
- Officer of the Order of Léopold II.
- Military Cross 1st class.
- Campaign Medal for foreign operations.
- NATO Medal for Kosovo.
- Ennobled a viscount in the Belgian nobility as a reward for his space achievements.
- Honorary doctorate from Hasselt University.
- Honorary doctorate from University of Antwerp.

Frank De Winne at Euro Space Day at Saarbrucken, Germany, 22 October 2014. (www.regionalverband-saarbruecken.de)

- Honorary doctorate from Gent University.
- Honorary doctorate from University of Liège.
- Honorary doctorate form University of Mons.
- Honorary doctorate from the University of Limburg.

8

Cervantes

Mission

ESA Mission Name:	Cervantes
Astronaut:	Pedro Francisco Duque
Mission Duration:	10 days, 1 hour, 37 minutes
Mission Sponsors:	ESA/CDTI
ISS Milestones:	23rd crewed mission to the ISS

Launch

Launch Date/Time:	18 October 2003, 05:38 UTC
Launch Site:	Pad 1, Baikonur Cosmodrome, Kazakhstan
Launch Vehicle:	Soyuz TMA
Launch Mission:	Soyuz TMA-3
Launch Vehicle Crew:	Alexandr Yuriyevich Kaleri (RKA), CDR
	Pedro Francisco Duque (ESA), Flight Engineer
	Colin Michael Foale (NASA), Flight Engineer

Docking

Soyuz TMA-3	
Docking Date/Time:	20 October 2003, 07:15 UTC
Docking Port:	Pirs
Soyuz TMA-2	
Undocking Date/Time:	27 October 2003, 23:17 UTC
Docking Port:	Zarya Nadir

J. O'Sullivan, *In the Footsteps of Columbus*, Springer Praxis Books,
DOI 10.1007/978-3-319-27562-8_8

Landing

Landing Date/Time:	28 October 2003, 02:40 UTC
Landing Site:	Near Arkalyk
Landing Vehicle:	Soyuz TMA
Landing Mission:	Soyuz TMA-2
Landing Vehicle Crew:	Yuri Ivanovich Malenchencko (RKA), CDR
	Pedro Francisco Duque (ESA), Flight Engineer
	Edward Tsang Lu (NASA), Flight Engineer

ISS Expeditions

ISS Expedition:	Expedition 7
ISS Crew:	Yuri Ivanovich Malenchencko (RKA), ISS-CDR
	Edward Tsang Lu (NASA), ISS-Flight Engineer
ISS Expedition:	Expedition 8
ISS Crew:	Colin Michael Foale (NASA), ISS-CDR
	Alexandr Yuriyevich Kaleri (RKA), ISS-Flight Engineer

8.1 THE ISS STORY SO FAR

A single Space Shuttle flew to the ISS between Soyuz TMA-1 and Soyuz TMA-3. This was STS-113, Assembly Flight 11A. It was a mirror image of STS-112, which had attached the starboard S1 truss to the central S0 truss, to give the ISS the beginning of its first 'wing'. STS-113 delivered the matching P1 truss and during three spacewalks this was installed portside in order to balance the structure. Also delivered was the second Crew and Equipment Translation Aid (CETA-2) and the Thermal Radiator Rotary Joint (TRRJ) for the P1 truss. As well as the assembly tasks, the Expedition 6 crew replaced the Expedition 5 crew. This mission would see a number of unfortunate milestones because of the loss of Columbia on the next Shuttle flight, STS-107, which was a 'solo' mission independent of the station. STS-123 marked the last time that a Russian cosmonaut flew on a Shuttle, and it would be 32 months before a Shuttle visited the ISS again.

STS-107 was a research mission carrying the SpaceHab Double Research Module and Israel's first astronaut. During the ascent a piece of thermal insulation foam detached from the External Tank (ET), struck the leading edge of the left wing and caused sufficient damage to compromise the structural and thermal integrity of the spacecraft. As a result, Columbia broke up on re-entry on 1 February 2003 and the crew were lost.

With the Shuttle fleet grounded pending an investigation, the allocation of ISS crews was changed. It would be maintained by two-person 'caretaker' crews launched on Soyuz spacecraft and they would operate the station with a reduced workload until the assembly flights could resume.

In April 2003 TMA-2 delivered Yuri Malenchenko and Ed Lu for Expedition 7. The Expedition 6 crew, who had arrived on STS-113 and were expecting to return on STS-114 in March, landed in the Soyuz TMA-1 'lifeboat' which had been delivered six months earlier.

The ISS with the P1 and S1 truss segments in place. Note the presence of the MBS and both CETA carts on the S1 truss (on the right side of this view). (NASA)

In other developments, on 15 October 2003 Yang Liwei was launched on Shenzhou 5, making him China's first taikonaut. This vehicle was similar in configuration to the Russian Soyuz, but somewhat larger.

8.2 PEDRO DUQUE

Early Career

Born on 14 March 1963 in Madrid, Spain, Pedro Duque graduated from the Escuela Técnica Superior de Ingenieros Aeronáuticos at the Universidad Politécnica in Madrid in 1986, having earned a degree in aeronautical engineering. That same year he joined the Grupo Mecánica del Vuelo and worked at ESA's European Space Operations Centre in Darmstadt, Germany, where he supported the Precise Orbit Determination Group. He was also a flight controller for the ERS-1 satellite and the European Retrievable Carrier (EURECA).

In May 1992 Pedro was selected to join the ESA astronaut corps, together with Maurizio Cheli, Jean-François Clervoy, Christer Fuglesang, Thomas Reiter and Marianne Merchez (the latter would never fly in space). He also trained at the Yuri Gagarin Cosmonaut Training Centre in Star City, near Moscow. In August 1993 he returned to Star City to train for the EuroMir94 mission and served as the prime Crew Interface Coordinator in the Mission Control Centre in Moscow, where he coordinated between the astronauts on Mir and the European scientists. Duque also served as a Crew Interface Coordinator for the STS-78 Life and Microgravity Spacelab mission in June 1996. Later in that year he was selected by NASA for Astronaut Group 16, called the Sardines, and flew as Mission Specialist on STS-95 in October/November 1998.

In 1999 Pedro returned to ESA and ESTEC in Noordwijk, the Netherlands, to support the Module Projects Division in what was then the Directorate of Manned Spaceflight and Microgravity, where he worked on the development of the Columbus laboratory and the Cupola module for the ISS.

Previous Mission

STS-95

Pedro Duque was a Mission Specialist for the STS-95 mission of Space Shuttle Discovery, from 29 October to 7 November 1998. That mission was renowned for the return to space of 77-year-old US Senator John Glenn, 36 years after his first spaceflight aboard the Mercury capsule 'Friendship 7' in February 1962, when he became the first American to orbit the Earth. Glenn had lobbied NASA for permission to fly on the Shuttle and perform a suite of experiments to investigate the effects of space travel and microgravity on an older person. His data could be compared to his original flight and the impressive database compiled from the annual medical checks that he received at the Johnson Space Center (as did all former astronauts). Over 80 experiments were conducted in the SpaceHab module, including experiments sponsored by ESA and JAXA; hence the international flavour of the crew.

The STS-95 crew with Pedro Duque second from the right in the back row. (NASA)

8.3 THE CERVANTES MISSION

Cervantes Mission Patches

The mission name was a tribute to the Spanish writer Miguel de Cervantes, known for his novel *Don Quixote*. In the mission patch, the stylised astronaut looked to the sky and extended his hand towards the stars that he hoped to reach one day. Cervantes' famous hero, Don Quixote, was riding across the sky towards the brightest star, which was the ISS.

The Soyuz TMA-3 crew with Pedro Duque on the right. (www.spacefacts.de)

A compact disk edition of *Don Quixote* was carried on the mission and was returned to Earth with the signatures of the crew.

By way of the Centro para el Desarrollo Tecnologico Industrial – the Centre for Development of Industrial Technology (CDTI) – the Spanish Ministry of Science and Technology sponsored Duque's flight within the framework of an agreement between ESA and Rosaviakosmos, the Russian Aviation and Space Agency. The patch included the logos of the four participating agencies: ESA; Energia, the Russian S. P. Korolev Rocket and Space Corporation; Rosaviakosmos; and the Spanish Ministry of Science and Technology.

The Soyuz TMA-3 patch was designed by Dutch artist Luc van den Abeelen and was in the shape of the descent module of the Soyuz spacecraft. The crew were represented by stylised cosmonauts in their Sokol suits with their national flags in evidence on their visors and with their hands joined in a display of solidarity. The ISS was shown in its planned final configuration orbiting a blue Earth.

Cervantes Mission Patch. (ESA)

Cervantes Mission Objectives

The Cervantes mission had four objectives:

- To exchange the Soyuz 'lifeboat' at the ISS.
- To carry out a programme of scientific and technical research organised by the European Space Agency and the Spanish Ministry of Science and Technology via the Centre for Development of Industrial Technology (CDTI). A number of experiments from the Odissea mission were to be repeated. Educational and promotional activities would be undertaken with the aim of bringing the European human space programme and research performed in space to a wider public, and to young people in particular.
- To increase operational experience aboard the ISS; from a European perspective the Cervantes mission was important because it would increase the experience of ESA's astronauts ahead of the launch of the Columbus laboratory module.
- To exchange the ISS Expedition crews, because Expedition 8 would fly up on Soyuz TMA-3 and Expedition 7 would return to Earth with Duque on Soyuz TMA-2.

Soyuz TMA-3 Mission Patch. (www.spacefacts.de)

Science

Duque was responsible for a total of 24 experiments. Most were to be performed aboard the Russian segment of the station but some would need the Microgravity Science Glovebox, the ESA facility in NASA's Destiny module that could provide a fully sterilised environment. Most of the experiments were new and sponsored by the Spanish government, but there were also several re-runs designed to obtain additional data for the Italian Marco Polo mission of April-May 2002 and the Belgian Odissea mission of October 2002.

The following experiments were to be carried out as part of the Cervantes mission:

- AGEING
- ROOT
- GENEEXPRESSION
- CARDIOCOG-2 (Odissea)
- NEUROCOG-2 (Odissea)
- SYMPATHO-2 (Odissea)

- BMI-2 (Marco Polo)
- MEDOPS
- MESSAGE-2 (Odissea)
- PROMISS-2 (Odissea)
- NANOSLAB-2 (Odissea)
- Crew restraint
- 3D Camera
- LSO-S (continuing LSO-B from Odissea)
- WINOGRAD
- CHONDRO
- VIDEO-2 (Odissea)
- APIS
- THEBAS
- ARISS-2 (Odissea).

AGEING

Objective: To study in more detail the mechanisms of the abnormal motility response encountered in microgravity by young flies (Drosophila melanogaster), with effects on the posterior ageing response of the flies. Three strains of fly with different phenotypes were to be used in four configurations. The strains were a long-lived strain, a short-lived strain, and a strain that showed an abnormal gravitropic response on the ground. Recently hatched flies of the three phenotypes would be exposed to the space environment.

Task: To record on video tape the in-flight motility of flies in the different experimental containers on the Russian segment of the ISS.

Equipment Used:

- Aquarius-B hardware, including a transport container-incubator to accommodate a set of AGE Biocontainers.
- AGE Biocontainers with a set of reservoirs for three different strains with different phenotypes of young flies.
- Biological transport container BIO-1 to carry the Biocontainers to the Russian segment of the ISS and return them to Earth.
- Temperature data logger and Sony camcorder DSR PD-150P.

Expected Results:

The AGE Biocontainers containing the young flies, along with video recordings of their motility and behaviour in the space environment.

ROOT

Objective: To study the changes that take place in the nuclear structure after exposure to microgravity during cell proliferation.

Task: Exposition of Arabidopsis thaliana root cells aboard the Russian segment of the ISS.

Equipment Used: Similar to AGEING.

Expected Results: The ROOT Biocontainers with seeds germinated in microgravity environment to be studied on Earth.

GENEEXPRESSION

Objective: To investigate the effects of the space environment upon the pattern of gene expression of Drosophila melanogaster pupae exposed to microgravity during their development. The sequencing of the Drosophila genome has made possible genomic and proteomic studies, especially those involving microarrays. A microarray analysis is able to identify the extent of gene expression changes that occur in microgravity using a key model system such as Drosophila.

Task: Exposition of Drosophila melanogaster pupae on the Russian segment of the ISS.

Equipment Used: Similar to AGEING, plus the Kriogem-3 refrigerator.

Expected Results: The GEN Biocontainers with the pupae exposed to the space environment.

MEDOPS

Objective: To identify variations in the carbon dioxide concentration at different locations in the US segment of the ISS.

Task: Measurement of carbon dioxide in places occupied during physical training and sleeping where, in the absence of adequate ventilation, elevated levels of carbon dioxide can occur in the microgravity environment.

Equipment Used: The carbon dioxide monitoring device of the Crew Health Care System (CheCS).

Expected Results: Data on carbon dioxide accumulation in various parts of the ISS.

CREW RESTRAINT

Objective: To test an astronaut restraint concept.

Tasks: Testing of the 'knee-block' concept using a device which restrains the body at knee level and allows the astronaut to work (such as at a laptop) over a prolonged period of time.

Equipment Used: The restraint device comprised the knee restraint and an interface beam. The restraint was designed to restrain the astronaut's feet and was attached on the interface beam, which was in turn mounted on the structural elements of the US segment of the ISS.

Expected Results: An assessment of the new astronaut restraint device.

3D CAMERA

Objective: To acquire practical skills in taking 3D images on board the ISS in order to determine the interest of professionals, mass media, and public in such imageries, and to collect data to be used in upgrading and improving the 3D camera for future utilisation of 3D technology on spacecraft during rendezvous operations.

Tasks:

- Testing of a 3D camera in the zero-g environment on board a Soyuz spacecraft and the ISS.
- Taking stereoscopic images of astronaut activities and hardware accommodation locations inside the ISS.
- Imagery of the Earth's surface.

Equipment Used:

- A stereoscopic 3D camera with twin-lens Nikon AIS and photoflash, 35-mm film, batteries and a universal bracket LIV for camera attachment.

Expected Results: Films returned to the ground with imagery results.

WINOGRAD

Objective: To study the effect of microgravity on the formation of a type of bacterial colony called a Winogradsky column. This is a colony that consists of different types of bacteria in which the waste products of one bacterium serve as the nutrients of the other. They need no other input than light for photosynthesis. Such ecosystems could be important for future long-duration spaceflight if bacteria could help to dispose of waste or to recycle air and water.

Task: Exposition of Winogradsky samples on board the Russian segment of the ISS.

Equipment Used: An upload container, a return container, an experimental unit, an illumination unit, a mockup and battery packs.

Expected Results: Exposed samples in the WINOGRAD return container.

CHONDRO

Objective: To test the experiment hardware as a method of achieving more stable cartilage growth.

Task: Cultivation of chondrocyte cells in microgravity conditions on the Russian segment of the ISS.

Equipment Used: The CHONDRO upload container, return container and battery pack.

Expected Results: Exposed samples in the CHONDRO return container.

APIS

Objective: A demonstration of the rotation of bodies that have different body mass distributions (i.e. spherical, cylindrical or ellipsoidal moment of inertia tensors) and the effects of mechanical energy dissipation under external or internal factors.

Tasks: Video recording demonstrations of rigid body rotation in zero-g conditions and processing of video materials for educational purposes.

Equipment Used: A hand grip, transparent hemispheres (two pieces), spin modules (three pieces), a Sony DSR PD-150P camcorder and video tape, plus a universal adapter bracket LIV for the camera.

Expected Results: Information recorded on video tapes.

THEBAS

Objective: To study the interaction between a container and solid spherical bodies within it when the system is not in a state of rest and is subjected to regular one-dimension oscillations in microgravity conditions. Control experiments using similar equipment would be carried out on the ground (with a deviation normal to the force of gravity) to obtain a qualitative assessment of the effect of gravity on such a system.

Tasks: Video recording for educational purposes of the behaviour of solid spherical bodies of different diameters in closed transparent containers.

Equipment Used: A pendulum, a set of transparent containers, a Sony DSR PD-150P camcorder and video tapes, plus a universal adapter bracket LIV for the camera.

Expected Results: Behaviour recorded on video tape.

Timeline

ESA Flight Schedule 2003

On 28 February 2003, ESA announced the postponement of the next two ESA missions to the ISS. In the wake of the tragic loss of Columbia on 1 February, the Soyuz flights would be required for crew transfers. The established routine was that the Expedition crews rode up and down by Space Shuttle, and that ESA (and other) astronauts visited for short periods during the 'taxi' flights that replaced the Soyuz 'lifeboat'. With the Shuttle fleet grounded, the ISS was reduced to two-person caretaker crews who would transfer by Soyuz. The ESA astronauts affected were Pedro Duque of Spain, whose April 2003 flight was postponed to October, and André Kuipers who didn't fly until April 2004.

"The agreement was made in the interest of a smooth continuation of Space Station operation and utilisation," explained Jörg Feustel-Büechl, ESA's Director of Human Spaceflight. "It should be seen as a sign of the close cooperation and solidarity among the International Space Station partners."

Swedish ESA astronaut Christer Fuglesang, whose Shuttle ride to the ISS had been scheduled for July 2003, was ultimately delayed until the STS-116 mission launched in December 2006.

Flight Day 1 – Saturday, 18 October 2003

The Cervantes mission got underway when Soyuz TMA-3 lifted off from Baikonur in Kazakhstan at 11.38 local time (07.38 Central European Summer Time).

Pedro Duque wrote in his ESA website blog, "A trip into space isn't something you can do every day, so I'm determined to use every moment to the fullest and to perform the various experiments as proficiently as possible. I am proud to be representing almost 40 million Spanish citizens."

Flight Day 3 – Monday, 20 October 2003

Soyuz TMA-3 docked with the Pirs port of the ISS at 09:16 CEST (07:16 UT) while approximately 400 km above Russian territory. At 12:14 CEST the newcomers, Pedro Duque, British-born NASA astronaut Michael Foale, and Russia's Alexander Kaleri entered the station.

Flight Day 4 – Tuesday, 21 October 2003

In the morning, Duque participated in the Sympatho, BMI and CARDIOCOG experiments. He also downlinked video of Soyuz TMA-3's approach to the ISS and a video of one his crewmates taking blood sample from him. In the afternoon he worked on the CARDIOCOG, NANOSLAB BMI and AGEING experiments.

At 10:30 CEST (08:30 GMT) Duque had a 15 minute conversation with José Maria Aznar, the Spanish Prime Minister, and Juan Costa, the Spanish Minister of Science and Technology. The link also included eight students from the School of Aeronautical Engineering of Madrid Polytechnic University.

Flight Day 5 – Wednesday, 22 October 2003

After downlinking some video material, the daily planning conference was held with the team in the Moscow Control Centre.

For the remainder of the morning Pedro worked on NEUROCOG, a neuroscience experiment to investigate human spatial perception. He also took some photos and shot some more video material. After lunch, which lasted from 13:15 to 14:15 CEST, he began using the 3D camera, which was an experiment in its own right. In addition to providing 3D images for illustrative purposes, the images would help in improving ISS simulators.

From about 19:30 CEST Pedro worked on PROMISS, an experiment on protein crystallisation. After dinner with the rest of the crew, he began work on the NANOSLAB experiment to study the formation of zeolites.

Flight Day 6 – Thursday, 23 October 2003

Writing in his ESA blog, Duque corrected an old urban legend that Americans used expensive space pens and the Russians used pencils.

> I'm writing these notes in the Soyuz with a cheap ballpoint pen.
>
> Why is that important? As it happens, I have been working in space programmes for seventeen years, eleven of these as an astronaut, and I've always believed, because that is what I've always been told, that normal ballpoint pens don't work in

> space. The ink doesn't fall, they said. Just try for a moment writing 'inverted' with a ballpoint pen and you will see I'm right, they said. During my first flight I took with me one of those very expensive ballpoint pens with a pressurised ink cartridge, the same as the other Shuttle astronauts. But the other day I was with my Soyuz instructor when he was preparing the books for the flight, and he was attaching a ballpoint pen with a string for us to write once we were in orbit. Seeing my astonishment, he told me the Russians have always used ballpoint pens.
>
> So I also took one of our ballpoint pens, courtesy of the European Space Agency – just in case Russian ballpoint pens are special – and here I am; it doesn't stop working and it does not 'spit' or anything. Sometimes being too cautious keeps you from trying, with the result that things are built more complex than necessary.

His first job of the morning was an interview with the Antena3 channel of Spanish TV. After that he worked on the Crew Restraint experiment and then had another TV interview with Tele-5. During the afternoon he worked on ROOT (plant molecular biology) and AGEING (flies). At about 16:20 he started working on the VIDEO-2 experiment to explain Newton's Laws of Motion to schoolchildren.

Pedro Duque during an interview with Spanish TV. (NASA)

Flight Day 7 – Friday, 24 October 2003

Duque awoke at approximately 07:00 CEST as usual, and after breakfast at 08:30 he devoted most of the morning to the NEUROCOG experiment and an interview with Spanish public television channel TVE. After lunch, he and the ISS Expedition 7 crew, Lu and Malenchenko, prepared for the return to Earth in Soyuz TMA-2.

Flight Day 8 – Saturday, 25 October 2003

From Duque's ESA blog:

> Imagine having to work in a 50 metre long laboratory with several lateral corridors. Everywhere in this lab there's equipment in which experiments are being performed.
>
> The work is planned by people from another country, who keep phoning up to learn details, to ask how things are going, or, more often, to order the operator to conduct the experiments in this way or that.
>
> There are several telephones in this lab, but none of them wireless and the operator needs to go to one of them to answer each call. It is true that being able to solve every doubt and to ask for advice sometimes makes the operator's job easier, and that he (or she) doesn't feel alone when problems arise.
>
> Ah! We're forgetting a little detail: The telephones have four lines, and the operator has to pay attention to the sound of the 'ring' to push the right button when answering, since each of the four groups that control the operator uses a different line. Oh, but it's true, we haven't said there are four groups, all of them very kind.
>
> The operator doesn't even pay attention anymore to the fact that the calls can come in two languages, none of them in one's mother tongue; nor to the fact that the operator spends the whole day meeting in the corridors the other four operators in the lab, each of them busy with a different task, be that an experiment or a repair to lab material. In such a lab one always has a feeling of rushing, of having to hurry everywhere. The day becomes tiring… and one feels like closing up, turning off the lights, and going home.
>
> Let's add another factor: It isn't possible to go home to sleep. Since the lab is in a remote region, one has to sleep inside it and eat inside it – pre-cooked food. There are even those who stay in the lab for up to a year. Every now and then one feels like going to see something else, have a walk, open the window for some fresh air, but that is not allowed either. In this lab the windows can never be opened, and the air gets recycled through filters.
>
> This is somewhat the feeling of the International Space Station, as experienced by a European performing experiments both in the US and the Russian segments, under the direction of people from the European Space Agency.
>
> Let's introduce now an additional factor: The lack of gravity. In the Station one does not walk nor run; one floats from one side to the other. The sensation is, of course, very interesting, and the lack of gravity is also the reason to invest so much

money in placing a lab here. It's true that it is possible to cross the corridor very quickly if in a hurry, only by pushing at one side and braking at the other. However, it is incredibly difficult to do this properly without touching the walls. And ah! As soon as one touches a wall at high speed and without control the laws of physics take revenge and one is sent rotating and kicking everything.

The people who have already spent six month in the Station can cross the 50 metres in little more than 15 seconds. My record after 3 days here (although I have experience from another flight) is 25 seconds – with a bump on the head. Even if it seems a game, sometimes you really have to fly from one side to the other, like when you've forgotten something at the other end (of course) and the time is precious. When there's some free time it is natural to train to move in a more controlled way, to avoid touching anything, especially the treacherous – and very hard – half-opened hatches. I personally estimate that I'd need a couple of weeks to fully adapt myself; to learn to move without having to think about every movement in advance.

Everyone knows that the very best part of this lab is the view. Alongside this is the pleasure of discovering that things get done despite the inconveniences.

Flight Day 9 – Sunday, 26 October 2003

Continuing Pedro's blog:

In a tidy house or office that has a clean floor, you can see immediately if someone has dropped something, a glasses case, for example. Lying on the floor, this object catches the eye, and everybody passing by will notice it. But in space a number of factors make it very difficult to find something that is lost.

First of all, of course, zero gravity. The other day I had a ballpoint pen clipped to my trousers when I brushed past something and lost it. I noticed immediately, so I turned to pick it up. Nothing. My pen was nowhere to be found. It had flown away, I didn't know in which direction, up, down. It could have been anywhere. As I began to accept the fact that I had lost it, I turned to proceed with my daily schedule and I saw it in front of me, flying in the same direction I was moving. When it flew away it must have bounced on something and kept going without waiting for me.

The Space Station is made up of modules in which the working area is more or less rectangular, but ahead and behind, up and down, and even right and left there are many adaptors with hatches. Consequently there are many corners in which things can hide.

Besides that: Which wall is the floor? And which is the ceiling? In many places the four walls of the 'tube' that make a module are all the same, making things difficult to find.

For example, I do quite a lot of work in a module that isn't in the main tube, but is attached to one side. You have to make a ninety-degree turn to enter that module from the others. Until then, everything is fine. When you enter, what is below you, you call 'the floor' and what is above is 'the ceiling'. But sometimes you might come from the other direction on your trip along the tube and make that turn.

Now you'll find that the ceiling is the floor and vice-versa. Because of this, if I have left the computer switched on and attached to the wall, when I go back I always have to make a full turn to find it.

And, finally, there are so many things attached to all four walls. Describing it like this must make it sound untidy, with everything in the way, but it's really a matter of necessity. The photo cameras cannot be put away because we take a lot of pictures of the work we do with the experiments, and of Earth as well. The cameras are stuck into place using Velcro, along with a variety of lenses, flashes, etc. Sometimes you need to turn three times to find the camera you're looking for.

Other items that are also in view are spare parts, food packages, bags full of clothes and so on. It is not that we use these things every day, but there's no more room in the cupboards. The Station is not finished yet, and there won't be enough storage space to put everything away until all the planned modules have been connected.

One of the things that I use most frequently is a notebook to take down information about the experiments. I carry it with me everywhere, I write down the precise hour at which I have changed the samples, the results, things to be aware of, etc. This notebook has to come back to Earth with all this data; it will help the scientists to rebuild exactly how everything has happened. During my first days this little notebook drove me crazy. I stuck a large piece of Velcro onto it to help keep it from flying around, but at the end of the day if it wasn't in the first place that I searched then it took me a while to find it. Now I have gotten used to leaving it in one of three different places, but at first it was a nuisance because I didn't know if I'd put it somewhere else or if it had loosened itself from the Velcro and was floating around – meaning it could be just about anywhere.

Once I made the famous turn carrying the notebook in my hand, plus some books and other items. When I reached the workplace I didn't have the notebook anymore. I knew I must have lost it when I slightly hit the corner, so I went back immediately. It was nowhere to be found. After some frantic minutes of fearing that I'd bitterly pay in delays by the end of the day, I found it in a little hidden corner. Thank heavens.

I forgot to mention one important factor in losing things: The air stream. Because everything floats, and because the air gets renewed and cleaned with fans and filters, anything that flies away tends to follow the direction of the air stream. The air stream actually helps more than it hinders, contrary to what you may expect. If you have lost something small you only have to wait a couple of hours and you'll probably find it at the filter where the air enters the circulation system. We're already used to checking there every now and again. All kind of interesting things can be found.

Flight Day 10 – Monday, 27 October 2003

From Pedro's blog:

All you know is that you've just woken up. Nothing, really nothing, squeezes you or pushes you. Nor do you have to make any effort to raise your arm and rub your eyes. In fact, strangely enough, your hand seems to want to get closer to your eye all of its own accord. You feel you have slept very deeply, and the world around you calmly comes into focus. A noise… engines? No, it's the fans – a light breeze.

You open your eyes to a weird light and laptop computers switched on all around you. Why are they up on the ceiling? Of course, you are waking up in a Space Station after a night inside your sleeping bag, with your arms floating in front of your face as usual and with your legs in that semi-curled position where the push and pull between the different muscles is evenly balanced. It's amazing how well you can sleep in here, thanks to how much they make you work and the softness of the mattress.

You check the time and it's a quarter to five. Soon, you think, the alarm clock will ring. You could sleep a few more minutes; it would do you good. The mist clears and the daily plan comes to your head. Let's see, biology, medicine, physical experiments ... and TV interviews. Then you suddenly remember the first interview is at 06:00 and that can't wait.

If you start an experiment 15 minutes late, you might be delayed for the rest of the day, but the orbits are a relentless clock and the Station will pass over the antennas that are to pick up the television signal at 06:00; not a second more, not a second less. Will you have time for breakfast after the connection?

Forget about sleeping a bit longer. Quickly get out of the sleeping bag and look for the precise plan of the day. Oops, too quickly! The sleeping bag and the plan that was hanging on the wall quickly move away. Without your reflexes fully awake, you don't manage to find a handle and you end up in the laptops, which fortunately are affixed to the wall and for the nth-time they cushion you. It doesn't matter, the laptops have the electronic version of your daily plan. In fact, after the television interview you will be straight into the experiment activities and you have to hurry.

You take your clean clothes, your soft-soled shoes, and the paper with the details of the interview. Flying towards the 'bathroom' you meet no one; everything is quite dark and you are the first up. The bathroom is just a point in the corridor with a large mirror and everything stuck to the wall with Velcro – combs, water bags, soap bags, no-rinse camping shampoo, etc. You find your towels, and wet one of them with the soap. You get cleaned, more or less; this isn't the day to be meticulous. You comb your hair, then get dressed… Now for breakfast.

Luckily your commander is up and, even if he hasn't had time to get dressed, he is switching on the Station's television system and setting up the camera. What does the note about the interview say? With the Spanish flag in the background, this television, this commentator, these are the possible questions.

Stretch the flag, which of course in zero gravity stubbornly persists in becoming all tangled up. This isn't the first time – all is under control and the background is ready in 2 minutes. It is 05:30.

You take some tea bags, a canned omelette and a bag containing bits of white bread. You begin filling the bags with lyophilised products and in the rush you burn your hand but it isn't serious because the low temperature avoids people getting badly burnt.

With the can-opener you uncover the omelette, which at first sight could be pâté or cheese. Fortunately the spoon hasn't flown away during the night, this time you won't have to recover it from a filter; you are not making beginner's mistakes anymore.

It is 05:48 and the lights have to be turned on. Your commander shouts out "balans bieloie!!" and you immediately grab a piece of paper and put it in front of the camera to help to define the white level for the camera; that's what your commander was asking you to do in Russian. The focusing process makes you leave your breakfast for a while.

At 05:58 you leave somewhere the last bits of omelette, half-eaten; most likely they will stain something. You squash the can to save room in the garbage. They are already calling through the radio in Moscow. Yes, we are ready (not true, you are still cleaning your face). Okay, you can start the interview.

When the interview ends, you share a burst of laughter with your mates because of the near miss. This relieves tension somewhat. You still have a full day of work ahead. Thank heavens it will end back in your sleeping bag, a pure feather mattress.

Flight Day 11 – Tuesday, 28 October 2003

From Pedro's blog:

This is the last full day that I will spend on the Station. There's still a lot to do, but the emphasis is now on finishing everything up and packing the results of the experiments correctly.

There are a number of experiments that make the most of the flight right up to the last minute, squeezing in the hours of zero gravity to study fluids or materials. Others must be either kept frozen or at a controlled temperature until just before we depart, to avoid losing data in the time they'll spend in the capsule. As we do not have fridges or heaters in there, they will be kept in a thermos instead.

The mood in the Station has changed. The outgoing crew have their minds more and more on Earth. They have already passed the 'baton' to the new crew. Some little things have already changed, even though I suppose the newly arrived crew will still need time to set everything up to their own satisfaction. Today I noticed that the items had already changed place in the bathroom area. It also surprised me when the new crew wanted to know where I had left everything, so that they could keep using them or put them away where they think is best.

I do not have first-hand knowledge of how it feels when you leave here after many, many days in orbit. I imagine that they must be ready to go back and eager to meet their families. They say the months they've spent in the Station have been very nice and they don't regret their coming to an end, but I can imagine that when they reach the ground happiness will be greater than nostalgia.

I have only met one colleague who, on the very same day of the landing, just as he was coming out of the plane bringing him from the Steppe, said, "I miss it, it was better in the Station." That was Valeri Polyakov in 1995, winner of the Principe de Asturias award, after the longest stay so far of an astronaut in space: 14 months. I don't know how Valeri feels about it now, I haven't thought about asking him. Maybe he'd say it was just an impulse and he changed his mind afterwards. But it is a privilege to work here for any amount of time and the experience is unforgettable.

In my case, after just 10 days, of course I could remain longer if the workload was reduced to equal that of my crewmates. My days would be calmer and I could enjoy the advantages. I feel a bit like a tourist who sees his final day approaching,

> and feels there are still so many things to do and to enjoy. Over the last few days I've been stealing a few minutes here and there to learn to move with ease and to look out of the window at Earth by day, Earth by night, the stars…
>
> I have been fortunate today. I've been able to see the fires in California, an amazing sight. I decided to take a moment to enter a module which is located out of the way and switch the lights off to see the night. I've seen something we can often see from up here, an electric storm in which the flashes of lightning propagate through the clouds. And in the middle of the show I saw something small but very beautiful: A shooting star below us. A nice farewell.

Soyuz TMA-2 landed near the town of Arkalyk in Kazakhstan at 08:40 local time, 03:40 Central European Time, with the Expedition 7 crew of Malenchenko and Lu having spent 185 days in space and Pedro Duque a mere 10 days.

"The experiment programme has been a complete success," said ESA Mission Manager, Aldo Petrivelli of the Cervantes mission, "and results have been obtained for all 22 experiments. These included two physical science experiments which utilised the European-built Microgravity Science Glovebox on board the ISS, four biological experiments, four human physiology experiments, and a number of educational experiments and technology demonstrations."

"We are both delighted and proud of the success of all aspects of the mission and thankful to our Russian partners," said Jörg Feustel-Büechl, ESA Director of Human Spaceflight. "This is the fourth such Soyuz mission to the ISS with an ESA astronaut, which strengthens the ties and experience we share with our colleagues at Rosaviakosmos. We look forward to the continuation of this cooperation. We are also pleased that the missions undertaken by ESA's astronauts are not only productive for the scientific community as a whole but for younger generations who will become Europe's scientists of tomorrow. Furthermore, the completion of this mission provides ESA with increased experience for future short and long-term missions, which will be useful in the operation and utilisation of Europe's Columbus Laboratory when it is launched to the ISS."

8.4 POSTSCRIPT

Subsequent Missions

At the time of writing, the Cervantes mission was Pedro Duque's last spaceflight.

Pedro Duque Today

Duque was seconded by ESA as Director of Operations of the Spanish User Support and operations Centre in Madrid, which is managed by the Instituto da Riva and Universidad Politécnica de Madrid. He managed the implementation and first operations of the Centre until 2006. In October of that year he took special leave from ESA to head an Earth observation project – leading Deimos Imaging S.L. (DMI) until 2011, initially as Managing Director and then as Executive President. DMI is a private company which operates a commercial Earth observation system with its own satellite and ground stations.

Pedro Duque being interviewed in May 2015. (www.cuatro.com)

Duque is a member of the Spanish Academy of Engineering and has the following honours:

- Order of Friendship awarded by Russian President Boris Yeltsin March 1995.
- Great Cross of Aeronautical Merit awarded by the King of Spain in February 1999.
- Principe de Asturias Prize for International Cooperation shared with three other astronauts.

After his special leave, Duque returned to ESA in October 2011 as member of ESA's astronaut corps and was placed in charge of the Flight Operations Office at the Columbus Control Centre near Munich, Germany, which supervises the mission flight control teams and supports the scientists and engineers who perform ground operations for the European module and its utilisation programme.

9

DELTA

Mission

ESA Mission Name:	DELTA
Astronaut:	André Kuipers
Mission Duration:	10 days, 20 hours, 52 minutes
Mission Sponsors:	ESA
ISS Milestones:	ISS 8S, 24th crewed mission to the ISS

Launch

Launch Date/Time:	19 April 2004, 03:19 UTC
Launch Site:	Pad 1, Baikonur Cosmodrome, Kazakhstan
Launch Vehicle:	Soyuz TMA
Launch Mission:	Soyuz TMA-4
Launch Vehicle Crew:	Gennadi Ivanovich Padalka (RKA), CDR
	André Kuipers (ESA), Flight Engineer
	Edward Michael 'Mike' Fincke (NASA), Flight Engineer

Docking

Soyuz TMA-4	
Docking Date/Time:	21 April 2004, 05:01 UTC
Docking Port:	Zarya Nadir
Soyuz TMA-3	
Undocking Date/Time:	29 April 2004, 20:52 UTC
Docking Port:	Pirs

J. O'Sullivan, *In the Footsteps of Columbus*, Springer Praxis Books,
DOI 10.1007/978-3-319-27562-8_9

Landing

Landing Date/Time:	30 April 2004, 00:12 UTC
Landing Site:	Near Arkalyk
Landing Vehicle:	Soyuz TMA
Landing Mission:	Soyuz TMA-3
Landing Vehicle Crew:	Alexandr Yuriyevich Kaleri (RKA), CDR
	André Kuipers (ESA), Flight Engineer
	Colin Michael Foale (NASA), Flight Engineer

ISS Expeditions

ISS Expedition:	Expedition 8
ISS Crew:	Colin Michael Foale (NASA), ISS-CDR
	Alexandr Yuriyevich Kaleri (RKA), ISS-Flight Engineer
ISS Expedition:	Expedition 9
ISS Crew:	Gennadi Ivanovich Padalka (RKA), CDR
	Edward Michael 'Mike' Fincke (NASA), Flight Engineer

9.1 THE ISS STORY SO FAR

There were no missions between Pedro Duque's flight on Soyuz TMA-3 and André Kuipers' flight on Soyuz TMA-4.

9.2 ANDRÉ KUIPERS

Early Career

André Kuipers was born on 5 October 1958 in Amsterdam, the Netherlands. In 1977 he completed the Van der Waals Lyceum high school in Amsterdam. He then enrolled at the University of Amsterdam and graduated in 1987 as a Doctor of Medicine. During his medical studies André worked in the Vestibular Department of the Academic Medical Centre in Amsterdam, where he was involved in research on the equilibrium system.

In 1987 and 1988, as an officer of the Royal Netherlands Air Force Medical Corps, he studied incidents caused by disorientation in pilots of high-performance aircraft. In 1989 and 1990 he worked for the Research and Development Department of the Netherlands Aerospace Medical Centre located in Soesterberg, where he was involved in research on the Space Adaptation Syndrome, contact lenses for pilots, the vestibular system of the inner ear, blood pressure and cerebral blood flow in both high-acceleration in a human centrifuge and in periods of microgravity on aircraft flying parabolic arcs.

He also performed medical examinations of pilots, monitored human centrifuge training, and taught pilots about the physiological aspects of flying.

From 1991 he was involved in the preparation, coordination, baseline data collection, and ground control of physiological ESA experiments for space missions. In particular, he was a project scientist for the Anthrorack human physiology facility that flew on the Spacelab D2 mission in 1993, and for two lung and bone physiology experiments that were undertaken during the EuroMir95 mission. He was then involved in the development of the Torque Velocity Dynamometer which flew on the LMS Spacelab mission in 1996, the Muscle Atrophy Research and Exercise System (MARES) on the ISS, and an electronic muscle stimulator (PEMS) for astronauts.

In July 1999 André joined the ESA astronaut corps and was assigned to ESTEC in Noordwijk, the Netherlands, to continue his work on microgravity experiments. Prior to initiating his training for his spaceflight, André supported a research programme in physiological adaptation to weightlessness in humans. He coordinated European experiments on lung function and blood pressure regulation using ESA's Advanced Respiratory Monitoring System (ARMS) on board the ill-fated STS-107 mission.

In 2002, after he completed the ESA Basic Astronaut Training Programme, Kuipers attended the Gagarin Cosmonaut Training Centre near Moscow, where his programme included the ISS systems, winter and water survival training, and spacewalk training.

In 2003 André was Crew Interface Coordinator at Mission Control in Moscow for both the Soyuz TMA-2 and Soyuz TMA-3 missions to the ISS, acting as backup for ESA astronaut Pedro Duque on the latter.

Previous Missions

N/A

9.3 THE DELTA MISSION

Mission Patches

DELTA was an acronym for the Dutch Expedition for Life Science, Technology and Atmospheric Research. It also referred to the Dutch Delta Works in the southwest of the Netherlands, the barrier which prevents the North Sea from threatening the large area of land formed by the Rhine-Meuse-Scheldt delta.

The logo for the mission was created by the Dutch design agency Crasborn BV. The planet was in red white and blue to represent the flag of the Netherlands. These same colours were employed in the word 'DELTA' in homage to the type of art from the De Stijl movement which advocated abstraction and simplicity with the reduction to primary colours with black and white using basic geometric forms.

The DELTA crew with André Kuipers on the right. (www.spacefacts.de)

DELTA Mission Patch. (ESA)

The three stars in the logo represented the Dutch national colour of Orange and the partners in the DELTA mission; European Space Agency (ESA), the Netherlands Government and Rosaviakosmos (the Russian Aviation and Space Agency, which had been the Russian Federal Space Agency, FKA, since 13 March 2004). The logo of ESA was joined by those of the two Dutch Ministries that funded the mission: the Ministry of Economic Affairs and the Ministry of Education, Culture and Science.

Soyuz TMA-4 Mission Patch. (www.spacefacts.de)

The Soyuz TMA-4 patch was designed by Dutch artist Luc van den Abeelen. He wanted the patch to express the pride of manned spaceflight and recapture some of the 'feel' of the old Apollo patches. The design featured the Soyuz TMA-4 spacecraft orbiting a blue Earth. The ISS was shown in its then-current configuration silhouetted against an orbital sunrise. A rather convoluted use of seven stars was used to commemorate the 43rd anniversary of Yuri Gagarin making mankind's first spaceflight (four on the left and three on the right) as well as the seven crew of Columbia. The three stars in the border represented the crew. Red, white and blue conveniently represented the flags of USA, Russia and the Netherlands, with the sequence being red, blue and white for one third of the way to represent Russia. The names of the crew were located above the Soyuz, with Kuipers' name in orange in honour of the Dutch national colour.

DELTA Mission Objectives

The DELTA mission was made possible, in part, by the Dutch Ministry of Economic Affairs and the Ministry of Education, Culture and Science. Both ministries wished to use the flight to emphasise the important role played by Europe and in particular the Netherlands in international space travel.

The DELTA mission had four objectives:

- Replace the ISS 'lifeboat' – As the orbital life of a Soyuz is six months, it must be replaced on a regular schedule. The DELTA crew would arrive in Soyuz TMA-4 and depart in Soyuz TMA-3.
- Change of ISS residents – Padalka and Fincke were to take over from Kaleri and Foale, with the latter pair returning to Earth with Kuipers.
- Science experiments – Kuipers' programme included about fifteen scientific experiments; for example a new type of light bulb for Philips and the Technische Universiteit Eindhoven. Some of the experiments would extend research carried out on earlier ESA missions.
- Education – The mission had a significant educational component. Kuipers was to grow cress at the same time as schoolchildren on Earth and for the VIDEO-3 project he would create a DVD for Dutch schools showing how he lived in space.

Science

The following scientific experiments were to be carried out during the DELTA mission:

- FLOW
- KAPPA
- ACTIN
- ICE-FIRST
- TUBUL
- CIRCA
- SYMPATHO-3 (continued from Odissea)
- MUSCLE
- Eye Tracking Device
- Motion Perception
- SUIT
- SAMPLE
- PROMISS-2 (continued from Odissea and Cervantes)
- ARGES
- HEAT
- Mouse Telemeter

- LSO-H (continued from Odissea and Cervantes)
- VIDEO-3 (continued from Odissea and Cervantes)
- ARISS-3 (continued from Odissea and Cervantes)
- SEEDS
- GRAPHOBOX
- Bug Energy.

FLOW

Objective: A study of cell mechanosensitivity (by the example of chicken osteocytes, bone cells) under near-weightless conditions.

Task: To study loading-induced osteoblast responses in microgravity conditions and their comparison with chicken osteoblasts and periosteal fibroblasts.

Equipment Used:

- Electronics units and memory modules.
- FLOW experiment containers (12 units) each one including a plunger box unit filled with a cell culture.
- KUBIK AMBER was an incubator (i.e. a thermally insulated container) with a set of mounting slots for containers with the insert centrifuge. The experiment was to be done in this incubator.
- KUBIK TOPAS was an incubator with a set of mounting slots for transporting the experiment containers to the Russian segment of the ISS and for stowage of containers after completion of the experiment until the return to Earth.
- FLOW Biokit to return the results of the experiment to Earth.

Expected Results: Incubated cell cultures in FLOW experiment containers and information recorded in memory modules.

KAPPA

Objective: Determination of the response of myeloid phagocytic cells on lipopolysaccharides of gram-negative molecules under microgravity conditions in activation of the transcription factor.

Task: To verify the hypothesis that microgravity inhibits activation of the transcription factor.

Equipment Used:

- Electronics units and memory modules.
- KAPPA experiment containers (4 units) each one including a plunger box unit filled with an embryonic calf serum.
- KUBIK AMBER incubator.
- KUBIK TOPAZ incubator.
- KAPPA Biokit to return the results of the experiment to Earth.

Expected Results: Incubated cell cultures in KAPPA experiment containers and information recorded in memory modules.

ACTIN

Objective: Investigate the effect of microgravity on the structure of actin microfilaments in mammalian cells, activated or not with growth factors.

Task: To incubate mammalian cells in microgravity conditions.

Equipment Used:

- Electronics units and memory modules.
- ACTIN experiment containers (8 units) each one including a plunger box unit filled with mouse fibroblasts.
- AQUARIUS incubator for storing the experiment containers aboard the ISS prior to starting the experiment.
- KUBIK AMBER incubator.
- KUBIK TOPAZ incubator.
- ACTIN Biokit for return of the experiment results to Earth.

Expected Results: Incubated mouse fibroblasts in ACTIN experiment containers.

ICE-FIRST

Objective: Study heritable changes caused by space radiation.

Task: To incubate strains of alive worms in space conditions.

Equipment Used:

- Electronics units and memory modules.
- ICE-FIRST experiment containers (8 units) each one including a plunger box unit filled with worm strains.
- AQUARIUS incubator.
- KUBIK AMBER incubator.
- KUBIK TOPAZ incubator.
- ICE-FIRST Biokit for return of the experiment results to Earth.

Expected Results: Incubated worm strains in ICE-FIRST experiment containers.

TUBUL

Objective: A study of the effects of microgravity on the cytoskeleton of individual walled plant cells.

Task: To cultivate plant cells in suspension culture (wild type Nicotiana tabaccum) and the chemical fixation of these plant cells at different time points under microgravity conditions.

Equipment Used:

- Electronics units and memory modules.
- TUBUL experiment containers (8 units) each one including a plunger box unit filled with plant cultures.
- TUBUL activation unit.
- AQUARIUS incubator.
- KUBIK AMBER incubator.
- KUBIK TOPAZ incubator.
- TUBUL Biokit for return of the experiment results to Earth.

Expected Results: Cultivated plant cells of wild type Nicotiana tabaccum in TUBUL experiment containers.

CIRCA

Objective: A study of space adaptation of the cardiovascular system to microgravity conditions.

Tasks: 24-hour monitoring of blood pressure, and measurement of finger blood pressure and ECG.

Equipment Used: A PORTAPRES to measure finger blood pressure, BMI-2 device to measure blood pressure, Cardioscience kit, CIRCA kit, blood pressure instrument kit, the EGE-2 experiment control system with switching unit and power supply, plus a DON-10 hard disk.

Expected Results: The hard disk and PCMCIA card containing blood pressure data and research results appropriate to the performed protocols.

MUSCLE

In weightlessness, astronauts often experience lower back pain. This was a surprise because on Earth back pain is associated with heavy spinal load, mainly a consequence of gravity. The hypothesis was that lower back pain may develop without compression of the vertebra. The explanation comes from the fact that the lower part of the vertebrae, the sacral bone, requires to be kept in position relative to the pelvic girdle (hip bones). And a deep 'muscle corset' plays an important role in this process, with the tonic postural muscles being activated when getting up in the morning and then deactivated when resting. It was hypothesised that this protective mechanism does not work in weightlessness. In space bones lose calcium and strength; hence the deep 'muscle corset' atrophies, leading to strain in certain ligaments, particularly in the lower region in the back, thereby causing lower back pain.

Objective: Assessment of astronaut deep 'muscle corset' atrophy in response to microgravity exposure.

Tasks: Assessment and recording of development of lower back pain in space.

Equipment Used: A set of questions for daily recording of subjective lower back pain experienced by the astronaut during a day.

Expected Results: Information recorded in the crew logbook.

ETD

Objective: Study of vestibular-oculomotor orientation in microgravity conditions.

Task: Measurement of eye and head motions by using a 3D eye-tracking device, recording and storage of received data.

Equipment Used: A system unit consisting of a dedicated computer that has a built-in power supply, a lightweight removable measuring unit for the head of the test subject that provides data and electrical interfaces for the system unit, an accessories kit, and a return container.

Expected Results: Data recorded on a hard disk.

Motion Perception

Objective: Assessment of the degree of vestibular adaptation to g-transitions.

Task: To gain insight in the process of vestibular adaptation by rating motion perception.

Equipment Used: Crew logbook with a set of questionnaires in which the astronaut would document his motion sensation and the nature of discomfort experienced in the course of his daily activity and associated with the space adaptation syndrome.

Expected Results: Information recorded in the crew logbook.

SUIT

Objectives: To study the role of tactile signals in the sensor system that determines spatial orientation, and develop a tactile support system for the astronaut that improves safety, performance and comfort.

Task: Evaluate a vibrotactile suit as a means of presenting information on orientation behaviour to an astronaut.

- A vibrotactile suit consists of small vibrating elements attached to the torso in a matrix pattern. The localised vibration of an element is directly mapped to the body coordinates, which makes it an intuitive and efficient way to present spatial infor mation such as directions. For activation of the vibrotactile array built into the suit, it uses information from gyroscopes possessing two degrees of freedom which are attached to the suit. The interaction with the suit is provided by using a control panel worn on the left arm of the participant. The SUIT experiment hardware is powered by a battery. The experiment data is recorded on replaceable flash cards.
- Each experiment session consists of two phases of activities: the first phase ('orientation') is a set of standard orientation tasks; in the second phase ('daily life') the participant performs his normal duties while wearing the suit.

- In the 'daily life' phase the astronaut works independently and performs some activities which are unrelated to the experiment. The suit registers the movements of the subject. It is necessary to obtain an understanding of both the positive and negative effects of the suit in an operational setting (while performing daily activities).

Equipment Used: Two SUIT Containers, ETD, SAMPLE, SUIT Pouch SUI-400, and a return pouch.

Expected Results: The data recorded on flash memory cards, together with a Dictaphone record of the test subject's answers and comments.

SAMPLE

Objectives: To evaluate microbial species that might inhabit the ISS, and to investigate the mechanism of microbial adaptation to microgravity.

Tasks: Microbial sampling of the Dutch astronaut and various surfaces on the Russian segment of the ISS for strains isolation and investigation of their adhesion abilities, plus exposure of Escherichia coli (commonly known as E. coli) on the Russian segment of the ISS during the VC-6 mission.

Equipment Used:

- SAMPLE Collection Kit, including sub-packs to take samples from the astronaut and various surfaces (computers, walls, personal cabins, technical equipment, etc.).
- SAMPLE Specimen Case filled with biological objects.
- KUBIK TOPAZ container, a thermally insulated container (incubator) for stowage of partial samples before the return to Earth.

Expected Results: Microbial samples taken on the Russian segment of the ISS and biological objects which were exposed to microgravity conditions.

ARGES

Objective: To identify the critical factors involved in the onset of helical instabilities in high-intensity discharge lamps. This experiment was to be performed in the Microgravity Science Glovebox (MSG) on the US segment of the ISS.

Task: Take emission spectra and data on helical instabilities in plasma in order to characterise radial de-mixing.

Equipment Used:

- ARGES Experiment Container (with damping frame) consisting of 20 lamps filled with xenon, mercury (up to 10 mg) and iodic salts of various metals.
- ARGES Kit with experiment accessories.
- ARGES Return Kit comprising mini-DV video tapes and pen drive data storage.

Expected Results: Information recorded on pen drive data storage and video tapes.

HEAT

Objectives: To characterise the heat transfer performances of a grooved heat pipe in microgravity and to validate the mathematical hydraulic models employed in designing new generations of heat pipe. It was to be performed in the Microgravity Science Glovebox (MSG) on the US segment of the ISS.

Tasks: Perform measurements in heat pipes in three functioning modes:

- Parallel heating/cooling (i.e. evaporation from below, condensing from below).
- Anti-parallel heating/cooling (i.e. evaporation from above, condensing from below).
- Mixed heating/cooling conditions (i.e. evaporation from below and above, condensing from below).

Equipment Used: An experiment box with a heat pipe in which liquid ammonia is used as a coolant, plus various experiment accessories.

Expected Results: Information recorded in the MSG computer and downlinked to Earth.

Mouse Telemeter

Objective: To test and calibrate small accelerometers developed by the Space Telemetry for Animal Research (STAR) programme. The calibration data could only be obtained in weightless conditions.

Task: Autonomous calibration of accelerometers in microgravity.

Equipment Used: The MOT Pouch containing a box with three orthogonal accelerometers, dedicated electronics and battery pack, and the memory card to record data about acceleration measurements in three orthogonal orientations.

Expected Results: Acceleration measurements in three orthogonal orientations recorded on the MOT memory card (which would be returned to Earth in the SUIT Return Kit).

SEEDS

Objective: To demonstrate that plants respond to gravity by a directed growth (on Earth) and exhibit a non-preferential or 'disoriented' growth direction when grown in the dark in microgravity.

Task: To demonstrate to young people (10–15 years) the influence of gravity on the growth of plants.

Equipment Used: A pouch providing folding rocket (3), plant growth chamber inserts (6), syringe with water (30 ml), Zip-lock bags (3), silica gel bag, filter paper with ordinary rocket lettuce (Rucola) plant seeds, plus a Sony DSR PD-150P camcorder and Mini-DV video tapes.

Expected Results: Video tape of seed growing in the microgravity environment, and some seeds not used in the experiment (which would be returned to Earth in the SUIT Return Kit).

GRAPHOBOX

Objective: Assess the effect of phototropism (growth towards a light source) and gravitropism (growth towards the gravitational vector) on the basic architecture of plants.

Task: To record the germination features of wild-type and mutant seeds of Arabidopsis thaliana in the dark and in blue (465 nm) low fluence light on board the Russian segment of the ISS.

Equipment Used:

- A launch container with GRAPHOBOX with agar dishes with Arabidopsis thaliana (2), ABS container (electronic device), and a temperature data logger.
- A Sony DSR PD-150P camcorder and a Nikon Coolpix 5400 camera.
- GRAPHOBOX return container.

Expected Results: Temperature values recorded by the temperature logger, as well as photo and video materials demonstrating plant growth in microgravity.

Bug Energy

Objective: To record the effects of microgravity on the output of bacterial fuel cells.

Task: To measure the current, voltage, and temperature in the experimental fuel cells throughout the course of the experiment in microgravity.

Equipment Used:

- A launch container with fuel cells (2) filled with biological and chemical materials, an electronic unit and a data logger.
- A Sony DSR PD-150P camcorder and a Nikon Coolpix 5400 camera.

Expected Results: Current, voltage and temperature records by the data logger, as well as photo- and video materials demonstrating the hardware operation.

Timeline

During his training for the DELTA mission, André Kuipers wrote a diary for the ESA website. This is a summary of his pre-launch experiences.

Training 31 November to 4 December 2003

This week Kuipers, along with his Soyuz TMA-4 crewmates and their backups, visited the European Space Research and Technology Centre (ESTEC) in Noordwijk, the Netherlands. Here they received training on the DELTA experiments. They also met the engineers and technicians who explained how they verify that the scientific equipment is sufficiently robust for the arduous journey into space; e.g. vibration testing, acoustic testing, and verification that the materials don't release dangerous vapours.

Kuipers found it amusing to be on the receiving end of medical and physiological testing, such as centrifuge testing. In his earlier career as a medical doctor specialising in aerospace medicine, he had been the one running these tests upon others. As Kuipers wrote in his training diary, "When I worked in Soesterberg, I was a regular test subject in the centrifuge there. This week I am back in it again for medical tests. I listen to some music whilst spinning around for an hour at three times the force of normal gravity. I am so used to it by now that I sometimes just fall asleep."

The cosmonauts, trainers and the doctors spent Wednesday evening together enjoying a meal in Amsterdam.

Training 5 December to 11 December 2003

While spending time at home in the Netherlands, Kuipers retrieved some personal items from his attic to display at an ESA Space Expo. These included the *Perry Rhodan* adventure books that inspired his love of space when a child, so he decided to carry one of them on his DELTA mission.

The issue of *Perry Rhodan* that André Kuipers took with him into space. (Wikipedia)

While in Soesterberg, Kuipers tested a vest that was developed by Nederlandse Organisatie voor Toegepast Natuurwetenschappelijk Onderzoek (the Netherlands Organisation for Applied Scientific Research, TNO). It contained vibrating elements to assist in orientating the wearer to prevent space sickness in weightlessness.

After a medical examination, Russian doctors expressed concern about the state of six of Kuipers' molar teeth. As there are no dentists on the ISS he was advised to have them crowned. This resulted in some hectic days in which he flew to Cologne so that a German air force dentist could fit six crowns before his return to Star City by way of Brussels.

Training 12 December to 18 December 2003

Over the weekend, the Soyuz TMA-4 crew of Kuipers, American Bill McArthur and Russian Valeri Tokarev visited Rostov, which was Tokarev's (and Valentina Tereshkova's) home town.

On Monday and Tuesday in Star City, Kuipers received further medical examinations and training on the laser rangefinder that would be required during a manual docking. Wednesday was spent with the film crew of the Dutch television programme *Jules Unlimited,* who made two shows featuring him. They filmed him in all aspects of his mission, including donning the spacesuit, experiencing g-force in the centrifuge, a tour of the ISS simulator and, of course, the staple of all media questions: the process of going to the toilet in space.

Training 25 December 2003 to 8 January 2004

During his last holiday prior to the spaceflight, Kuipers enjoyed Christmas at home in the Netherlands with his family before they headed to Kiruna, Sweden, 300 miles above the Arctic Circle, for the New Year. Since skiing was deemed too risky this close to the launch, his activities focused on sledding and learning about the Sami lifestyle.

On Monday, Kuipers attended a New Year reception at the Ministry of Economic Affairs as guest of honour to Minister Laurens Jan Brinkhorst.

Training 8 January to 14 January 2004

Kuipers' crewmate Bill McArthur was replaced by his American colleague Leroy Chiao on the crews for Soyuz TMA-4 and ISS Expedition 9. (A decision that would be revisited.) Owing to a medical issue McArthur had to remain grounded for six months; he would eventually fly on Soyuz TMA-7 to command Expedition 12 in October 2005.

The crew, including Chiao this time, continued to train for every emergency in the Soyuz simulator including fire, loss of pressure and ultimately evacuation. Kuipers requested and received extra tuition from master simulator trainer Igor Ivanovic on Sunday.

Monday saw video training, and on Tuesday the ESTEC trainers arrived in Star City to review the experiments that would occupy 80% of Kuipers' time aboard the station. The scientist responsible for the Eye Tracking experiment even arrived in person.

On Wednesday Kuipers joined NASA astronauts to watch CNN as President Bush announced his Vision for Space Exploration that was intended to expand the human presence deep into space. The optimism that this announcement generated would be short lived however, since the programme was never fully funded and was eventually cancelled by President Obama and replaced by a considerably less adventurous proposal.

Training 15 January to 21 January 2004

This week's training focused on the database that contained the ISS inventory. Kuipers learned that every item aboard was logged, with the data including its mass and location. With crews changing routinely every 6 months, it was vital that the new crew be able to find anything they required. If a crewmember moved something, they had to update the database. Although his DELTA experiments and personal equipment had not yet been entered into the database, Kuipers could find items which were used by his colleagues Pedro Duque and Frank De Winne during their Cervantes and Odissea missions. He even found references to their bodily wastes that were to be stowed aboard a Progress cargo ship to burn up during re-entry!

The continuing language training included speaking in Russian, singing Russian songs, reading newspapers, and watching videos. On this occasion, Kuipers translated Russian tourist videos about Amsterdam.

Although Kuipers hadn't been able to participate in potentially risky skiing when on vacation in Sweden, the cosmonaut trainees were encouraged to cross-country ski around Star City for fitness, as well as making use of the gymnasium, pool, and sports hall. There was even an opportunity for some 'navigation practice' when Kuipers used a GPS to find his way around the lakes.

Like all ESA astronauts before him, Kuipers visited the factory in Tomlino, near Moscow, where the spacesuits were made. A mould of his body had been made during his time backing up Duque, but when he was retested it was decided to alter the sleeves because his fingertips were touching the ends of the glove.

Training 22 January to 28 January 2004

Kuipers trained with the TNO SUIT experiment; i.e. the vest that had 56 mobile-phone-type vibrating elements embedded in its fabric. Kuipers wrote in his blog, "The vest is intended to aid astronauts with their sense of direction. Once you have calibrated the equipment by the position of the floor, and then you turn sideways for example, you feel vibrations running from bottom to top against your side. Or if you are hanging upside-down, you feel the elements vibrating against your shoulders. TNO wishes to know whether the vest – which actually functions as an extra sensory organ – is useful in getting your bearings in the Space Station, where you naturally do not know what is 'up' and what is 'down'. It is possible that feeling where 'down' is will also help to counter space sickness and help the astronauts to sleep better. On the other hand, the vest could also be used on Earth. By pilots, for example, or by fire fighters to help them find their way through thick smoke."

Kuipers was a tourist in Moscow that weekend, when his girlfriend Helen and his parents came to visit. He took them to see the Kremlin, Red Square, the Metro with its mosaics, and Gorky Park.

Back at Star City, Kuipers tested, tasted, and chose his meals for the duration of his visit to the ISS. The menu contained pork, pasta, goulash, and mincemeat with onions. The breakfast choices included yoghurt, muesli bars, all sorts of fruit juices and other beverages. He wrote, "We passed all forty dishes round between us. We had to take a forkful, taste it, and then move on to the next dish. A kind of feast that leaves you feeling rather full. In my opinion, the goulash tasted best. A type of yoghurt which had nuts in it was very good too. And I do love fruit juices. There was some kind of cereal dish that I found quite inedible; it had no taste at all. There were one or two other things that I definitely didn't want on my menu, so I only gave them a mark of three or four."

On Wednesday, Kuipers had a good opportunity to compare weightless training in parabolic flight as provided by ESA and Roscosmos. Previously, as a doctor and researcher he had flown in the Airbus where the training was to gather experimental data. Here, in the much bigger Ilyushin 76 cargo aircraft, the training was operational for astronauts.

He wrote, "Together with my colleagues, I trained for the heavier work. In the Airbus everything is bolted down and the biggest item you are allowed to float is a pen or a bottle of water. Now we had to handle 50-kilogram weights to find out what it is like to move big racks in space. Although there is no weight in the Space Station, the mass remains the same. If you give something a push, it floats on until you stop it. In the Space Station, a heavy instrument rack can do a lot of damage unless you slow down such a big mass in good time. That can be pretty dangerous, so you have to learn how to do it properly. I also had to don a spacesuit during a parabola. It is really difficult to do, because you have nothing to support you. If you push one way, you fly off in the other direction. The suit was specially made for these training exercises, and unfortunately it was a size too small for me. It was a tight squeeze and it took a lot of effort to get it on. After we landed, I sent my girlfriend a text message. It was along the lines of: 'That's it for now.' The next time that I am weightless it will be for 864,000 seconds instead of just twenty."

Training 29 January to 4 February 2004

The Soyuz on which Kuipers was to launch was to deliver the Expedition 9 crew to the ISS, but with the Space Shuttle fleet grounded after the Columbia tragedy the crew roster was still in a state of flux. As it was considered too short a time in which to integrate Chiao and Tokarev as a team, the pair were replaced by the backup crew of Gennadi Padalka and Mike Fincke. Padalka would command the Soyuz flight and be commander of the ISS during Expedition 9. He had previously spent time on Mir and he would return to command the ISS again three more times; at the time of writing (2015) he has spent a record 879 days in space. This was NASA astronaut Fincke's first spaceflight, and he was the flight engineer. He would command the ISS on his next mission in 2008 and return to it as a Mission Specialist on STS-134 in 2011, which was to be the penultimate flight of the Space Shuttle Program.

This week Kuipers' training switched from the cramped Soyuz simulator to the full-sized Zvezda and Zarya models in Star City's simulator hall.

On Friday, he celebrated New Year at the Dutch Embassy, and on Sunday spent a more sombre evening commemorating the Columbia crew at the American Embassy.

Kuipers wrote, "It was also a personal commemoration for me. I had been involved in the medical research and the training of these astronauts, and I showed them around Amsterdam when they were in the Netherlands. Dave Brown was a good friend of mine. He even sent me an email from on board the Shuttle. I have printed his email on a photo of Columbia's crew, and I will take it with me on my own journey into space."

Training 5 February to 11 February 2004

After a weekend in the Netherlands, Kuipers headed to Houston to train on the US systems of the ISS, on Monday morning reporting to Building 4S at the Johnson Space Center. He had been to JSC before as an ESA researcher and even as backup crewmember for Soyuz TMA-3, but he wrote about this trip in his blog, "4S is a magical building where all the astronauts work. They meet together every Monday morning. It is a kind of scene that you normally only see in films, and there I was sat amongst them as a prime crewmember."

After a NASA welcome, the crew watched a video provided by Michael Foale on the ISS in which he showed where the equipment for the HEAT and ARGES experiments for the DELTA mission were stored. Training started with initial familiarisation with the American segment of the ISS. A television crew for the Dutch show *Netwerk* filmed footage for their broadcast on Sunday evening, and NASA's own TV station conducted interviews.

During a simulation on Wednesday Padalka, Fincke and Kuipers repeated the Russian emergency scenarios of depressurisation, fire and evacuation, this time from the American segment.

Training 12 February to 18 February 2004

Kuipers' second week at JSC focused on the ESA Microgravity Science Glovebox (MSG), which was an enclosed box from which hazardous gases couldn't escape. The operator manipulated the contents by inserting their arms into sealed rubber gloves.

Earlier in the week, he trained with the communication and computer systems. As well as official communication with Houston and Moscow, an astronaut can send and receive personal emails as long as the addresses are pre-approved and set up.

To complement the training he had received in Star City, Kuipers was trained on the photographic and video equipment on the American side of the station.

After formal meetings of all kinds, Kuipers caught up with some colleagues from ESA to get the "inside stories" and "tips and tricks" of living and working aboard the ISS. He met Léopold Eyharts, for example, who flew to Mir in 1998, and Roberto Vittori, who made a similar mission to the ISS in 2002. Kuipers learned a valuable lesson concerning wet towels. "Those wet towels that you wash with in the Zarya module: What do you do with them when you've finished? These were matters that had only been touched on in training. I found out that the towels, for example, have to be hung out to dry first, because water is very precious and you want to recycle it via the closed system on board." Both men would revisit the ISS: Eyharts as a member of Expedition 16, and Vittori twice more, once on a Soyuz and again on the Space Shuttle.

The busy week continued with a presentation on the variety of medical equipment on the ISS. As Kuipers wrote, "There is an awful lot of it, [ranging] from a defibrillator to injections and from dental equipment to stitches. A complete hospital in space, you might say. If necessary, we have to be able to pull a tooth, deal with an asthma attack, or carry out resuscitation. The equipment will probably never be used (and we certainly hope so), but it is excellent nevertheless."

The last activity on Friday was a visit to Mission Control, and in the evening there was a party to celebrate the fact that the present crew had been on the ISS for 100 days. Kuipers spent the weekend with his visiting daughters and dining out with his old crewmates Chiao, McArthur and Tokarev.

Finally there was a meal involving Kuipers' new crewmates Fincke and Padalka, the backup crew of Salizhan Sharipov, Leroy Chiao and Gerhard Thiele, plus all their families. Kuipers counted eight nationalities: American, Russian, Dutch, German, Chinese, Swedish, and even Indian and Uzbek. In writing of this evening, he said, "Fantastic! And very typical of the international cooperation that we see in the International Space Station."

Training 19 February to 25 February 2004

After spending a weekend with his parents in the Netherlands, Kuipers had a meeting with scientists from the STATUS programme of the University of Groningen who sampled the bacteria on his body. On board the ISS he was to take samples from "various places, such as handles, keyboards, headsets, the corners between panels, the sleeping quarters, the areas around warm lamps, humid areas and (of course) the toilet. All of these are places where you would expect to find a lot of bacteria. I will also grow bacteria cultures in space; a box that I will take up with me will contain prepared slides. The box will remain closed throughout the mission and after I return, the DNA of the bacteria will be studied in Groningen. The aim is to see how their characteristics change in conditions of weightlessness and to observe the extent to which they cling differently to surfaces when there is no gravity."

Back in Star City, training continued on the ESTEC experiments including CIRCA which required measuring blood pressure. On two occasions aboard the station he was required to wear blood pressure monitors for 24 hour periods during which stressful and calm conditions would be monitored.

Kuipers was back in the Soyuz simulator practising manual re-entry. He compared this task to an enjoyable computer game. "My best score was a deviation of zero kilometres: Right on the spot. My worst attempt took me about 25 km away from it. The computer itself has a margin of error of 19 km, so I am quite happy with my performance. What's more, if the computer handles the re-entry we will have to put up with a g-force of 5. If I do it myself, it is usually about three. I have already asked my instructors to let me to take care of the job..."

Training 26 February to 3 March 2004

If something goes awry during the return to Earth, the Soyuz capsule can land anywhere within 51.6 degrees of the equator, that being the inclination of the ISS orbit. This takes in a vast territory, which includes bodies of water and inaccessible mountain ranges. In addition to practising for a splashdown on water, Soyuz crews learn winter survival. Using only what is available in the capsule, they have to construct shelter, prepare food and await rescue. Kuipers and his crewmates were allowed just one day for this training, and they took care not to get frostbite so close to the start of the mission.

Writing of his preparation for the CIRCA experiment Kuipers said, "I had to be up early the next morning. At seven in the morning, I had to kit myself out in all kinds of

André Kuipers in the Soyuz simulator. (ESA)

equipment. Actually CIRCA consists of three experiments in one. Readings are taken on the ground prior to and after the flight, to provide comparative data for the readings I'll obtain during the flight. This is known as baseline data collection. So on Monday and Wednesday, throughout the day and the night, I wore a band around my arm and blood pressure monitors on my fingers. Every 15 minutes during the day and every half hour throughout the night, the cuff on my arm was briefly inflated to measure my blood pressure. The little cuffs on my fingers measured continuously. During the day, measurements were taken under a variety of circumstances. For example, I had to perform stress tests such as doing arithmetical calculations, or simply relax, or concentrate on my breathing while either standing or lying down. It is not much fun, especially at night. It feels as if someone is gripping your arm every half hour. Exactly the same will happen when I am in space; I hope that it will not disturb my sleep too much."

Kuipers and Padalka trained using the laser rangefinder which would be required in the event of a manual docking. In this scenario, Kuipers would go into the habitation compartment of the Soyuz and aim the device through a forward-facing window to measure the changing distance to the ISS. Padalka would use this information when adjusting the speed of approach; he had already performed a manual docking with the Mir space station.

During this week, Kuipers received his first instructions on the SEEDS experiment that he was to conduct on the ISS in parallel with Dutch children in their classrooms.

Training 4 March to 10 March 2004

To Follow up De Winne's educational video on liquids in weightlessness and Duque's similar film on Newton's laws of physics, Kuipers prepared for the space video that he was to make showing how the human body is affected by weightlessness.

For the first time the crew rehearsed the landing procedure together, fully kitted out in spacesuits. This was actually done twice, the first time with Kuipers occupying the left-hand seat that is reserved for the flight engineer and then again with Fincke in that place.

Since Yuri Gagarin, every cosmonaut has urinated against the rear wheel of the bus on the way to the launch pad. This had to be rehearsed as well, since it is apparently difficult to do when wearing a spacesuit!

At a lunch with various embassy staff at the ESA office in Moscow on Friday afternoon, Kuipers was thrown in at the deep end when he was expected to address the assembly and tell them about the mission and the training. He spent the weekend with his two daughters. It was the last time he would see them before they met at the Baikonur Cosmodrome, but by then he would be under quarantine.

After publicity photographs complete with national flags and wearing spacesuits adorned by the DELTA and Soyuz TMA-4 patches, the training resumed with Kuipers learning the radio techniques that he would use when continuing the tradition of talking to school children via 'ham' radio while in space. He also communicated with Dutch children via an internet chat session on Tuesday. Once again the old question on using the toilet in space arose. Kuipers wrote, "I was completely overwhelmed. In less than an hour, thousands of questions were fired at me. It was really fun, trying to answer as many of them as possible and as fast as I could. There were some great questions among them. For example, whether I can see stars and planets from the Space Station (yes, and stars do not twinkle because there is no atmosphere). There were the inevitable questions about how I go to the toilet (an airflow ensures everything ends up in the disposal system) and sleep (well buckled-in). But there were also questions about how fast you actually fly through space (28,000 km/hour) and what you have to do if you have used up all your oxygen (make some more from water using electricity). It was a good session, even though I was unfortunately unable to answer all the questions. On my return from space, we will do it again."

After the medical grounding of his colleague Bill McArthur, Kuipers nervously entered a 4 hour examination with an optician, an ENT specialist, a surgeon, an internist, and a neurologist. He once again passed with flying colours.

Training 11 March to 17 March 2004

On Thursday, Kuipers was fortunate to be at Star City to celebrate the 70th birthday of Yuri Gagarin. Following a formal ceremony at the Gagarin statue there was a concert, then a buffet for cosmonauts past and present. Kuipers met and spoke with Aleksei Leonov, the first man to perform a spacewalk.

Kuipers went langlauf skiing with Mike Fincke before the last snows thawed, and went for a meal with Dutch friends over the weekend. Since it was Russian election day on Friday, they stayed out of Moscow at the request of Star City security advisors.

Shep's bar, created by Expedition 1 commander and ex-Navy Seal, Bill Shepherd was the venue for Fincke's 37th birthday celebrations. Kuipers said of the unofficial astronaut club, "There are benches, a margarita mixer, a jukebox and a piano. The walls are covered with photographs. Astronaut signatures cover the bar. You can play table tennis or billiards, and if you crawl through a hole in the wall you're in the gym! Of course, there are no bar staff; if you want something, you simply take it from the fridge. Otherwise, it is like a friendly pub. I go there now and again, if there is a good film on, for example, or if there are special celebrations on one of the public holidays. Last Halloween, they had a theme party there: The men dressed as women and the women as men. I wore an enormous blonde wig. The theme resulted in all kinds of comical scenes."

Pedro Duque and Frank De Winne visited on Monday and Tuesday, again imparting some nuggets of wisdom from their time on board the ISS. Kuipers recalled their advice, "I need to take a 'bum bag' with me to store the compact flash cards from the cameras I will sleep best in the Soyuz with my legs towards the docking entrance. And I should make the most of the time during the first two days to look out the window because, after that, time will be limited. Actually, that is something that I should not do too much. Pedro told me that he had got slightly sunburned during his flight; the sunlight is much more intense because the ultraviolet light is not filtered by Earth's atmosphere. The stars are best seen from the Soyuz or the Pirs airlock on the station because you can turn off all the lights in these. In addition, they both told me that the landing, especially, will be quite an experience. You hear a few big bangs as the various parts detach during re-entry. There are another couple of shocks when the parachutes open and, of course, when you hit the ground. The worst things about it, they say, are the huge oscillations under the parachutes and the uncomfortable position if the capsule comes to rest on its side."

On Tuesday, the results of the recent medical tests were all positive.

For the last time as a complete crew, they rehearsed the landing procedures in the simulator before the technical exam.

Training 18 March to 24 March 2004

Kuipers received a lot of attention from the Dutch media as the launch date grew near. This week in Star City he was accompanied by newspaper and magazine journalists, by radio broadcasters, and by TV crews from NOS, RTL and SBS6.

On Friday he visited the Mission Control Centre (TsUP) in the Moscow suburb of Korolyev. Then he spent the weekend with his girlfriend and prepared his personal kit, including the music which he would take. Music was delivered on CD and all discs had to be checked for viruses that could disrupt the station's systems. "Music is very important to me, especially in combination with the view. Then I can just daydream and enjoy everything. Not just in space, by the way; I do the same on Earth. In the mountains, or visiting ancient ruins where no one else is about… I listen to powerful, swelling music: Glorious!"

More examinations followed on Monday. During the laser rangefinder measurement there was an equipment failure which Kuipers suspected was part of the test, but when he noticed that a cable had come lose he plugged it in again and the test continued. As he wrote, "As you can see, real problems can arise even in simulations."

There was an exam in the centrifuge on Tuesday. It was a landing simulation and he made the best landing of his entire training course by achieving an 'excellent' mark five times in a row. "For me, it is just like a computer game. You try to get the highest possible score. The only difference is that you are really being thrown about."

Kuipers described the "standard day in the Space Station" training on Wednesday. "That involved us doing everything, for an entire day, that we would do as a crew up there. For me, it meant carrying out a lot of experiments. In the morning, I was busy with the 'amateur radio'. Then I ran through the procedures for the educational activity VIDEO-3 and collected bacteria from all over the station using cotton buds. After lunch, I did the Eye Tracking Device (ETD) experiment, which involves measuring the interaction of the eyes with the internal balance mechanism. I'd just finished this when (of course) something went wrong. After all, it was a training exercise. We had all laid bets on whether it would be pressure loss or a fire. It was pressure loss. At a time like that, you have to check the Soyuz first; if the leak is not there, you have to comb through the entire station. You always keep the Soyuz behind you, because it is your 'lifeboat'. In an emergency, you have to be able to use it to evacuate the station. The leak turned out to be at the coupling between the Russian modules, Zvezda and Zarya, which is a very awkward place. We had to seal off the Zvezda because there is no Soyuz attached to it. Also known as the 'service module', this the main section of the station. It has all the monitoring and air purification equipment, plus the water and the toilet. As carrying on without a service module is pretty difficult, in the end we abandoned the ISS."

Training 25 March to 31 March 2004

This was the week of final examinations, and a failure now could cost a cosmonaut his or her seat on the flight.

After saying goodbye to their trainers, Kuipers and his ESA backup Gerhard Thiele spent several days studying into the small hours. His ESA and NASA colleagues assisted by preparing meals and allowing them uninterrupted revision time. Monday was the first exam in the Space Station simulator. While his colleagues on the Expedition crew were tested on station maintenance tasks, Kuipers had to run through a number of experiments for which he had rehearsed, namely the CIRCA/BMI, KUBIK and Aquarius equipment (including the part that was to be done in the Soyuz). Of course there was a simulated leak in the middle of the test to be addressed. The results were announced immediately and happily everyone passed the ISS exam.

On Tuesday it was the exam in the Soyuz simulator, which for Kuipers was the more difficult of these two days:

> As First Engineer, I am jointly responsible for the spacecraft. For this reason, 80% of my training is geared towards emergency procedures and safety measures. A lot of different procedures were covered in this exam. In our spacesuits we took our seats in the simulator. The start, preparation, and launch ran according to plan.
>
> Then the problems started. First the radio let us down, so I had to switch over to the reserve system. One of the angular acceleration meters also broke down, whereupon it automatically switched over to the second system. For the rest of the flight I really had to pay attention because I was now working all the time with the reserve system. Then both radar systems broke down, which meant we knew for certain that the commander would have to dock with the Space Station manually. The docking operation appeared to go well, but the antenna that makes first contact with the Space Station was stuck, preventing us from completing the docking.
>
> Then to make matters worse, a fire started in our capsule which was not put out by shutting off the electrical systems. We extinguished the fire by letting all the air escape from the Soyuz. As a result of this, the mission had to be called off and we had to return to Earth immediately. Employing an emergency procedure, we had to separate and take up the right position.
>
> After that we tried to start an automatic emergency landing programme, known as programme number five. But that also refused to work and so we had to do everything manually. I had to start up the engine exactly at the right time to slow down and make the return to Earth. The acceleration meter appeared not to be functioning correctly, so the computer was unable to work out when the engine was supposed to cut out again. The fuel consumption was normal and we knew how long the engine had to burn, so we shut it down ourselves at the right time. The separation of the landing

capsule from the living quarters and from the engine compartment of the spacecraft, and the steep, stable re-entry through the atmosphere went normally. Finally we saw to our amazement that the braking forces had increased to more than 17g! But fortunately this appeared to be a fault in the simulator computer...

If this examination had been for real, then it would have been a short mission. Just two days. The training was still more difficult than the exam. In fact, the purpose of the training is that you pass the exams and then the real flight is more straightforward than the training.

After that hard day we emerged from the simulator tired, stiff, and sweating. After changing our clothes we had to appear before the exam commission and the specialists. The commission was unable to find a fault with anything in our exam. The entire crew had worked faultlessly. Reason for a celebration.

After successfully passing the exams, the crew adhered to tradition. The next day is, according to Russian tradition, reserved for more official proceedings. First, a meeting of the great commission in the white hall in the main building where the results of the examinations are discussed. Then a press conference attended by members of the Dutch media. After that they signed the visitors' book kept in Yuri Gagarin's restored room as part of the Star City museum. Then the crew travelled to Moscow to meet Anatoly Perminov, the new head of Roscosmos.

At the Kremlin Wall in Red Square they laid flowers at the memorials to Yuri Gagarin (the first man in space), Sergei Korolev (the famous Chief Designer) and the fallen heroes Vladimir Komarov (Soyuz 1), and Georgy Dobrovolsky, Vladislav Volkov and Viktor Patsayev (Soyuz 11).

Training 1 April to 11 April 2004

After all the exams, press conferences and official tasks the crew were able to enjoy a few days of rest and relaxation. Kuipers had four days in Star City with his girlfriend Helen. He attended a symposium in Moscow in honour of 15 years of collaboration between the ESA and Roscosmos. After Helen went back to the Netherlands, Kuipers reviewed the communications and operational plans with his backup, Gerhard Thiele and ESA astronaut Reinhold Ewald who would lead the ESA ground teams during the mission.

The final preparations that week included further medical tests, rehearsing experiments, baseline eye movement measurements, sessions on the tilt table (to simulate how the head feels in the state of microgravity), centrifuge training, and organising his personal belongings for the flight, whose total mass was limited to 1.5 kg!

Then it was off to the Baikonur Cosmodrome in Kazakhstan, with the prime and backup crews in separate aircraft so that a single accident would not endanger the mission. After being greeted by the reception committee, they were driven (in two buses) to the Cosmonaut Hotel.

As Kuipers recalled, "After signing yet more photos, I went for a run that evening. In the grounds of the hotel there is a small avenue with trees planted by the first cosmonauts.

Just beyond the trees is the boundary of Baikonur. You can get to the river through a hole in the fence, and if you wanted you could run endlessly through the Steppe." Luckily he stayed on site.

Flight Day 1 – Monday, 19 April 2004

At 09.19 local time (05.19 Central European Time) Soyuz TMA-4 lifted off from Baikonur with the ESA DELTA mission cosmonaut André Kuipers and the ISS Expedition 9 crew of Gennadi Padalka the Roscosmos commander, and Mike Fincke the NASA flight engineer. Nine minutes later they had achieved orbit and Kuipers became the second Dutch citizen in space, after Wubbo Ockels in 1985.

"It is always satisfying to see a mission under way after observing all the hard work that has gone into its preparation," said Jörg Feustel-Büechl, ESA's Director of Human Spaceflight. "The DELTA mission is a very diverse programme, with relevance to the lives of people old and young. It is a very positive sign to witness the degree of enthusiasm that has come into the mission from Dutch research and education institutions and industry, and I am pleased that, together with the scientific programme, children also get to play an active part during the mission."

Also present at the launch was the Minister for Economic Affairs, Laurens Jan Brinkhorst and His Royal Highness Prince Johan Friso of the Netherlands, recently appointed as a director at TNO Space. The Prince, who studied Aerospace Engineering at Delft University of Technology, was attending in a professional capacity. "I am convinced that space activities advance innovation in a wide variety of fields," he said. "Especially human spaceflight like now with the DELTA mission and André Kuipers. This mission will result in new scientific and technological understanding, which will in the mid- and longer term always contribute to the well-being of all of us here on Earth."

Flight Day 3 – Wednesday, 21 April 2004

Soyuz TMA-4 docked with the nadir port of the Zarya module at 07:01 Central European Time. After the standard post-docking checks, the hatches were opened at 08:27.

Kuipers' first day on the station saw him transfer biological samples between the European-built KUBIK incubators for the variety of biological experiments: ACTIN, ICE-first, FLOW, KAPPA and TUBUL. He also conducted the human physiology experiments CIRCA, Motion Perception (MOP), and MUSCLE, and took samples for the microbiology experiment SAMPLE. In addition he installed the two university-backed experiments, Bug Energy and GRAPHOBOX.

Flight Day 4 – Thursday, 22 April 2004

The SEEDS experiment was kicked off on this day by Maria van der Hoeven, the Dutch Minister for Education, Culture and Science. In a broadcast from De Klaverweide, a junior school in Noordwijk, a group of children planted rucola seeds on Earth while Kuipers did the same in space.

"This is real scientific research," said Minister van der Hoeven. "You have to discover something for yourself, which is very different to learning from a textbook." The schoolchildren who planted their seeds under the watchful eye of the Minister were very excited about the experiment. "This is a once in a lifetime experience. And it is really fun to do."

The kick-off for SEEDS was broadcast during *Space News*, a special edition of *het Jeugdjournaal* – a Dutch news programme for children. The children could see how Kuipers prepared his own 'growing rocket' via a live video link-up with the Space Station. As Kuipers pointed out, "Everything floats here in space. That is why I have used some glue to hold the seeds in place. Watering the seeds is also a bit tricky, so I am using a special syringe."

When asked whether he was strapped into his seat, Kuipers floated towards the ceiling and joked, "That is not exactly the case."

Flight Day 5 – Friday, 23 April 2004

Kuipers performed the FLOW and KAPPA experiments in the Soyuz during the journey towards the station. FLOW looked at whether bone cells have a decreased sensitivity to stress in weightlessness. The purpose of KAPPA was to study the influence of gravity on NFkB, a protein that plays a crucial role in inflammation and immunity in the body.

Aboard the ISS, only 20% of the HEAT experiment was completed during this time period due to sub-optimal interfacing between the experiment and a cooling plate inside the Microgravity Science Glovebox. As a result of the higher than expected temperatures the experiment had not progressed as planned. Another headache for Kuipers was the centrifuge in the AMBER KUBIK incubator; it was meant to replicate 1g but was not working, so Kuipers attempted a repair.

Flight Day 6 – Saturday, 24 April 2004

The Space Expo was in full swing in Noordwijk, the Netherlands, and the 18 students who won ESA's art and essay competition called 'Zeg het ISS' (Talk ISS) were ready and waiting for Kuipers to make contact from space. This was part of the Amateur Radio on the International Space Station in Europe (ARISS-Europe) programme that had been ongoing since Frank De Winne's visit to the station. These 18 students and over 80 classmates were gathered in the Ariane room. While waiting for Kuipers to fly into radio range ESA astronaut Reinhold Ewald entertained the group with tales of his 1997 mission to Mir. And then they all heard the call, "Hello, this is André Kuipers from on board ISS. I am ready for your questions."

Cynthia asked about time in space. Kuipers explained that they used Greenwich Mean Time (GMT) on the ISS, and although there was a clock on board the astronauts also wore watches. Julia wanted to know whether it was possible to breathe normally in space. Kuipers said that space was a vacuum, but in the station the air was normal for the Earth's surface. Rik asked how Kuipers was getting on with his crewmates. He replied, "We have a lot of fun together. We talk Russian and English to each other. We get on well." Just after the 18th question had been answered, the signal from space slowly faded away. The radio chat was finished. "That's a pity," said Sandy.

Flight Day 8 – Monday, 26 April 2004

Today, Kuipers successfully tested the energy-saving plasma lamps which were developed by Philips and Eindhoven Technical University. Gerrit Kroesen from the latter said, "I am very pleased that we were able to conduct this experiment on a human spaceflight. It would not have worked without the tireless André Kuipers."

After a shaky start to the ARGES experiment conducted in the Microgravity Science Glovebox in the American Destiny laboratory, proceedings were quickly back on track. By the end of the weekend 100% scientific success was achieved. The full results of the plasma lamp tests were returned to Earth with Kuipers on Soyuz TMA-3. Kroesen said of the results, "This experiment brings us one step closer to selling these super energy-saving lamps to the consumer."

After failing to get the centrifuge in the AMBER KUBIK incubator working, Kuipers did not have a control set of data representing 1g. The control data would have to be gathered on Earth with all the other conditions except the space environment being identical, so it was a partial success.

Flight Day 9 – Tuesday, 27 April 2004

Ahead of a special *Space News* bulletin in which Kuipers was to reveal how his SEEDS experiment went, the children at De Klaverweide junior school had already opened their own 'growing rocket'. Inside the rocket were two chambers, one with light and one without. The seeds in the dark chamber had grown taller, but were yellow in colour. In the light chamber, the seedlings were shorter but of a healthier green colour with wider stems.

The children concluded that on Earth it didn't matter whether the plants had light or not, they still grew upwards. They speculated about the colour and shape of the seedlings that had grown in space. Perhaps they had not grown at all, or they might have grown in circles, or perhaps there would not be any difference to those grown on Earth. To settle the issue, Kuipers opened his 'growing rocket' on a live video downlink. "The plants in the dark chamber have grown in all directions. Upside down, left, right. They are all over the place," he reported. In space the seedlings in the light chamber had grown larger and were greener, and they had all grown towards the light.

The lesson from the SEEDS experiment was that either light or gravity will inform a plant of the direction in which it should grow. In the absence of both, it is confused. The experiment showed the schoolchildren how scientific research is done and also made a contribution to future space travel. If humans are to go on longer space missions, say to the Moon or even to Mars, they will know the best way to grow their food. According to this experiment, they should be sure to take enough light bulbs along for the journey.

The SEEDS experiment was organised by Space Research Organisation Netherlands (SRON) and was financed by the Dutch Ministry for Education, Culture and Science.

Flight Day 10 – Wednesday, 28 April 2004

On his last full day on the station (until his return for the Promisse mission in 2011) Kuipers wrapped up the remaining experiments and packed his equipment for departure. He and Padalka conducted the SUIT tests during the last few days. The vests operated flawlessly and the experiment was conducted according to the pre-defined procedures. The CIRCA experiment that monitored the pattern of blood pressure and heart rate in a weightless environment over a 24-hr period was also completed.

And Kuipers collected bacterial samples throughout the station for the SAMPLE experiment. In a weightless environment some bacteria, including those which cause infections in humans, grow faster than on the ground and they are more resistant to antibiotics.

Flight Day 11 – Thursday, 29 April 2004

At 20:52 UT, Soyuz TMA-3, which had delivered ESA's Pedro Duque to the ISS in October 2013, undocked from the ISS. Aboard were André Kuipers and the Expedition 8 crew of Alexander Kaleri and Michael Foale. Padalka and Fincke remained aboard the ISS to start the six month Expedition 9.

A bespoke procedure for re-entry was developed by TsUP and used by the crew on this occasion. After docking in October, a leak was discovered in one of the two helium systems. Helium was used to pressurise fuel and oxidiser during the de-orbit burn. While the leak was isolated during the 6 months docked to the ISS, the nominal re-entry procedure called for an open connection between both of the systems. As this could not be permitted in this case, the opening of the valves was delayed to the last moment. After the Soyuz had moved away to a distance of 19 km from the ISS, a de-orbit burn lasting 261 seconds slowed it sufficiently to begin the descent to Earth. Only when the burn was complete did the crew know that a full duration burn was possible.

The re-entry module containing the crew was the only component to possess a heat shield to withstand the high temperatures of re-entry. The jettisoned service and orbital modules burned up in the atmosphere.

Flight Day 12 – Friday, 30 April 2004

Kuipers monitored the automatic re-entry, ready to take over if anything didn't operate correctly, but his comprehensive training wasn't needed and at 07:12 local time the capsule landed near Arkalyk in Kazakhstan after a return flight of just over 3 hours.

Additional – 7 September 2004

Due to a loose connection in the heating plate, Kuipers was unable to complete the HEAT experiment during his DELTA mission on the ISS. The experiment, which was to study how heat was transferred in space, was part of a programme researching thermal control in spacecraft. ESA asked NASA if the ISS crew could repair the apparatus and complete the experiment. Mike Fincke successfully finished the experiment in early September.

Marc Heppener, ESA's head of ISS scientific research said, "HEAT was an important experiment in the package of experiments during the DELTA mission, and I am very pleased that it has now been successfully completed. This is another important accomplishment for the DELTA mission. Overall, therefore, we can now say that 85% of the experiment programme has been successful."

Kuipers was also delighted, "It really was disappointing there was no time to solve the problem in April, so I am pleased that Mike has been able to work on it. Thanks Mike!"

9.4 POSTSCRIPT

Subsequent Missions

See the Promisse mission.

André Kuipers Today

After his DELTA mission Kuipers was assigned post-flight activities and other duties at the EAC and ESTEC. He supported ESA payload development, parabolic flight campaigns and healthcare spin-offs, as well as offering ground-support for the missions of other ESA astronauts. André also qualified as a Eurocom, communicating with astronauts from the Columbus Control Centre in Munich, Germany. In 2005 he was assigned as backup for the first Canadian ISS increment, for which he received additional training on the US and Russian segments, including operating robotic systems.

From 2007, André trained as backup to ESA astronaut Frank De Winne for Europe's second long-duration spaceflight to the ISS. He received user, operator, and specialist level training on all modules of the station, training on experiments in ESA's Columbus laboratory, and training on the Automated Transfer Vehicle (ATV). From May 2009 he was Eurocom at the Columbus Control Centre in support of the six month Oasiss mission by De Winne. Thereafter, Kuipers began training for his next mission to the ISS. Besides qualifying as Flight Engineer for the Soyuz TMA-M series, he was certified to use the SSRMS robotic arm, to perform EVAs, and to berth the SpaceX Dragon spacecraft.

During his first mission in 2004, André decided to be an ambassador for sustainability, science and technology, education and some charity organisations. His activities are coordinated through the André Kuipers Foundation. As an ambassador, he supports: Wereld Natuur Fonds, WE Foundation, Stichting Hoogvliegers and Emma Kinderziekenhuis.

Other organisations with which Kuipers is involved are:

- Techniekpact, a partnership between Dutch government, industry and education sector to enthuse young people in science and technology.
- Member of the Royal Palace Committee.
- Member of the Governmental Sports Council.
- Royal Netherlands Meteorological Institute (KNMI, Supervisory Board).
- Science Museum Nemo (Advisory Board).
- Platform Beta Techniek (Supervisory Board).
- Dutch Aerospace Fund.
- Outdoor Medicine Association.

André Kuipers receives the Order of Friendship from Russia's President Vladimir Putin on 3 July 2013. (ESA)

- Natuurwijs (Foundation for nature education).
- Dutch Air Rally.
- Historical Science Museum, Hofwijck.

As well as membership of ad hoc organisational committees and juries for science competitions.

He is a member of the following organisations:

- Member of the Dutch Aviation Medicine Society.
- Member of the Dutch Association for Spaceflight.
- Member of the Royal Dutch Society for Aviation.

He has received the following honours:

- Officer of the Order of Orange-Nassau, 2004.
- Honorary Citizen of Ouder-Amstel, 2004.
- Knight of the Order of the Netherlands Lion, 2012.
- Honorary Doctorate of the University of Amsterdam, 2012.
- Honorary citizen of Haarlemmermeer, 2012.
- Andreas Medal of the City of Amsterdam, 2013.
- Recipient of the Russian Order of Friendship, 2013.
- King of Arms at the inauguration of King Willem-Alexander, 2013.

10

Eneide

Mission

ESA Mission Name:	Eneide
Astronaut:	Roberto Vittori
Mission Duration:	9 days, 21 hours, 21 minutes
Mission Sponsors:	ESA
ISS Milestones:	ISS 10S, 26th crewed mission to the ISS

Launch

Launch Date/Time:	15 April 2005, 00:46 UTC
Launch Site:	Pad 1, Baikonur Cosmodrome, Kazakhstan
Launch Vehicle:	Soyuz TMA
Launch Mission:	Soyuz TMA-6
Launch Vehicle Crew:	Sergei Konstantinovich Krikalev (RKA), CDR
	Roberto Vittori (ESA), Flight Engineer
	John Lynch Phillips (NASA) Flight Engineer

Docking

Soyuz TMA-6	
Docking Date/Time:	17 April 2005, 02:20 UTC
Docking Port:	Pirs
Soyuz TMA-5	
Undocking Date/Time:	24 April 2005, 18:44 UTC
Docking Port:	Zarya Nadir

J. O'Sullivan, *In the Footsteps of Columbus*, Springer Praxis Books,
DOI 10.1007/978-3-319-27562-8_10

Landing

Landing Date/Time:	24 April 2005, 22:08 UTC
Landing Site:	Near Arkalyk
Landing Vehicle:	Soyuz TMA
Landing Mission:	Soyuz TMA-5
Landing Vehicle Crew:	Salizhan Shakirovich Sharipov (RKA), CDR Roberto Vittori (ESA), Flight Engineer Leroy Chiao (NASA), Flight Engineer

ISS Expeditions

ISS Expedition:	Expedition 10
ISS Crew:	Leroy Chiao (NASA), ISS-CDR Salizhan Shakirovich Sharipov (RKA), ISS-Flight Engineer
ISS Expedition:	Expedition 11
ISS Crew:	Sergei Konstantinovich Krikalev (RKA), ISS-CDR John Lynch Phillips (NASA) ISS-Flight Engineer

10.1 THE ISS STORY SO FAR

There had been only one flight since the DELTA mission by André Kuipers. In October 2004 Soyuz TMA-5 delivered Expedition 10 to the ISS along with Yuri Shargin, who was the Russian Military Space Force's first cosmonaut; he replaced a Russian businessman 'space tourist' who had failed a medical.

10.2 ROBERTO VITTORI

Early Career

See the Marco Polo mission.

Previous Mission

See the Marco Polo mission.

10.3 THE ENEIDE MISSION

Eneide Mission Patches

The name stood for Esperimento di Navigazione per Evento Italiano Dimostrativo di EGNOS, and it was chosen in tribute to Virgil's Aeneas who travelled from Troy to Italy and became an ancestor of Romulus. Although the patch included what appeared to be a spacewalking astronaut, no spacewalks were planned for the mission. The word 'Eneide' was prominent, flowing in the Italian colours of red, white and green. The five stars represented the participating organisations, whose names appeared on the border: the European Space Agency (ESA); the Russian Federal Space Agency, Roscosmos; the Italian industrial group, Finmeccania; the Ministero Difesa, the Italian Ministry of Defence; and the Region of Lazio.

The Soyuz TMA-6 crew with Roberto Vittori on the left. (www.spacefacts.de)

The Soyuz TMA-6 patch was designed by Russian artist Alex Panchenko, who evoked the early graduation badges of Soviet Academies and early space patches with the diamond shape. He says he designed the patch on the back of a napkin in 15 minutes along with Sergei Krikalev while visiting a Starbucks in Santa Monica. The patch showed a Soyuz orbiting over Russia. Three stars represented the crew. Their names were in the border, with the abbreviations for the space agencies Roscosmos, NASA and ESA.

Eneide Mission Objectives

Roberto Vittori's second mission to the ISS was again on board a Soyuz spacecraft, but this time the updated TMA model. He had previously visited the station on Soyuz TM-34. Joining him on this new launch was the Expedition 11 crew of the vastly experienced Sergei Krikalev of Roscosmos and John Phillips of NASA. Krikalev had already flown on Soyuz and Space Shuttle vehicles and spent time as long term crew on Mir as well as the first crew of the ISS. In fact, because he had also been part of the STS-88 crew who delivered the Unity module this would be his third trip to the ISS. Phillips, although older, was not as experienced. Nevertheless he was also returning to the station, having been a

Eneide Mission Patch. (www.spacefacts.de)

Mission Specialist on STS-100 which delivered Canadarm2, among other things. In due course, Phillips would return on STS-119 and assist in the installation of the final segment of the integrated truss structure.

Eneide was an ESA mission sponsored by the Italian Ministry of Defence and the Lazio Region of that country, and supported by Finmeccanica and the Roman Chamber of Commerce (CCIAA) under the terms of an agreement between ESA and Roscosmos.

It had four main objectives:

- To exchange the ISS 'lifeboat' by replacing Soyuz TMA-5 with Soyuz TMA-6.
- To retrieve the Expedition 10 crew of Leroy Chiao and Salizhan Sharipov and replace them for the next six month increment with Sergei Krikalev and John Phillips of Expedition 11. Chiao and Sharipov would return to Earth with Vittori.
- To perform an experimental programme of scientific interest. Vittori was to spend 40 hours on these activities. Most of the experiments were developed by Italian researchers and fabricated by local industry and research institutions.
- To expand the experience of the European astronaut corps aboard the ISS in readiness for the installation of the ESA Columbus laboratory.

Soyuz TMA-6 Mission Patch. (www.spacefacts.de)

Science

The following experiments were to be carried out as part of the Eneide mission:

- CRISP-2
- BEANS
- SEEDLINGS
- FRTL-5
- MICROSPACE
- VINO
- HPA
- NGF
- VSV

- ETD-I
- FTS
- MOP-I
- HBM
- GOAL
- ENEIDE
- LAZIO
- EST
- E-NOSE
- SPQR
- ASIA
- BOP
- ESD
- ARISS-4 (continued from Odissea, Cervantes and DELTA).

CRISP-2

Objective: To study the effects of microgravity on the proliferation of neurons in crickets in which egg fertilization occurs in space, and also the effects of gravity deprivation on gravity-related behaviour in crickets.

The female crickets would be allowed to deposit eggs in the respective egg collectors during two periods of the flight. This requirement considered that embryos were exposed for different periods to microgravity. After the first period of egg deposition, the embryos would be developed for 8 days in microgravity. During this period the neuronal proliferation under 1g conditions would be completed. After the second period of egg deposition the embryos would remain in microgravity for just 3 days, and neuronal proliferation would resume after return to Earth.

Task: Incubation of animals on the Russian segment of the ISS.

Equipment Used:

- Experimental Cricket Containers CRI (2) with laboratory crickets, i.e. female crickets Acheta domesticus, seven female crickets in each container. Each Cricket Container accommodated three Egg Collectors and one water bottle.
- KUBIK – AMBER – thermally insulated container (incubator) with a set of slots for Cricket Containers.
- A Sony DSR PD-150P camcorder and Mini-DV video tapes.

Expected Results: Incubated embryos of crickets in Cricket Containers and video data recorded during the experiment showing walking abilities and health conditions of the female crickets.

BEANS

Objective: This experiment sought to involve students in the implementation of a space mission and to increase their understanding of the space environment and the potential for applying space technology. They were to investigate the conditions leading to the germination of seeds and observe in parallel the early stages of the plant life cycle on Earth and in space.

Task: Germinate plants (beans) on the Russian segment of the ISS.

Equipment Used: The SBS container was a sealed plastic bag with a paper towel, six bean seeds and the 20 ml water container. The experiment also used a Sony DSR PD-150P camcorder, a Nikon D1X camera and 256 kB memory card.

Expected Results: Germinated seeds returned to Earth with a memory card containing photographic and video data of the germination process.

SEEDLINGS

Objective: To study the effect of microgravity on germination, growth, and nutrition properties of vegetable sprouts.

Tasks: Assess potential in-space production of vegetable cultures as fresh ready-for-use food products, and assess effect of microgravity on germination, growth and nutrition properties of vegetable cultures.

Equipment Used: Two SED containers, each a sealed plastic bag with the paper towel with attached seeds and the 20 ml water container. The experiment also used a Sony DSR PD-150P camcorder, a Nikon D1X camera and 256 kB memory card.

Expected Results: Germinated seeds returned to Earth with a memory card containing photographic and video data of the germination process.

FRTL-5

Objective: To assess the effects of microgravity and the radiation in the space environment on normal, differentiated in-vitro cultures of Fisher rat thyroid cells. This unique in-vitro test system would allow the use of cells in a quiescent state (non-proliferating) which could be kept almost indefinitely without culture medium exchanges or any manipulations.

Task: Exposure of animal cell samples on the Russian segment of the ISS.

Equipment Used: A container with 10 culture flasks presenting sealed culture bottles, with each flask containing rat thyroid cell strains (FRTL5) in proliferative and quiescent states (i.e. five samples per state). The experiment also used the AQUARIUS-B incubator for cell exposure.

Expected Results: Exposed rat thyroid cell strains returned to Earth.

MICROSPACE

Objective: To study the response of representative non-pathogenic micro-organisms to the spaceflight environment.

Tasks: Exposure of micro-organisms on the Russian segment of the ISS.

Equipment Used: Three identical pouches of microbial cultures were to be used, each of four subunits. Each subunit was sealed in a plastic bag and had three Sigma tubes with 8 cryotubes (with lyophilised cultures). There was also a temperature data logger. Two dosimeters were placed in the first subunit of Pouch-3. A Camera Nikon D1X was used.

Expected Results: Lyophilised cultures exposed returned to Earth.

VINO

Objective: To verify the ability of vine tendril grafts from vineyards of the Toscana area to survive and grow after the exposure to the space environment.

Task: Passive exposure of vine tendril grafts on the Russian segment of the ISS.

Equipment Used: The VINO Container accommodating the vine tendril grafts.

Expected Results: Samples of vine tendril grafts exposed to the space environment and returned to Earth.

HPA

Objective: Experimental research on the performances of the human upper limb in weightlessness.

Tasks: To investigate motor coordination while reaching for and grasping objects, and assess muscle fatigue during the execution of sustained handgrip and pinch force.

Equipment Used:

- Posture Acquisition Glove (PAG).
- Wrist Electronic Box (WEB) containing the Inertial Tracking System (ITS) composed of three accelerometers and three gyroscopes to measure both the bending angles of the fingers and the kinematical parameters of the wrist (acceleration, speed and orientation).
- Handgrip Dynamometer (HGD).
- Pinch Force Dynamometer (PFD).
- Floppy disks to record and store experiment results.

Expected Results: Files of experiment data recorded on floppy disks.

NGF

Objective: To assess alterations of nerve growth factor (NGF) and other neurotrophins influencing a response to pre-flight, in-flight, and post-flight stresses on the Soyuz spacecraft.

Task: Collection of saliva samples by the astronaut during the VC-8 mission.

Equipment Used: Saliva-A.NGF Kit.

Expected Results: Twelve collected saliva samples placed in two Saliva-A.NGF Kits.

VSV

Objective: To study the contribution of visceral receptors to the detection of the subjective sense of vertical (defined as the astronaut's body z-axis in microgravity) in an environment which rules out possible bias due to visual and gravito-inertial forces.

Task: Assess the contribution of visceral receptors to the sense of subjective vertical.

Equipment Used: The Chibis Lower Body Negative Pressure device, the GAMMA-1M apparatus, a Subjective Vertical Analyser, a fixing system, a Nikon D1X camera and a memory card.

Expected Results: Downlinked telemetry information, data documented in the Appendix to the crew procedures, and digital imagery on the memory card.

ETD-1

Objective: To further investigate vestibular-ocular-motor orientation in microgravity conditions.

Task: Measure eye and head motions using an eye-tracking device.

Equipment Used: A system unit consisting of a dedicated computer that has a built-in power supply, a lightweight removable measuring unit for the head of the test subject that provides data and electrical interfaces for the system unit, an accessories kit, and a return container.

Expected Results: A hard disk containing data from the eye tracking device.

FTS

Objectives: To identify food items from the Lazio region (Italy) that could be incorporated into the diet of the ISS crew, to demonstrate that traditional foods from the Lazio region are tasty and nutritious and don't lose their quality in spaceflight conditions, and to increase the variety of food items available for the station's crew.

Tasks: Taste new product items and assess any organoleptical alterations when changing a food.

Equipment Used: A tray with food items (both typical and traditional) from the Lazio region of Italy.

Expected Results: Pages of the crew procedures with a filled-in questionnaire to assess organoleptical tests of food products in spaceflight, as well their suitability for use, plus digital imagery on a memory card.

MOP-I

Objective: To investigate vestibular adaptation to a change of gravity.

Task: Assess the degree of vestibular adaptation by defining an individual's perception of motion.

Equipment Used: Questionnaire with a set of forms.

Expected Results: Questionnaires filled in by the astronaut during a 10 day mission describing how his sense of balance was affected by head motions, and in particular the presence or absence of discomfort during daily life associated with space adaptation syndrome.

HBM

Objective: Development of 'smart' clothing for astronauts, capable of checking their life functions and using devices to allow free motion on board.

Task: Measurements to monitor heart beat rate.

Equipment Used: The HBM belt with sensors connected to an analogue-to-digital interface, an IBM A31p ThinkPad laptop, a Sony DSR PD-150P camcorder, and a memory card.

Expected Results: A PCMCIA card with data obtained during the course of measuring the astronaut's heart beat rate, plus video tape of the experiment run.

GOAL

Objective: To increase the comfort and efficiency of astronauts by improving their psychological and physiological well-being in terms of enhanced garment wearability, aesthetics, thermal stability, and bodily hygiene on board.

Task: Comparative analysis between GOAL clothing items and the usual garments on board with the intention of verifying the clothing system efficiency.

Used Clothing: T-Shirt developed using new fabrics and design.

Support Facilities: A Nikon D1X camera and memory card.

Expected Results: A memory card with imagery of the astronaut performing routine planned activities while wearing the GOAL clothing.

ENEIDE

Objective: Measurement and verification of GPS and EGNOS (European Geostationary Navigation Overlay Service) signals in low Earth orbit.

Tasks: To verify the performances obtainable using a GPS/EGNOS receiver, and perform a PVT measurement using a combined GPS/EGNOS receiver during different phases of the mission and compare the results with the equivalent data from the spacecraft.

Equipment Used:

- The receiver FM inside Soyuz TMA-6 was an integrated GNSS receiver for space applications which could determine position and time using GPS signals. In addition to real-time navigation data with conventional accuracy, it could measure pseudo-ranges and integrated carrier phase.

- The wide beam antenna FM installed externally on Soyuz TMA-6 was an airborne antenna with a dual-band L1-L2 that had a hemispherical coverage pattern. The gain was greater than –4dBic for elevations exceeding 5 degrees and right hand circular polarisation.
- Receiver FM, wide beam antenna FM and miscellaneous cables.
- IBM A31p ThinkPad laptop (with cables) and PCMCIA card.

Expected Results: A PCMCIA card containing data that was received automatically during the orbital phase of the Soyuz TMA-6 flight to the ISS and during a mated flight with the Russian segment of the station.

LAZIO

Objective: A detailed investigation of the radiation environment in space, its interaction with Earth's magnetosphere and its effects on human physiology.

Tasks: To record, identify, and measure flows of electrically charged particles with a determination of their direction; to measure the intensity and variation of the magnetic field present in the ISS; to report phosphenes, where an astronaut sees a flash of light without any light actually having entered the eye; and to verify the ability of different shielding materials to reducing the radiation environment.

Equipment Used:

- Main Electronic Box (MEB) with Magnetometer Box (MB) and Magnetometer Head (MH).
- LAZIO Kit 1 consisting of the Increment 11 Kit, dosimeter tiles, voice cassettes, and a variety of memory cards.
- LAZIO Kit 2 consisting of light shielding mask, joystick, AIWA tape recorder, batteries and a spare fuse.
- Kit LAZIO-MB.
- Magnetometer head kit.
- AST spectrometer.
- Nikon D1X camera.

To achieve the first goal it was planned to use a high performance cosmic ray detector (MEB). To meet the second goal a high precision low frequency magnetometer would be used, consisting of the EGLE magnetometer box (MB) and EGLE magnetometer head (MH). LAZIO Kit 2 would be used to verify the hypothesis that the observed frequency of light flashes decreases with permanence in space. Light flashes had first been reported by Apollo astronauts on lunar missions, when they were outside the Earth's magnetosphere and more susceptible to cosmic rays. The current investigations continued the experiment started by Roberto Vittori on his previous visit to the ISS (Marco Polo, VC-3 in 2002). The assessment of the ability of different kinds of shielding to protect against the cosmic rays would use the AST spectrometer that had been delivered to the ISS for the Marco Polo mission.

Expected Results: Information recorded on PCMCIA cards (11 items), a CF memory card, voice tapes (5 items) and dosimeters (2 items).

EST

Objective: To verify the functioning of particular electronic equipment in the space environment.

Task: To test industrial grade electronic devices on board the Russian segment of the ISS.

Equipment Used: An electronic module with a set of industrial electronic equipment (microcontroller, SRAM, FRAM, FLASH, EEPROM memory, FPGA, reader devices, pressure sensors, impact sensors, temperature and humidity sensors) and associated accessories, plus a Sony DSR PD-150P camcorder and Nikon D1X camera.

Expected Results: Samples of industrial electronic equipment exposed to spaceflight conditions, and information on hardware operations recorded in various forms.

E-NOSE

Objective: To demonstrate the ability of an electronic nose to detect odours.

Task: Test the ability of E-Nose to detect the presence of even small amount of gases released by the on-board equipment.

Equipment Used: ENM electronic unit, reference probe tubes (2) and cables (3), IBM A31p ThinkPad laptop, Sony DSR PD-150P camcorder and Nikon D1X camera.

Expected Results: Data regarding air quality in different areas of the ISS, received automatically and recorded on a PCMCIA card and on video tape.

SPQR

Objective: To assess whether the imaging system on board the ISS could operate with special ground-based optics and image processing aids to detect possible damage to its external surfaces.

Task: At a time coordinated with Mission Control in Moscow, a laser beam would be aimed from the ground at the ISS. On hitting a cube cone reflector placed inside a window of the Zvezda module the laser would return to its source to be processed by the receiving station.

Equipment Used: A glass cone reflector with a bracket, a window shield and a safety bag, plus a Nikon D1X camera and memory card.

Expected Results: Laser reflected signals received on the ground, and return to Earth of the memory cards with photos of the hardware mounting on the window at the time of the experiment.

ASIA

Objective & Task: To assess the sensitivity of a circuit board of electronic components to the radiation in the space environment on the Russian segment of the ISS.

Equipment Used: A pouch consisting of the ASIA unit, an electronics board that would be carried in a metallic container (a minicrate PEER) to protect it against damage during its transportation and storage before and after the exposure exercise, a memory card to store the experiment data, plus a Nikon D1X camera and memory card.

Expected Results: The electronics board (in the protective container) would be returned to Earth after its exposure to the spaceflight environment inside the ISS, plus images of the hardware.

BOP

Objective: To investigate the molecular mechanisms regulating osteoblast physiology in microgravity.

Task: Culturing of differentiated human osteoblast cell line MG-63 on the Russian segment of the ISS.

Equipment Used: The main unit to house a set of culture chambers; Zip-lock bags with empty syringes, syringes with medium, syringes with PBS, and syringes with Lysing buffer; a bag with adapter; a bag for waste; various batteries, video tapes and memory cards; the culture chambers for the main unit; the Aquarius-B Incubator to cultivate cells in space; and the apparatus to return the samples to Earth. The experiment would also use a Sony DSR PD-150P camcorder and a Nikon D1X camera.

Expected Results: Human osteoblast cells cultivated in the space environment, plus a memory card and a video tape recording of the experiment.

ESD

Objective & Task: To make a video to familiarise students with the differences between the ground and space environments, in this case by demonstrating the electrostatic self-assembly of two different types of macroscopic components or spheres in weightlessness on board the Russian segment of the ISS. The experiment was to document the behaviour of 500 spheres made of polymethylmethacrylate (PMMA) and polytetrafluoroethylene (PTFE) inside transparent polycarbonate cube containers. When agitated, the spheres would acquire opposite electrical charges and in weightlessness they would self-assemble into ordered 3D molecular structures.

Equipment Used: The Electrostatic Self-Assembly Demonstration pouch with cube containers (each containing a set of spheres made of different materials), plus a Sony DSR PD-150P camcorder.

Expected Results: Video tape recordings.

Timeline

Flight Day 1 – Friday, 15 April 2005

Roberto Vittori lifted off from Baikonur Cosmodrome in Kazakhstan in Soyuz TMA-6 at 06:46 local time (02:46 CEST, 00:46 UT) for the Eneide mission to the ISS.

"I am very pleased to see this mission successfully on its way," announced Daniel Sacotte, ESA's Director of Human Spaceflight, Microgravity and Exploration Programmes. "The launch of Eneide is the culmination of a great deal of effort by all parties in Italy, Russia and the ESA centres involved in its preparation. I know this effort will continue throughout the mission, and beyond. This cooperation will bring benefits to European citizens of all ages through the extensive scientific, educational, and technological programme that the mission will carry out, and I hope this will fire the imaginations of the children who will become our scientists and engineers of tomorrow."

Flight Day 3 – Sunday, 17 April 2005

After a two day pursuit, Soyuz TMA-6 docked with the Pirs module of the Russian segment of the ISS at 6:20 Moscow Time (4:20 CEST). The hatches were opened just over 2 hours later and Leroy Chiao, the ISS commander, welcomed the newcomers aboard with bread and salt in accordance with Russian tradition.

Daniel Sacotte was the first to offer congratulations on behalf of Europe, telling Vittori, "I know you have already performed a navigation experiment during your flight to ISS."

When asked how he was feeling, Vittori said, "I'm very proud that the Italian and European flags have flown in space for the second time. Thank you to everyone that has made this mission possible."

Flight Day 4 – Monday, 18 April 2005

As a colonel in the Italian Air Force, Vittori participated in a video conference with the Interforce Operational Centre of the Italian Ministry of Defence. He spoke with Admiral Giampaolo di Paola, Chief of Staff of Italian Defence, and General Leonardo Tricarico, Chief of Staff of the Italian Air Force. The Italian Air Force had forged strong links, via ASI, with ESA and Roscosmos. As part of this, the Italian Air Force Flight Test Centre (CSV) in Pratica di Mare, near Rome, had collaborated with the Medical Division of the Yuri Gagarin Cosmonaut Training Centre. This had resulted in the first international students training at Star City in order to qualify as Space Flight Surgeons.

Flight Day 5 – Tuesday, 19 April 2005

Vittori worked on the Bone Proteomics experiment (BOP) into how osteoblast cells produce bone in microgravity. As humans have shown a loss in bone mass of 1% per month, this is a serious problem for spacefarers.

Adalberto Costessi was the Principal Investigator for the experiment. He won the Success 2002 competition for European university students. This competition was organised by the ISS Utilisation Strategy and Education Office within ESA's Directorate of Human Spaceflight, Microgravity and Exploration. As he explained, "I have been at ESTEC, ESA's research and technology centre in the Netherlands since June 2004 as my prize for winning the contest. It has been an amazing experience with lots of very hard work. In the past, experiments have shown the number of osteoblast cells that reach full maturation is decreased under weightless conditions compared to Earth. Also, the mature cells produce much less bone material. We don't yet know why. We hope that BOP will allow us to understand better how osteoblast cells work. We hope to help identify new methods for treating bone diseases – not only for astronauts, but also for the millions of people affected by osteoporosis here on Earth."

Flight Day 6 – Wednesday, 20 April 2005

On this day the Amateur Radio on the ISS (ARISS) programme was active again. Each ESA astronaut since Frank De Winne had communicated with school children via 'ham' radio. This time 200 children – winners of the 'mISSione possibile' competition – visited ESA's European Space Research Institute (ESRIN) near Rome. As the station was not over Italy at the time, the radio link was established with a NASA facility in Greenbelt, Maryland, and the signal relayed to ESRIN by telephone line.

Flight Day 7 – Thursday, 21 April 2005

The ENEIDE Esperimento di Navigazione per Evento Italiano Dimostrativo di EGNOS (Navigation experiment for an Italian EGNOS Demonstrative Event) was conducted by Vittori in order to test the operation of the European Geostationary Navigation Overlay Service. EGNOS uses data provided by GPS for civil use, which can identify a vehicle's position to a precision of within 20 metres. EGNOS can improve that resolution to just a few metres, enabling it to be employed for critical applications such as by civilian air traffic. "ENEIDE seems to have performed optimally," noted Vittori, after the final experimental session. "The receiver found the satellites very quickly, and the software worked perfectly."

The VSV experiment was led by Colonel Enrico Tomao, who was in charge of flight and space medicine at CSV. Vittori conducted the experiment to "analyse the visceral receptor performance within an environment which rules out possible bias due to visual and gravitational inputs." It was known that orientation is mainly derived from inputs provided by specialised sensors located in the eye, the inner ear, joints and muscles. Recently new inputs had been revealed. Some receptors in the kidneys and the thorax are also sensitive to shifts in blood mass, and they therefore contribute to an individual's perception of what constitutes vertical. Since this finding was so recent, there was little data on the effect it would have in an environment such as the ISS. The first phase of the experiment was carried out at the Gagarin Cosmonaut Training Centre. On 21 April, Vittori underwent a second session on board the ISS. This lasted an hour and used the Lower Body Negative Pressure unit which drew fluids, mainly blood, toward the feet, thus stimulating the visceral receptors. The final phase of the experiment would be carried out after re-entry to see the effects, if any, of prolonged exposure to microgravity.

Flight Day 8 – Friday, 22 April 2005

Throughout the mission, Vittori supported the LAZIO experiment. Its Principal Investigator, Roberto Battiston, was director of the Perugia section of the Italian National Institute of Nuclear Physics. "The data gathered during the mission will be analysed in the following months," Battiston said. "We will try to identify evidence of a connection between variations in the particle flux and seismic phenomena registered at ground stations during the same period." The aim was to test a 20-year-old hypothesis that a high emission of low frequency electromagnetic waves originate from an area where an earthquake is about to occur. The initial observations had been made on board the Soviet Salyut stations. Should the hypothesis prove correct, then by measuring the intensity and variations in the charged particles around Earth it would be possible to predict an earthquake several hours before it started, and identify the area in which it would occur.

Vittori also worked with the Alteino instrument from the Marco Polo mission 2002. In this case, he wore a helmet to block any light and whenever he saw a light flash he pushed a button. The experiment sought a correlation between the occurrence of the light flashes and the flux of cosmic rays which were being measured at the same time. As Battiston pointed out, "Light flashes could be due to light emitted by the charged particles and revealed by the retina, or to effects related to the stimulation of nerve cells by heavy ions. The preliminary data obtained on the Marco Polo mission have encouraged us to repeat the experiment."

Flight Day 9 – Saturday, 23 April 2005

The Agrospace Experiment Suite (AES) was similar to the SEEDS experiment conducted by Kuipers on his mission and it involved growing beans and rocket plants in space to test whether production of food in space was sustainable. As Principal Investigator Marco Casucci explained, "It's an experiment which could be important in the future, for long-duration space missions. If we're able to demonstrate that astronauts can grow rocket sprouts, then we're a step closer to improving their diet and their lives in orbit."

The Space Beans for Students experiment was an educational programme where students were to grow the same kind of beans on Earth as Vittori did on board the ISS, and then the results would be compared.

Flight Day 11 - Monday, 25 April 2005

The hatches between the Soyuz TMA-5 spacecraft and the ISS were closed at 17:34 CEST on Sunday 24 April, with Vittori, Sharipov and Chiao aboard. At 20:44 it undocked and drew away. The capsule re-entered the atmosphere at 23:44, the main parachute opened at 23:53 and the landing occurred near the town of Arkalyk in Kazakhstan at 00:07 CEST (04:07 local time).

10.4 POSTSCRIPT

Subsequent Missions

See the DAMA mission.

Roberto Vittori Today

See the Marco Polo mission.

11

Astrolab

Mission

ESA Mission Name:	Astrolab
Astronaut:	Thomas Arthur Reiter
Mission Duration:	171 days, 3 hours, 54 minutes
Mission Sponsors:	ESA
ISS Milestones:	ISS ULF1.1, 30th crewed mission to the ISS

Launch

Launch Date/Time:	4 July 2006, 18:37 UTC
Launch Site:	Pad 39-B, Kennedy Space Center
Launch Vehicle:	Space Shuttle Discovery (OV-103)
Launch Mission:	STS-121
Launch Vehicle Crew:	Steven Wayne Lindsey (NASA), CDR
	Mark Edward Kelly (NASA), PLT
	Michael Edward Fossum (NASA), MSP1
	Lisa Marie Caputo Nowak (NASA), MSP2
	Piers John Sellers (NASA), MSP3
	Stephanie Diana Wilson (NASA), MSP4
	Thomas Arthur Reiter (ESA), MSP5

Docking

STS-121

Docking Date/Time:	6 July 2006, 14:52 UTC
Docking Port:	PMA-2, Destiny Forward

J. O'Sullivan, *In the Footsteps of Columbus*, Springer Praxis Books,
DOI 10.1007/978-3-319-27562-8_11

STS-116
Undocking Date/Time: 19 December 2006, 22:10 UTC
Docking Port: PMA-2, Destiny Forward

Landing

Landing Date/Time:	22 December 2006, 22:32 UTC
Landing Site:	Runway 15, Shuttle Landing Facility, Kennedy Space Center
Landing Vehicle:	Space Shuttle Discovery (OV-103)
Landing Mission:	STS-116
Landing Vehicle Crew:	Mark Lewis 'Roman' Polansky (NASA), CDR William Anthony 'Bill' Oefelein (NASA), PLT Nicholas James MacDonald Patrick (NASA), MSP1 Robert Lee, Jr. 'Beamer' Curbeam (NASA), MSP2 Arne Christer Fuglesang (ESA), MSP3 Joan Elizabeth Miller Higginbotham (NASA), MSP4 Thomas Arthur Reiter (ESA), MSP5

ISS Expeditions

ISS Expedition:	Expedition 13
ISS Crew:	Pavel Vladimirovich Vinogradov (RKA), ISS-CDR Jeffrey Nels Williams (NASA), ISS-Flight Engineer 1 Thomas Arthur Reiter (ESA), ISS-Flight Engineer 2

ISS Expedition:	Expedition 14
ISS Crew:	Michael Eladio 'LA' Lopez-Alegria (NASA), ISS-CDR Mikhail Vladislavovich Tyurin (RKA), ISS-Flight Engineer 1 Thomas Arthur Reiter (ESA), ISS-Flight Engineer 2 replaced by Sunita Lyn 'Suni' Williams (NASA), ISS-Flight Engineer 2

11.1 THE ISS STORY SO FAR

Since Roberto Vittori's Eneide mission in April 2005, the Space Shuttle had resumed operations and there had been two Soyuz launches to the ISS.

STS-114 marked the Return to Flight in July 2005. In addition to delivering cargo and removing accumulated waste using the MPLM Raffaello, a new set of safety procedures were implemented to verify that no damage to the orbiter's thermal protection tiles and carbon-carbon sections had occurred during the ascent into space that could pose a risk when re-entering the atmosphere. An unprecedented array of ground and air based high definition and high speed cameras recorded the launch, and this footage was analysed before re-entry was approved. If re-entry had been ruled out, the crew would have been transferred to the ISS to await rescue and the orbiter would have been abandoned. Impact sensors on the leading edges of the wings had been installed to record damage to the carbon-carbon sections. In space, the RMS lifted the new Orbiter Boom Sensor System (OBSS) which contained cameras and a Laser Dynamic Range Imager (LDRI). This was lowered beneath the orbiter to scan the nose and the leading edges of the wings. And shortly prior to docking, at a stationary point directly below the ISS the orbiter made the Rendezvous Pitch Manoeuvre, a 'backflip' at 0.75 degrees per second to enable the station crew to take high resolution photographs of the tiles and send them to NASA for analysis. The damage detected on this mission wasn't sufficient to prevent the vehicle from returning to Earth, but it grounded the fleet for another year.

In October 2005 Soyuz TMA-7 delivered the Expedition 12 crew to the ISS, together with 'space tourist' Greg Olsen, whose mission had been delayed from TMA-6 owing to health issues. It was the last in a series of flights that had been sold by Roscosmos to NASA to carry American astronauts. A new deal was signed in order to continue this arrangement.

Meanwhile, on 12 October 2005 Shenzhou 6 launched China's first two-person crew into orbit. As previously, when the descent module returned to Earth, the orbital module continued to operate under ground control before being de-orbited in April 2006. This is unlike the orbital module of the Russian Soyuz spacecraft, where the orbital module is jettisoned to fall into the atmosphere. The Chinese used the unmanned modules to help in planning the construction of their own space station.

Soyuz TMA-8 delivered the Expedition 13 crew to the ISS along with Brazilian astronaut Marcos Pontes. Ten days later Pontes returned to Earth with the Expedition 12 crew on board Soyuz TMA-7.

11.2 THOMAS REITER

Early Career

Thomas Reiter was born in Frankfurt on 23 May 1958. Upon graduating from Goethe-High School in Neu-Isenburg in 1977 he joined the Luftwaffe and studied at the Bundeswehr (German Armed Forces) University in Munich, from which he gained a master's in aerospace technology in 1982.

After completing his military jet training at Sheppard Air Force Base in Texas, Reiter was based at Oldenburg, Germany. He was involved in the development of computerised mission planning systems and became a flight operations officer and deputy squadron commander. In 1990 he received test pilot training in Manching, Germany, and then worked on test projects. After converting to the Tornado for operational flying, he completed advanced test pilot training at the Empire Test Pilot School (ETPS) at Boscombe Down in England.

Thomas Reiter. (ESA)

In 1992, Reiter joined the ESA European astronaut corps. His first mission was EuroMir95 to the Soviet Mir space station. Training at Star City, Moscow, commenced in August 1993. It included Mir operations and maintenance, Mir EVAs in Soviet suits, and training as a flight engineer for the Soyuz spacecraft. During his 179 days aboard Mir he carried out two EVAs and became the first German to perform a spacewalk. Back on Earth, he trained on the Soyuz TM type and gained the Soyuz Return Commander certificate that qualified him to command a three-person Soyuz capsule during its return from space. Reiter also helped to develop the European Robotic Arm (ERA) that was intended for use on the ISS.

From September 1997 to March 1999, Reiter returned temporarily to the Luftwaffe as operational group commander of a Tornado wing. Then he returned to the European Astronaut Centre at Cologne, Germany, where he worked on the Automated Transfer

Vehicle (ATV) and ERA programmes. He also continued training at the Gagarin Cosmonaut Training Centre until March 2000 in preparation for ISS missions. In April 2001, Reiter was assigned to the first ISS advanced training class to prepare for the inaugural European long-term mission. Within the Directorate of Human Spaceflight and Exploration he worked on the Columbus programme. In September 2004 he was assigned to the Expedition 13/14 roster.

Previous Mission

Soyuz TM-22 EuroMir95

In September 1995 Thomas Reiter flew to Mir on board Soyuz TM-22. The intention was for Reiter to join the Mir Expedition 20 crew as part of his EuroMir95 programme for 135 days. This was extended to 179 days as a result of the financial difficulties being experienced by the Russians at the time which delayed the construction of the Soyuz rocket and spacecraft for the next crew. EuroMir95 consisted of 41 biomedical, technical, and materials processing experiments. One EVA for Reiter had always been intended, and a second was added after the mission was extended.

During the first EVA in October 1995 Reiter worked with Sergei Avdeyev to install the European Science Exposure Facility (ESEF) and to exchange sampling cassettes on a

The Soyuz TM-22 crew with Thomas Reiter in the centre. (www.spacefacts.de)

Russian experiment that measured orbital debris. For an EVA in December, he remained in Soyuz TM-22 ready to depart for Earth if anything should happen to his crewmates outside the station. For the third EVA in February 1996 Reiter and Yuri Gidzenko received training by radio owing to the ad hoc nature of the tasks.

After the successful EuroMir94 and EuroMir95 missions, ESA would readily have funded a third mission with Reiter's backup, Christer Fuglesang, but attention was turning to the new ISS, in which Europe would have a bigger part to play.

11.3 THE ASTROLAB MISSION

Astrolab Mission Patches

The ASTROLAB mission patch was designed by German artist Detlev van Ravenswaay. The outer circle comprised 24 sections to represent the ancient astronomical computer called an astrolabe. The black, red and yellow border of the inner circle matched the German flag in honour of Thomas Reiter.

The ISS was shown in its then current configuration, against a field of stars. The three bright stars to the left symbolised the Expedition 13 crew and the 17 smaller stars were Member States of ESA.

The STS-121 crew with Thomas Reiter second from the right. (NASA)

The Expedition 13 crew with Reiter on the left. (NASA)

The Astrolab mission honoured the German geographer, astronomer and mariner Martin Behaim, who made major improvements to the astrolabe. And by referencing ASTROnauts and LABoratory it highlighted Reiter's role as ESA's first long-duration ISS crewmember.

The STS-121 patch showed the Space Shuttle docked with the ISS, which was in its then current configuration. The Sun and the astronaut symbol depicted orbital dawn. The names of the crew were displayed around the border.

The Expedition 13 patch was designed be Alex Panchenko, and evolved as the crewing changed. The initial version didn't have either Reiter's name or the German flag. Then there was a patch with Reiter's name but no flag. The final version had both. The ISS was shown in its configuration at the start of the six month expedition, with trailing elements from the national flags representing the three crewmembers. The Expedition 13 crew said, "The dynamic trajectory of the space station against the background of Earth, Mars and the Moon symbolises the vision for human space exploration beyond Earth orbit and the critical role that the ISS plays in the fulfilment of that vision."

The Expedition 14 crew with Reiter on the left. (NASA)

The Expedition 14 patch was designed by mission commander Michael 'LA' Lopez-Alegria. He wanted a simple patch without names. The trajectory represented the future path to the Moon and to Mars. The five stars honoured the men and women who lost their lives either training for or actually flying on Apollo 1, Soyuz 1, Soyuz 11 and the Space Shuttles Challenger and Columbia.

Astrolab Mission Patch. (ESA)

Astrolab Mission Objectives

The Astrolab mission was historic for a number of reasons. Starting with the launch of Thomas Reiter on STS-121, continuing through ISS Expeditions 13 and 14 and ending with the landing of STS-116 it would be the first ESA long-term mission to the ISS.

It had many objectives:

- First ESA astronaut member of an ISS Expedition – Reiter became a member of Expedition 13 shortly after Discovery docked with the ISS and stayed on to join Expedition 14 in September 2006 when Vinogradov and Williams were replaced by Lopez-Alegria and Tyurin. As the first ESA member of a main ISS Expedition, he was also the first non-US, non-Russian, long-term crewmember.
- First ESA astronaut to perform a spacewalk from the ISS.
- First long-duration European science programme on the ISS – Astrolab marked the first time that a European scientific programme was assembled for a long-duration

STS-121 Mission Patch. (NASA)

mission, including experiments in human physiology, biology, physics, radiation dosimetry, various industrial experiments, technology demonstrations and educational projects.

- Delivery of European experiment facilities – Space Shuttle Discovery delivered three ESA experiment facilities to the ISS:
  - Minus Eighty Laboratory Freezer (MELFI)
  - European Modular Cultivation System
  - Percutaneous Electrical Muscle Stimulator.
- Resumption of three-member Expedition crewing – Since the Columbia tragedy, the ISS had been reduced to two-person caretaker crews and Assembly Flights were put on hold. Reiter's arrival saw the crew return to three and the recommencement of the construction of the truss structure.
- First long-duration ISS mission to be managed by a European Centre – The control centre for the Astrolab mission was at the new Columbus Control Centre in

Expedition 13 Mission Patch. (NASA)

Oberpfaffenhofen, Munich, Germany. Astrolab gave this centre vital operational experience in advance of the delivery of the Columbus module to the station by STS-122 in February 2008.

Science

The following experiments were to be carried out as part of the Astrolab mission:

- CARD
- CARDIOCOG-2
- Chromosome-2
- CULT
- Eye Tracking Device
- Lymphocytes (L) and Granulocytes (G)
- Immuno
- NOA-1
- NOA-2
- Epstein-Barr

Expedition 14 Mission Patch. (NASA)

- Renal Stone
- BASE
- Leukin
- Ying
- ALTCRISS
- Matroshka-2
- Under the Background Influence
- PK-3+
- ERB 3D Video Camera
- Special Event Meal
- Skin Care
- GTS-2
- ARISS
- Lessons from Space
- E-learning Session
- Oil Emulsion.

Human Physiology

CARD

Salt intake can increase cardiovascular measurements such as the total volume of blood pumped from the heart over a given period. This experiment would increase salt intake to increase blood volume in order to assess the effects on blood pressure, heart rate, cardiac output and the neuroendocrine system.

It was to involve six subjects across multiple Expeditions and use the Pulmonary Function System and the MELFI freezer. The results were expected to provide insight into the mechanisms underlying certain cardiovascular problems on Earth, in particular heart failure.

CARDIOCOG-2

This experiment was to study the effects of weightlessness on the cardiovascular system, as well as stress and the cognitive and physiological reactions of an astronaut.

On four occasions over the course of a mission, an astronaut would undertake a half-hour protocol of normal and controlled breathing, together with a stress test. Cardiac activity, respiration, and blood pressure were to be measured continuously during this activity using Cardioscience equipment on the ISS. This would be compared against ECG, blood pressure, respiration, and ultrasound measurements obtained during pre-flight tests.

The experiment was a continuation of CARDIOCOG and was to be performed by three additional long-term subjects. The aim was to increase understanding of the common clinical problem known as orthostatic intolerance, which was a proneness for fainting.

Chromosome-2

While in space, astronauts are always exposed to different types of ionising radiation. This experiment was to assess the genetic impact of these radiations by studying chromosome changes and sensitivity to radiation in white blood cells. It was to involve eight subjects: four from short-duration flights and four Expedition crewmembers.

CULT

This experiment was to study cultural aspects and leadership styles of ISS crewmembers. Data from questionnaires would be analysed to investigate dynamics as a function of mission duration. Research on ground personnel would be carried out in parallel. The results would provide recommendations on how to interact with multinational crews. In particular the objective was to provide recommendations for communication between the ground and space. This experiment would require eight subjects over multiple Expedition crews.

Eye Tracking Device

Our sense of balance is strongly interconnected with our eyes, and understanding how weightlessness affects this relationship ought to increase our knowledge of 'space sickness' in the space environment and also terrestrial conditions such as vertigo and nausea.

In space our eyes can rotate around three axes, whereas normally only two are used. The name of the coordinate framework that describes the movement of the eyes in the head is known as Listing's plane. This experiment was to investigate Listing's plane under different gravity conditions using the Eye Tracking Device to record horizontal, vertical and rotational eye movements, and measure head movement.

This experiment would require eight subjects from long-duration missions and eight from short-duration missions. It began with ESA astronaut André Kuipers during the DELTA mission in April 2004.

Lymphocytes (L) and Granulocytes (G)
This required taking blood samples to study white blood cells Lymphocytes (L) and Granulocytes (G).

Immuno
This experiment was to determine changes in stress and immune responses during and after a stay on the ISS. It required sampling saliva, blood and urine to check for hormones associated with stress and for analysing white blood cells. One focus would be the adaptation of energy metabolism, which can affect immune response. A better understanding of the linkage between stress and the functioning of the immune system would have relevance for people on the ground. The experiment would involve a total of six subjects over multiple Expedition crews.

NOA 1
Research had demonstrated that more nitric oxide in our breath is an early and accurate sign of airway inflammation, especially in asthma but also after inhalation of dust. This experiment was to exploit an improved technique for analysing nitric oxide in expired air to study the physiological reactions of humans in weightlessness.

Dust never settles in weightlessness, so it seemed highly likely that there would be an increased exposure of the human airways to inhaled particles in space. The crewmembers were to carry out a simple inhalation-exhalation procedure on a bi-weekly basis throughout their stay on the station. An increased level of expired nitric oxide, compared to preflight, would indicate airway inflammation. The data would be stored on a credit-card-sized memory unit. This experiment would require eight subjects over multiple Expedition crews, beginning with Expedition 12.

The Platon device that was developed for spaceflight served a dual role, because it could also be used to improve the treatment of asthma by allowing monitoring of patients at home.

NOA 2
In scuba divers the presence of gas bubbles in the bloodstream due to decompression is well-known, and bubbles can occur after dives even when there are no symptoms of decompression sickness. The occurrence of decompression sickness in astronauts

following a spacewalk is not known. Employing decompression techniques on Earth produces symptoms in about 6% of cases. This suggests a much higher frequency of gas bubbles without clear symptoms of decompression sickness. A non-invasive and simple technique for assessing current decompression techniques prior to and after a spacewalk would be beneficial.

In this experiment, astronauts would perform a simple inhalation-exhalation procedure (as in the NOA 1 protocol) as late as possible before starting the standard spacewalk preparations, and again as soon as possible after finishing the spacewalk. An increase in expired nitric oxide would indicate the presence of gas emboli, suggesting a requirement for adaptation of existing procedures.

NASA experiments

In addition to the European experiment programme in human physiology, Thomas Reiter took part in two experiments on behalf of NASA scientists: Epstein-Barr and Renal Stone.

Epstein-Barr

This experiment covered the reactivation of the latent Epstein-Barr virus with which about 90% of the adult population is infected. Blood and urine samples were to be collected before and after the mission in order to assess the functioning of the immune system.

Renal Stone

Exposure to weightlessness could increase the risk of renal stone development during and immediately after spaceflight. This experiment would test the use of potassium citrate as a countermeasure to reduce the risk of renal stone formation in space. On Earth, potassium citrate is a proven means of minimising calcium-containing renal stone development.

Biology

BASE

With the Bacterial Adaptation to Space Environments (BASE) experiment, scientists set out to study how bacteria cope with and adapt to aspects of the space environment; in particular weightlessness, cosmic radiation, electric and magnetic fields, and space vibrations. The results would indicate how such adaptations might influence their potential to contaminate and biodeteriorate the space habitat, their potential to endanger crew health, and their function in waste recycling and food production systems. In addition, several model bacteria were to be cultured to observe their physiology, gene expression, gene rearrangement, and gene transfer when subjected to the space environment.

LEUKIN

The purpose of this experiment was to study the signal transduction pathway of the activation of T-lymphocytes. The focus was on the role of the IL-2 receptor and on the determination of its genetic expression. The hypothesis was that a lack of expression of IL-2 R is the major cause of the loss of activation in re-suspended cells in weightlessness. Improving our knowledge of the mechanisms by which spaceflight changes the functioning of immune cell could assist in the development of better preventative and/or corrective measures for immune suppression during long-term space missions.

YING

The Yeast-In-No-Gravity experiment was to investigate the influence of weightlessness on so-called Flo processes – in particular the cell-surface interaction on solid media and the cell-cell interaction in liquid media in Saccharomyces cerevisiae yeast cells. Weightlessness has a direct effect on the physiology of yeast cells due to an altered gravitational micro-environment, and in the case of yeast cell cultivation in liquid media the changed shear environment in microgravity also has an effect. The goal was to obtain a detailed insight into the roles of gravity and shear stress in the formation of organised cell structures such as yeast flocs, biofilms and filaments. The results would of interest for both fundamental science and industry as well as the medical field.

Radiation

ALTCRISS

ALTCRISS (Alteino Long-Term monitoring of Cosmic Rays on the International Space Station) was an ESA experiment to study the effect of shielding against cosmic rays in two complementary ways. The Alteino detector monitored differences in the flow of cosmic rays with regard to the position and orientation of the device and also with regard to different shielding materials placed over the particle acceptance windows. It had previously been operated by Roberto Vittori, during his Marco Polo and Eneide missions. It consisted of a cosmic ray detector (AST/Sileye-3) and an Electroencephalograph (EEG), but the EEG was not necessary for the Altcriss experiment. The data was to be used to better understand the radiation environment aboard the ISS and how to provide efficient shielding against it.

Matroshka 2

The ESA Matroshka facility was installed on the external surface of the ISS on 27 February 2004 in order to study the radiation levels experienced by astronauts on spacewalks. It consisted of a human shape (head and torso) known as the Phantom which was equipped with active and passive radiation dosimeters. It was placed in a carbon-fibre, reinforced plastic container to simulate a spacesuit. The facility was brought back inside on 18 August 2005. Passive sensors were removed and returned to Earth with Expedition 11. The facility was to be placed outside again in September 2006 in order to collect further data.

Under the Background Influence

This was an advanced technology demonstration which involved measuring the background radiation inside the modules of the ISS using a new type of sensor. The goal was to verify the radiation models used to predict internal radiation levels, and if necessary correct them. The new radiation sensor was made of an alloy of cadmium, zinc and tellurium. Its advantage was that it was compact and did not require cryogenic cooling. The experiment would demonstrate this specific sensor technology for the first time in space.

If successful, it could act as a precursor for an instrument in ESA's Atmospheric-Space Interaction Monitor (ASIM), which was intended to be placed on the exterior of the European Columbus laboratory.

Understanding space radiation, its interaction with the station, and its impact on the human body were key factors in safely planning long-duration flights.

Physics

PK-3+

Plasma is the most disordered state of matter, being composed of electrically charged electrons and ions. A 'complex plasma' enriched with micro-particles possesses special properties which facilitate fundamental investigations in weightless conditions. In addition to improving our understanding of fundamental physics this research has applications across many scientific disciplines such as plasma processing and fluid dynamics.

The PK-3+ facility was to investigate complex plasmas in weightlessness over a broad range of fundamental parameters. It was built under the responsibility of the German Aerospace Centre, DLR and superseded the PK-3 (PKE-Nefedov) facility that had been in use on the ISS since March 2001.

Technology

ERB - 3D video camera

The 3D video camera known as the Erasmus Recording Binocular was to be tested in weightlessness on board the ISS and used to accurately map its interior. This would involve using images from three cameras: the ERB 3-D video camera, a Sony PD-150 video camera, and a Nikon 3-D still camera. It would improve models available on the ground, as well as improve the fidelity of the ISS 3D virtual reality simulator at the Erasmus Centre located at ESTEC, which was ESA's research and technology centre in Noordwijk, the Netherlands. Of special interest was filming subjects and objects moving to and from the camera and objects which protrude from a surface, such as cables on experiment racks.

Special event meal

The French space agency (CNES) sponsored a project to provide the ISS crew with high quality food for celebratory meals such as New Year, the arrival of a new crew, and birthdays. The ability to break the monotony of standard daily food would have a positive psychological effect, especially on a long-duration mission.

Skin Care

This human physiology experiment was to characterise a number of parameters of human skin such as hydration grade and transepidermal water loss in weightlessness. It would test the applicability of the space environment as a model for the ageing of skin. It involved non-invasive medical equipment, in particular video imaging of the skin surface.

GTS-2

Global Transmission Services 2 was a technology experiment to test, validate and demonstrate radio transmission techniques for the synchronisation of Earth-based clocks and watches from the ISS. The GTS data services, based on a unique coding scheme, could eventually lead to commercial functions such as blocking stolen cars or credit cards directly from space.

After theoretical investigations and practical tests had suggested measures to remedy the weaker-than-expected strength of the transmitted signal earlier experienced by receivers on Earth, the GTS system was reactivated on 5 December 2005.

Education

ARISS

For the Astrolab mission, the objectives of the Amateur Radio on the ISS (ARISS) programme were to build, develop and maintain the amateur radio activities on the station and in particular to provide real-time radio links during which primary schools in Germany and Switzerland could put questions to ESA astronaut Thomas Reiter. Among the children chosen were the winners of national competitions set up by ESA's Education Office. The ground stations were provided by local amateur radio clubs.

Lesson from Space

The theme of the fourth DVD produced for the 'Lessons from Space' series was robotics. This would be made with support and input from ESA astronauts, robotic experts, and teachers and their pupils in Germany, Sweden, Switzerland, and the United Kingdom. It would feature footage of demonstrations from the Astrolab mission.

E-learning session

An e-learning session was scheduled as an 'Earth-based' lecture during which a live link-up was to be made with Thomas Reiter on the ISS. The lecture was to be presented to European university students following the EuMAS Masters Programme in Aeronautics and Space by Dr. Hubertus Thomas of the Max-Planck-Institute in Garching, Germany. The scientific theme was plasma crystals and complex plasmas, with reference to the PK-3+ plasma crystal experiment that was to be performed during the Astrolab mission.

Oil Emulsion

This experiment was to be carried out on Earth by school pupils in the 11–14 year age range, and by Thomas Reiter in space. The space part of the experiment would be filmed and downlinked. The aim was to highlight that an oil and water emulsion behaves differently in weightlessness and on Earth. A sealed container with two immiscible fluids (clear oil and ink-coloured water) was to be shaken until the two fluids began to mix. Their weightless behaviour would then be filmed at specific times over a period of two weeks, then shown to schools by a children's programme on German public television. The different kinds of segregation that occur during the experiment, in space and on Earth, would be explained by teachers. The experiment was developed by DLR as a cooperative exercise between the German and European space agencies.

Timeline

Flight Day 1 – Tuesday, 4 July 2006

The Space Shuttle Discovery lifted off at 1:38 p.m. CDT on 4 July 2006 for STS-121, bound for the ISS to enable Thomas Reiter to join the Expedition 13 crew.

As part of the post-Columbia regime, there was greater vigilance and monitoring of the orbiter's heat shield tiles. In addition to ground based cameras monitoring the ascent through to the jettisoning of the SRBs, the orbiter crew took photographs and video of the ET after this separated in space, the Orbiter Boom Sensor System inspected the vehicle soon thereafter.

Having spent 179 days on board Mir a decade earlier, Reiter was starting his second long-duration mission.

Daniel Sacotte, the ESA Director of Human Spaceflight said, "Astrolab is a milestone for human spaceflight in Europe. Thomas is just the first to go and more will follow. Even if our astronauts will not be permanently on the Space Station over the coming years, we must prepare for having them on board very often and for long durations. This is really the beginning of a long-term European human presence in space."

ESA Director General Jean-Jacques Dordain added, "Europe is a visible and reliable partner in one of the most complex projects ever carried out in space – with Thomas on board this year, and with the launches of Columbus and the first ATV servicing mission next year. Thanks to the commitments of its partners, in particular NASA and the Russian space agency, as well as the commitments of its Member States, ESA can actively prepare for utilisation of the Space Station and – looking beyond that – for its contribution to the exploration of the solar system."

Flight Day 2 – Wednesday, 5 July 2006

As was traditional, the Shuttle crew were awakened to the sound of music chosen for crewmembers by their families and friends. In this case it was *Lift Every Voice and Sing* by the New Galveston Chorale, from Stephanie Wilson's home town.

This was officially the second and final Return to Flight mission, rectifying the problems that had been observed by STS-114, so Mission Control promptly set about analysing the videos of the launch to determine whether the Shuttle was in a safe state for re-entry. As on STS-114, the Laser Dynamic Range Imager, Laser Camera System, and Intensified Television Camera of the OBSS inspected the carbon-carbon panels of the wing leading edges and nose cap. This 6.5 hour inspection found nothing amiss. Then the TV cameras on the Shuttle's robot arm scanned the exterior of the crew cabin for an hour.

During this time, Thomas Reiter prepared the cargo in the mid-deck for transfer to the ISS. Sellers and Fossum installed the centreline camera in Discovery's docking mechanism and checked out their EVA suits.

Flight Day 3 – Thursday, 6 July 2006

In honour of Reiter's son Daniel, the wakeup call was Elton John's *Daniel*. After the famous backflip manoeuvre to enable the station crew to photograph its thermal protection tiles, Discovery docked at PMA-2 on the front of the Destiny module at 9:52 a.m. CDT. The hatches were opened at 11:30 a.m., and Reiter promptly installed his seat liner in the Soyuz spacecraft to officially become a member of the Expedition 13 crew.

Sellers and Fossum transferred their EVA spacesuits to the Quest airlock. There was some serious robot arm activity when Nowak, Wilson and Williams lifted the OBSS using the station's Canadarm2 and passed it to Discovery's robotic arm. This was in preparation for the first spacewalk, when Sellers and Fossum were to mount the end of the OBSS/RMS combination to see whether the 100-foot-long combined arm was a stable work platform.

Flight Day 4 – Friday, 7 July 2006

Lisa Nowak's children chose *Good Day Sunshine* by the Beatles as Friday's wakeup call, which was heard by the crew at 2:14 a.m. CDT. Expedition 13 were awakened 30 minutes later to their standard wake up tone.[9]

Stephanie Wilson operated Canadarm2 to transfer the Leonardo Multi-Purpose Logistics Module from the payload bay of the Shuttle to the Unity node.

As a result of the OBSS survey of Discovery's heat shield, NASA ordered a further inspection of the nose cap because the heat shield blanket around the nose cone and two 'gap fillers' seemed to be protruding from the underside.

As Fossum and Sellers prepared their spacesuits and tools for Saturday's EVA, media obligations saw Lindsey and Kelly talking to CBS, Fox, ABC, and NPR in one session, and the Expedition crew talking with CNN, CBS News, and the Associated Press in another.

After reviewing the rate at which the orbiter's consumables were being used, Mission Control told the crew that the mission would be extended by an additional day to 13 days. This would provide time for the third spacewalk whose execution had been made conditional on there being sufficient resources available.

Flight Day 5 – Saturday, 8 July 2006

To celebrate Fossum's upcoming first spacewalk his family chose *God of Wonders* by Marc Byrd and Steve Hindalong as the wakeup call.

---

[9] Lisa Novak would achieve notoriety in 2007 when, on 5 February, she was arrested for an attempted kidnap and battery of Air Force Captain Colleen Shipman. Novak was having an affair with fellow astronaut William Oefelein, who was in turn having a relationship with Shipman. Novak's assignment to NASA was curtailed on 7 March 2007. In November 2009 she pleaded guilty in the civil court and was sentenced to a year's probation. In July 2011 she was demoted from Navy Captain to Commander and discharged in "other than honourable conditions".

In performing the first spacewalk of the mission, Sellers and Fossum installed a 'blade blocker' in the zenith umbilical cable for the Mobile Transporter to prevent an unintentional severing of the cable (as occurred to the matching nadir cable in December 2005). They then carried our several tests of the OBSS attached to the Shuttle's robotic arm in order to assess its suitability as an EVA work platform. Variously with Sellers and then with both Sellers and Fossum, they performed simulations of a variety of work movements while sensors in the foot restraint recorded the forces.

Meanwhile on board the ISS, Expedition 13 crewmembers Vinogradov and Reiter unloaded cargo from the MPLM, including a new oxygen generator for the Destiny laboratory module and the Minus Eighty Laboratory Freezer (MELFI).

Mission Control gave a clean bill of health to the orbiter's thermal protection system, apart from a protruding gap filler near the ET umbilical doors.

Flight Day 6 – Sunday, 9 July 2006

The Discovery crew was awakened with *I Have a Dream* by ABBA, requested for Mark Kelly by his children.

On this day, a new heat exchanger for the Common Cabin Air Assembly, a component of the ISS environmental control system to collect condensate from the air, and a spare US spacewalking suit and emergency jet pack were transferred from the MPLM to the station. Sunday also saw the filling of the MPLM with cargo for return to Earth, including experimental results and samples, and miscellaneous waste.

Meanwhile Sellers and Fossum prepared for their second EVA, scheduled for Monday. The two crews took part in a joint press conference with journalists at NASA and ESA centres. Before the end of the day John Shannon, the deputy Shuttle Program Manager and chairman of the STS-121 Mission Management Team, reported that Discovery's heat shield was free from any damage and was cleared for its return to Earth. Under new procedures, if an orbiter was unable to make a safe return, the crew would wait on the ISS until a rescue mission could be launched.

Flight Day 7 – Monday, 10 July 2006

On being awakened by a rendition of *Clocks* by Coldplay, Sellers was informed, "That song was from Mandy and the kids, and they hope you enjoy your EVA today."

During a 6 hour and 47 minute EVA, Sellers and Fossum carried out two tasks. First they replaced the Trailing Umbilical System on the nadir of the Mobile Transporter because it had been inadvertently severed in December. The replacement, matching the zenith example installed on the first EVA, had no blade. During this work, the robot arm of the Shuttle transferred a new Pump Module from the payload bay to External Stowage Platform 2, located on the forward side of the Quest airlock. Then Sellers and Fossum interrupted their Mobile Transporter work to secure this apparatus to the storage platform.

Expedition 13's Vinogradov, Williams and Reiter continued their tasks of transferring equipment and supplies from the MPLM to the station and loading unneeded cargo into that module for return to Earth.

Flight Day 8 – Tuesday, 11 July 2006

Expedition 13 continued loading and unloading the MPLM, including a new window and the window seals for the Microgravity Science Glovebox, then Reiter went to install these new items. Sellers and Fossum carried out preparations for their third EVA, prior to joining their crewmates to attend a press conference with Associated Press and USA Today.

Flight Day 9 – Wednesday, 12 July 2006

The third spacewalk of the mission got underway at 6:20 a.m. CDT, and focused upon procedures for inspecting and repairing damage to the heat shield. First, Sellers was swung on Canadarm2 controlled by Nowak and Wilson to the starboard wing, where he used an infrared camera to supply video of the carbon-carbon panels on the leading edge. Meanwhile in the payload bay, Fossum carried out the Non-Oxide Adhesive Experimental (NOAX) experiment involving test repairs to twelve sample carbon-carbon panels. NOAX was a pre-ceramic polymer sealant which contained carbon-silicon carbide powder. They also completed three gouge repairs and two crack repairs with NOAX. An additional task which was added by the mission controllers was to install a fixed grapple bar to an ammonia tank on the S1 segment of the truss.

Vinogradov, Williams and Reiter packed the MPLM with returning experiment samples, unneeded hardware and rubbish.

Flight Day 10 – Thursday, 13 July 2006

This was a well-deserved day off for Discovery's crew. The Expedition 13 astronauts held interviews with ESA astronaut Pedro Duque and students at the Columbus Control Centre in Cologne, Germany, and with Russian media at the Mission Control Centre in Moscow.

Texas Agricultural and Mechanical (A&M) University alumnus Mike Fossum spoke with Texas Governor Rick Perry who said, "Aggies are all up in great arms to have the first Aggie in space. You are making some history, Michael."

In addition, Fossum and Nowak conducted interviews with MSNBC and Fox News Live.

Flight Day 11 – Friday, 14 July 2006

The crew awoke to *The Aggie War Hymn* performed by the Fighting Texas Aggie Band, played for Mike Fossum. In addition to interviews with CNN, ABC, NBC and CBS, Discovery's Steve Lindsey and ISS Flight Engineer Thomas Reiter deactivated the MPLM and closed the hatches, then Nowak and Wilson used the Shuttle's robotic arm to transfer the module back to the payload bay.

Flight Day 12 – Saturday, 15 July 2006

Former Arizona State Senator and future US Congresswoman Gabrielle Giffords chose U2's *Beautiful Day* as the wakeup call for her husband Mark Kelly.[10]

---

[10] Gabrielle Giffords would be shot in an attempted assassination on 8 January 2011 in Casas Adobes, Arizona, while speaking at an event. Nineteen people were injured and six were killed. Although she returned to Congress, she resigned in 2011 in order to concentrate on her recovery.

Discovery's crew bade farewell to the ISS crew, including their former crewmate Thomas Reiter, and the hatches were closed at 3:15 a.m. CDT. After departing the station, the Shuttle crew used the robotic arm and boom sensors to thoroughly inspect the starboard wing and the nose cap heat shield, making certain there was no damage from orbital debris. A similar survey of the port wing had been made the previous day. After the nose cap survey, the boom was berthed on the starboard sill of the payload bay and the arm was stowed on the opposite sill.

Flight Day 13 – Sunday, 16 July 2006

As Discovery's crew prepared for de-orbit and landing, stowing gear and securing the vehicle for the trip home, the Expedition 13 astronauts enjoyed an off-duty day after the intense activity of the joint programme.

Flight Day 14 – Monday, 17 July 2006

Discovery touched down on Runway 15 at the Shuttle Landing Facility in Florida at 8:15 a.m. CDT.

Expedition 13 Week Ending Friday, 21 July 2006

Expedition 13, now expanded to three crewmembers, got down to work. The newly installed Minus Eighty Laboratory Freezer was activated. It would permit biological and human research experiment samples to be stored pending return to Earth for analysis.

The new oxygen generation system was checked out this week to prevent its internal valves from sticking over long periods. This was to augment the Russian Elektron oxygen-generation system and enable the ISS to expand to a crew of six. They also flushed the cooling loops in the Quest airlock in preparation for the upcoming spacewalk by Reiter and Williams.

On 20 July 2006, Reiter hosted a conversation with the ESA Space Operations Centre (ESOC) in Darmstadt, Germany, where dignitaries were gathered, including German Chancellor Angela Merkel, Minister-President of Hesse (where ESOC was located) Roland Koch, and the ESA Director General Jean-Jacques Dordain.

"Space exerts not only a great fascination but also provides via scientific experiments great value, in particular for research in the health and medical sector," observed Chancellor Merkel. "This visit to ESOC has demonstrated again that ambitious space programmes cannot be done by one country alone; international cooperation is of the utmost importance. Therefore, Germany will remain a constant and reliable partner in the European space programme."

ESA Director General Dordain said, "It is a great honour and a pleasure to welcome Chancellor Angela Merkel and Hessen Prime Minister Roland Koch to ESOC for this in-flight call. Their visit demonstrates deep support for Thomas Reiter's ISS mission as a key element of Europe's human spaceflight programme, and is particularly significant for this ESA operations centre in Germany."

Expedition 13 Week Ending Friday, 28 July 2006

On Tuesday, Russian flight controllers commanded the engines of the docked Progress vehicle to fire to boost the orbit of the ISS to 219 by 203 statute miles in preparation for the arrival of STS-115 and Soyuz TMA-9, both of which were scheduled for September.

Expedition 13 Week Ending Friday, 3 August 2006

On 1 August, Reiter participated in the first of three ARISS amateur radio sessions. On this occasion he spoke with around 150 children at the ESA Space Camp held in Patras, Greece. Present were the Greek Minister of Education and Religious Affairs, Marietta Giannakou, the Head of the European Astronaut Centre, Michel Tognini (a former astronaut and veteran of Soyuz, Mir, and Shuttle missions), and ESA astronaut Reinhold Ewald (Soyuz and Mir veteran).

Thomas Reiter during a spacewalk. (NASA)

On Thursday, Williams and Reiter conducted an EVA from the Quest joint airlock wearing NASA spacesuits. The first task was to install the Floating Potential Measurement Unit. This was to measure the electrical potential of the station to provide information needed to develop procedures to minimise the risk of electrical current jumping ('arcing') from a conductor to a ground as the ISS was expanded. Next they installed two Materials on International Space Station Experiment (MISSE) packages. These suitcase-style containers were left open in order to evaluate the long-term effects of space exposure on a variety of materials samples as part of an investigation to identify the best materials for use in future spacecraft. These were positioned on the exterior of Quest; MISSE-3 on one of the high-pressure tanks that surround the crew lock and MISSE-4 on the outboard end.

The two astronauts then worked individually. Williams installed a controller for a thermal radiator rotary joint on the S1 truss. Meanwhile Reiter replaced an external computer. Williams then installed a starboard jumper and spool positioning device (SPD) on the S1 truss. Reiter inspected a radiator beam valve module SPD site where one device was already installed, prior to installing another one. He then installed an SPD on a port cooling line jumper. The jumper units were intended to improve the flow of ammonia through the radiators once the coolant was inserted.

Williams began the setup for the final major scheduled task, which was to test an infrared camera designed to detect damage to the carbon-carbon protection of a Shuttle. This camera would highlight damage by the difference in temperature between clean and damaged material. While Reiter used the camera to inspect appropriate test samples, Williams moved on to one of the additional tasks.

The first of the additional tasks was to install a light to assist future spacewalkers on the truss's railway handcart. Williams then removed a malfunctioning GPS antenna. After Reiter finished the infrared camera experiment, he installed a vacuum system valve on the Destiny module for future scientific experiments. Williams then moved two articulating portable foot restraints to prepare for spacewalks by the STS-115 crew and then he took pictures of a scratch on the airlock hatch. Reiter wrapped up by inspecting apparatus attached to the PMA that joined the Zarya module to the Unity node.

Throughout the spacewalk, the astronauts photographed the worksites after completing their tasks and finally took some pictures of each other prior to re-entering the airlock. The EVA ended at 3:58 p.m. EDT, having lasted 5 hours 54 minutes.

Expedition 13 Week Ending Friday, 11 August 2006

The Expedition 13 crew spent two days this week carrying out maintenance on the treadmill vibration isolation system in the Zvezda habitat module. It was important for crew health and muscle mass that the astronauts exercised daily and the treadmill had gyroscopes to stabilise itself and hence prevent the vibrations caused by vigorous exercise activities from being transmitted to the station's structure, lest that disturb sensitive experiments.

The crew also devoted much of the week to packing items that were to be transferred to the Shuttle Atlantis when it docked in the coming weeks. Mission Control remotely moved Canadarm2 to position cameras to view the markings on the exterior of the station that would be used as a graphical computer alignment aid when new components were attached to the station by the robotic arm. That aid, called the Space Vision System, would be used during Atlantis' flight to help with the attachment of the new truss section.

On Friday, 4 August, Reiter set a new record for the number of days spent in space by a European astronaut, surpassing the 209 days of Jean-Pierre Haigneré.

ESA Director General Jean-Jacques Dordain congratulated Reiter. "At the end of your mission you will have spent one year in space. With this outstanding expertise and experience you, as our 'highest flying' colleague, symbolise Europe's commitment to space," said Dordain. "We are extremely proud of your achievements and wish you the best of luck in your remaining time up there."

Expedition 13 Week Ending Friday, 18 August 2006

Early during the week, Williams worked with flight controllers and the Dynamic Onboard Ubiquitous Graphics (DOUG) programme that simulated the robotic arm's operation in the training role. He then moved Canadarm2 from its position on the Destiny module to a power and data grapple fixture on the Mobile Transporter in order to inspect the end of the P1 truss, where the next Shuttle mission was to attach a new segment. The crew also reviewed the installation procedures for this new segment, and Williams carried out spacesuit maintenance.

Vinogradov and Reiter worked on the Russian-German Plasma Crystal experiment that examined the behaviour of tiny particles when excited by high-frequency radio signals in a vacuum chamber; it functioned by itself most of the time.

They also activated the Earthkam experiment which allowed students to request photographs from space of specific locations on Earth via email.

Expedition 13 Week Ending Friday, 25 August 2006

The ISS crew continued to pack items which were to return to Earth on Atlantis. They reviewed joint operations plans. They trained to take pictures of the tiles and carbon-carbon shielding on the belly of the Shuttle during its backflip prior to making its final approach to the station.

Flight controllers tested the operation of a NASA air scrubbing system in advance of Atlantis's arrival. This Carbon Dioxide Removal Assembly (CDRA) was turned on for an extended period of time to test its ability to remove carbon dioxide from the air. It augmented the Russian air scrubber, Vozdukh, which was turned off during the test.

On Wednesday, flight controllers also fired the engines of the docked Progress to raise the orbit of the station by 2.5 miles in readiness for the arrival of Atlantis and, shortly thereafter, Expedition 14's arrival on Soyuz TMA-9.

Williams replaced filters in the cooling system and conducted the Dust and Aerosol Measurement Feasibility Test (DAFT); the latter being a test of the P-Trak air quality meter that counted ultra-fine dust particles in the microgravity environment inside the station.

The crew continued with setting up and checking out of the European Modular Cultivation System (EMCS) that had been delivered by STS-121. This contained a centrifuge capable of subjecting a wide range of small plant and animal experiments to partial gravity conditions. The first experiment was the Analysis of a Novel Sensory Mechanism in Root Phototropism (TROPI) which studied mustard seeds to identify the genes that were responsible for successful plant growth in weightlessness.

Vinogradov and Reiter participated in ESA science experiments which tested the response of the cardiovascular system to long-duration exposure to microgravity.

Expedition 13 Week Ending Friday, 1 September 2006

With the STS-115 Atlantis launch postponed due to Tropical Storm Ernesto, the ISS crew had more time this week to pack cargo for return to Earth.

Over a 3 day period Williams worked on an experiment into the anomalous long-term effects on astronauts' central nervous systems. It involved tracking cosmic radiation and monitoring both brain activity and visual perceptions. He also worked with the Capillary Flow Effects experiment. And he tested the new seal on the Microgravity Science Glovebox in the Destiny module.

Expedition 13 Week Ending Friday, 15 September 2006

On 9 September the Expedition 13 crew of Vinogradov, Williams and Reiter watched the launch of Atlantis via a live television transmission from Houston. Its main payload was the combined P3/P4 truss segments.

During 10 September, as Atlantis made its way to the station in a series of orbital manoeuvres, the ISS crew prepared for its arrival. They readied the cameras that were to photograph the orbiter's heat shield elements, and PMA-2 on the front of the Destiny module where the Shuttle was to dock. On 11 September, Atlantis performed the now routine backflip to allow the station's crew to photograph its heath shield. After docking, the hatches opened at 7:30 a.m.

Getting straight to work, STS-115's Ferguson and Burbank extracted the P3/P4 truss from the payload bay and handed it to Canadarm2, operated by Expedition 13's Williams and the Shuttle's MacLean. MacLean thereby became the first Canadian astronaut to operate Canadarm2 in space.

On 12 September, during the mission's first spacewalk, Joe Tanner and Heide Stefanyshyn-Piper began the task of connecting the truss. After getting ahead of the timeline, Mission Control allocated them some 'get ahead' tasks that had originally been assigned to the second spacewalk. At the end of the EVA, they connected power cables on the truss, released the launch restraints on the Solar Array Blanket Box and on the Beta Gimbal Assembly, and configured the Solar Alpha Rotary Joint which would enable the arrays to track the Sun. Furthermore, they removed two circuit interrupt devices in preparation for the upcoming STS-116 mission.

On the second spacewalk of STS-115, on 13 September, Burbank and MacLean continued with activation of the Solar Alpha Rotary Joint (SARJ). After the spacewalk, Mission Control tested the SARJ but when they commanded the second Drive Lock Assembly DLA-2 to rotate the SARJ 360 degrees there was no response. After a workaround on 14 September, Mission Control successfully finished the SARJ checkout. As the 73-metre-long solar arrays were unfolded, several of the panels became stuck but, as had occurred when the P6 solar arrays were installed by STS-97 in December 2000, the problem was soon fixed. The new solar arrays were fully unfolded at 7:44 a.m. CDT, but were not delivering power; that would have to await STS-116, which was to wire the arrays to the station along with the new cooling systems.

On 15 September, Tanner and Stefanyshyn-Piper completed the third and final spacewalk of this mission, powering up a cooling radiator for the new solar arrays, replacing an S-Band radio antenna, and installing insulation for another communications antenna.

Expedition 13/14 Week Ending Friday, 22 September 2006

On 17 September, the crew of Atlantis bade farewell to Expedition 13 and undocked from the station. Meanwhile the Expedition 14 crew of Commander Michael Lopez-Alegria, Flight Engineer Mikhail Tyurin and visiting spaceflight participant Anousheh Ansari were one day away from lifting off from Baikonur.[11]

As Soyuz TMA-9 carried out its rendezvous operations, the station crew dealt with an emergency. The Elektron oxygen generator had been offline for the Shuttle's visit. As Vinogradov was bringing it back online, the system shut down immediately. After a number of fruitless attempts at restarting the unit, Vinogradov noticed smoke. The fire alarm was set off to shut down the ventilation systems and prevent the smoke from spreading between modules, and the crew were told to wear surgical masks, goggles and gloves. They collected any leaking liquid (potassium hydroxide was used in the Elektron) and station operations had returned to normal within an hour. The overheating on starting up the unit had melted a rubber seal.

The next day a Progress freighter was undocked from the station and de-orbited to burn up on re-entry.

On 19 September, as Soyuz TMA-9 was on its way to the ISS with three people on board, Atlantis was homebound with six. Although this wasn't a record,[12] having a dozen people in space aboard three spacecraft was an excellent opportunity to organise a conference link.

"It's a little crowded in the sky today," said Williams from the station. "We look forward to having you guys on board," he told the Soyuz crew. "We'll see you back on Earth sometime soon," Atlantis's Commander Brent Jett told Expedition 13.

As Russian engineers continued to investigate the malfunction in the Elektron, single-use oxygen canisters were being 'burned' to replenish the cabin air.

The hatches to Soyuz TMA-9 were opened at 3:34 a.m. CDT on Wednesday, 20 September, and Michael Lopez-Alegria, Mikhail Tyurin and Anousheh Ansari were welcomed on board the ISS by Expedition 13.

As Atlantis touched down in Florida on 21 September, the handover of duties continued on the station.

Expedition 14 Week Ending Friday, 29 September 2006

Various members of the joint crew took part in media events during the busy week of handover. The entire crew spoke with Russian media on Monday, and Williams and Lopez-Alegria spoke with CBS News and with AP Television. On Tuesday, Williams and Lopez-Alegria were interviewed by both CNN Espanol and the *Houston Chronicle*.

---

[11] Anousheh Ansari was the Iranian-born American telecommunications entrepreneur who sponsored the $10 million Ansari X Prize that was won when SpaceShipOne reached space twice in two weeks during suborbital flights in 2004.

[12] In March 1995 Soyuz TM-21 was delivering three cosmonauts (including NASA's Norman Thagard) to the Mir space station while Space Shuttle Endeavour was flying the STS-67 mission. The three Expedition 17 crew on Mir made it thirteen people in space for the first time.

The official change of command ceremony occurred on 27 September when 'LA' Lopez-Alegria became the station commander, taking over from Pavel Vinogradov. Thomas Reiter transferred from Expedition 13 to Expedition 14.

Then Vinogradov and Williams undocked Soyuz TMA-8 and returned to Earth with spaceflight participant Anousheh Ansari, who had spent nearly 11 days on the station. They landed at 8:13 p.m. CDT some 50 miles northeast of Arkalyk in Kazakhstan.

Expedition 14 Week Ending Friday, 6 October 2006

As Lopez-Alegria and Tyurin familiarised themselves with the station, the Expedition 14 research got underway and samples of the Passive Observatories for Experimental Microbial Systems in Micro-G (POEMS) experiment were put in the MELFI freezer. Lopez-Alegria started the most comprehensive in-flight study yet attempted of human physiological changes during long-duration spaceflight. This required him to monitor all his food, drinks, vitamins, and minerals as well as to take blood and urine samples.

On 2 October the crew set up a live linkup with the 2006 International Astronautical Congress in Valencia, Spain. Reiter welcomed everyone in English, Spanish-born Lopez-Alegria added a greeting in Spanish and Tyurin did likewise in Russian.

All three men took part in the monthly fitness test, exercising on a stationary bicycle and measuring their heart rates and blood pressures. As part of an ESA experiment, Reiter also monitored his oxygen uptake. Lopez-Alegria and Tyurin also checked out the emergency medical equipment and supplies; a check that was performed early in each crew's tour.

Tyurin replaced more components in the faulty Elektron's control panel during the week, but to no avail. His colleagues in Moscow promised to send up the necessary replacement parts.

During this week the station operated with three of its four Control Moment Gyroscopes (CMG) online. CMG-3 was taken offline to enable vibrations that had been noticed during a thruster firing in September to be investigated.

Expedition 14 Week Ending Friday, 20 October 2006

This week Lopez-Alegria and Tyurin practised the manual docking procedures for the next Progress cargo ship. In addition to the regular complement of food, fuel and supplies, it carried spare parts for the Elektron oxygen-generator.

Lopez-Alegria took his second batch of samples for the Nutrition Experiment. Reiter started the Analysis of a Novel Sensory Mechanism in Root Phototropism (TROPI) experiment in the European Modular Cultivation System. It involved growing seeds under differing light frequencies and gravity conditions in a centrifuge to study the direction that their roots and shoots grew, to determine which genes were responsible for successful plant growth.

Ground controllers tested the CMG-3 unit by spinning it to 500 rpm and then monitoring how it decelerated using the Microgravity Acceleration Measurement System to identify any irregularities.

On Friday Lopez-Alegria replaced parts in the Carbon Dioxide Removal System, restoring it to a fully operational state. There had been a blockage in an air flow regulator valve. Once the new valve and a new filter were installed, the second of its two adsorbent beds was brought back online.

On 10 October the station was left without a crew for over an hour owing to the need to move the Soyuz TMA-9 spacecraft from the aft port of the Zvezda module to the nadir port of Zarya in order to clear the way for the next Progress freighter.

Prior to entering the Soyuz spacecraft the crew left the ISS systems in standby mode in case they were unable to re-enter the station and had to return to Earth in the Soyuz, leaving the ISS vacant. All three crewmembers had to perform this short flight because no one could be left on board the station without a 'lifeboat' being present.

On 12 October ESA Council Chairman Professor Sigmar Wittig and ESA Director General Jean-Jacques Dordain congratulated Reiter on his 100th day on board the ISS. In a link to the ESA HQ in Paris, Reiter thanked the ESA Council for their message, "It is definitely a pleasure to be up here. In retrospect these 100 days went by like in the blink of an eye. The Space Station is really a fascinating building."

Alluding to the fact that the ESA Council had agreed to release further funds for the ISS, Dordain joked, "OK Thomas, since now we have a budget you will have some food for the rest of your flight."

Expedition 14 Week Ending Thursday, 26 October 2006

In preparation for the next Shuttle mission, this week flight controllers tested the Thermal Radiator Rotary Joints (TRRJ) on the S1 and P1 truss segments. This was because the STS-116 crew were to activate the external thermal loops. Once the TRRJs were able to revolve, they would dissipate heat from the avionics contained within the truss segments.

The new Progress supply ship was launched from Baikonur and on Thursday it docked to the aft port of Zvezda. Controllers delayed the final latching for three and a half hours while they checked whether an antenna had retracted as commanded. During the time that this ship was docked but not securely latched, the station's orientation was allowed to drift in order to prevent any motions from disturbing the softly docked craft. As this meant the station's solar panels were not ideally oriented, some non-critical apparatus was put on standby to save power. Once the hard docking was achieved, attitude control was restored and power generation was returned to normal. The freighter delivered 1,918 pounds of thruster propellant, 110 pounds of oxygen, and almost 2,800 pounds of spare parts, experiment hardware and life support components, including parts for the Russian Elektron oxygen-generation system.

Ground controllers reviewed the S-band communications system – this had begun to experience faults over the previous week. The crew measured sound levels on the station, which could be noisy with ventilation fans running. Reiter continued his work with the plant growth experiments. Lopez-Alegria made log entries for a sleep experiment.

On Wednesday Reiter was the guest lecturer for the European Masters in Aeronautics and Space Technology (EuMAS) programme based at the Technical University of Munich. Some students who were visiting the Columbus Control Centre in Oberpfaffenhofen,

Germany, were to interact with him but the others at universities in Spain, France and Germany simply viewed the lecture. Reiter showed the operation of the PK3+ experiment apparatus that he had recently assembled, plus a demonstration of the phenomenon of electrostatic charging in weightlessness.

Expedition 14 Week Ending Friday, 3 November 2006

On Monday, Tyurin installed the new valves and cables in the Elektron oxygen-generation unit and successfully reactivated it. Lopez-Alegria moved Canadarm2 to the Mobile Transporter, then set up cameras for the Earth Knowledge Acquired by Middle School Students (EarthKAM) experiment.

Reiter harvested the second sample of seeds from the TROPI experiment and stored them in the MELFI freezer pending their return to Earth. He also continued the CARD study of the relationship between salt intake and the cardiovascular system when exposed to the microgravity environment.

Expedition 14 Week Ending Friday, 9 November 2006

Throughout the week the crew prepared the Pirs docking compartment for the upcoming spacewalk by Lopez-Alegria and Tyurin. Reiter spent much of his time preparing material for carriage on STS-116, when he returned home in December.

Lopez-Alegria collected his third set of blood and urine samples for the Nutrition experiment and stored the samples in the MELFI freezer. He also voted in the US presidential election by sending an encrypted computer ballot to Mission Control in Houston for forwarding to the county clerk's office.[13]

On 7 November Reiter participated in a video link, this time with the 'ISS Research Technology from Europe' Industry Day at the ESTEC facility in Noordwijk, the Netherlands.

Expedition 14 Week Ending Friday, 17 November 2006

On Wednesday, Lopez-Alegria appeared in the first live high-definition television broadcast from a vehicle in space. NHK Television in Japan and the Discovery HD Theatre made this broadcast using Space Video Gateway technology. Hitherto HD video was physically returned to Earth for broadcast.

Meanwhile, engineers on the ground continued to test CMG-3, which was now to be replaced by the STS-118 mission.

Reiter continued his research programme by working on CASPER, which was to help astronauts sleep better during long-duration missions and also on the Alteino Long-Term Monitoring of Cosmic Rays (ALTCRISS) experiment that studied the effects of shielding on cosmic rays.

On Friday, Lopez-Alegria and Tyurin tried on their Russian Orlan spacesuits in order to test their systems in preparation for the upcoming spacewalk.

---

[13] The first in-space voting was by David Wolf aboard the Mir space station in 1997.

Expedition 14 Week Ending Friday, 23 November 2006

At 6:17 p.m. CST on 22 November, Lopez-Alegria and Tyurin started the first ISS spacewalk of the Expedition 14 mission by exiting the airlock in the Pirs docking compartment.

In an event sponsored by the Canadian company Element 21, Tyurin hit a golf ball using a gold plated six iron. The shot did not go precisely as intended. Lest this be seen as creating orbital debris, the 3 gram ball (a regular golf ball weighs 48 grams) was expected to re-enter the atmosphere after only a few days and burn up.

The spacewalkers examined the problematic antenna on the recently arrived Progress ship. It had indeed failed to retract. When they found that they couldn't retract it manually they took photographs to help engineers on the ground decide what to do. They also repositioning a communications antenna to be used in docking the ESA Automated Transfer Vehicle, the first of which was expected in 2008. Finally Lopez-Alegria and Tyurin installed the BTN-Neutron device that would measure the neutron particles in solar flares that reach low Earth orbit.

On 24 November, Reiter conducted an audio-only conversation with his ESA astronaut colleague Paolo Nespoli, who was at the European Space Research Institute (ESRIN) in Frascati, near Rome. In later comments, Nespoli spoke about his forthcoming flight to the ISS on STS-120, "I am very much looking forward to my first spaceflight. I have already started an intensive preparation phase which is very challenging and exciting. With the launch of Node 2, Europe will see the launch of its first [ISS module]. It is an important step because Node 2 will serve as a connecting element for the European Columbus laboratory, the US laboratory Destiny and the Japanese laboratory Kibo. It will also be the attachment point for the Japanese H-II transfer vehicle. It will carry a docking adapter for the Space Shuttle and serve as an attachment point for the MPLMs – the Italian-built cargo modules which are carried to and from the Station inside the Space Shuttle."

Expedition 14 Week Ending Friday, 1 December 2006

The Expedition 14 crew reviewed the STS-116 mission plans this week. They prepared the station's Quest airlock, spacesuits, and tools for three spacewalks planned during the Shuttle visit that was to deliver the P5 truss segment and then return Thomas Reiter to Earth. They packed up apparatus that was to return to Earth aboard the Shuttle, including Reiter's personal items.

Flight controllers fired the engines of the Progress ship to boost the orbit of the station, ready for the new Shuttle, but the planned firing of 18 minutes 22 seconds ended after just 3 minutes due to an orientation change of the station as the thrusters were fired. It was established that the recently added truss and solar arrays had altered the mass and balance for the station, hence a software modification would be required prior to attempting another boost. Flight controllers also investigated a fault which occurred while testing new software for the Solar Alpha Rotary Joint – a circuit breaker had opened during the test. It was reset and software modifications were planned.

Renowned French chef Alain Ducasse had worked with ESA and CNES to develop gourmet food that could be served to astronauts in space on special occasions. On 26 November Expedition 14 was able to sample this gourmet menu:

Main dishes:
Effiloché de volaille en Parmentier
Shredded chicken Parmentier
Dos d'espadon façon Riviera
Riviera style swordfish
Volaille épicée, sauté de légumes à la Thaï
Spicy chicken with stir-fried Thäi vegetables
Cailles rôties au Madiran
Quails roasted in Madrian wine
Magret de canard confit, condiment aux câpres
Duck breast 'confit', with capers

Side dishes:
Carottes de sable au goût d'orange et coriAndré
Sand carrots with a hint of orange and coriander
Céleri rave en délicate purée à la noix de muscade
A light puree of celery with a hint of nutmeg
Caponata
Tomato, aubergine and olive dip

Desserts:
Gâteau de semoule de blé fine aux abricots secs
Semolina cake with dried apricots
Morceaux de pommes fondantes
Apple fondant pieces
Far de l'espace
Space 'far' (a Brittany tart)

Thomas Reiter said, "It was absolutely delicious. It was a really nice treat for a Sunday evening." He added, "Food is really something which gives us a break. It is something where we find some joy and we're really trying to take some time for our meals. We all agreedthat we are enjoying this food, but we have no doubt that it would taste much better if we had some wine with it as well!"

The next week saw the arrival of ESA's Christer Fuglesang on STS-116, so for the remainder of Thomas Reiter's time on the ISS and his return to Earth on Discovery see the chapter on the Celsius Mission.

11.4 POSTSCRIPT

Subsequent Missions

Astrolab was Thomas Reiter's final spaceflight.

Thomas Reiter, ESA Director for Human Spaceflight and Operations, at the Berlin Air and Space Show on 22 May 2014. (ESA)

Thomas Reiter Today

Following his active career as an astronaut, on 8 August 2007 Thomas Reiter became a member of the Executive Board of the German Aerospace Centre (DLR) with responsibility for Space Research and Technology.

On 1 April 2011 Reiter became the Director of ESA's new Directorate of Human Spaceflight and Operations (D/HSO). It included management of the European International Space Station elements, ESA's Automated Transfer Vehicle and the European Astronaut Centre (EAC), in addition to ESA's unmanned missions and ground-based mission infrastructure.

Reiter has logged just over 350 days in space (the most by any non-American or non-Russian) and has been awarded the following honours:

- Order of Friendship from Russia (1996).
- Order of Merit of the Federal Republic of Germany (2007).
- Bavarian Medal Europe (2008).
- Honorary doctorate of the Faculty of Aeronautics and Astronautics at the Bundeswehr. University Munich (2010).
- Order of Friendship (Russia, 1996).
- Medal 'For Merit in Space Exploration' from Russia (2011).
- Honorary membership in the Danish Astronautical Society.

12

Celsius

Mission

ESA Mission Name:	Celsius
Astronaut:	Arne Christer Fuglesang
Mission Duration:	12 days, 20 hours, 45 minutes
Mission Sponsors:	ESA
ISS Milestones:	ISS 12A.1, 33rd crewed mission to the ISS

Launch

Launch Date/Time:	10 December 2006, 01:47 UTC
Launch Site:	Pad 39-B, Kennedy Space Center
Launch Vehicle:	Space Shuttle Discovery (OV-103)
Launch Mission:	STS-116
Launch Vehicle Crew:	Mark Lewis 'Roman' Polansky (NASA), CDR
	William Anthony 'Bill' Oefelein (NASA), PLT
	Nicholas James MacDonald Patrick (NASA), MSP1
	Robert Lee, Jr. 'Beamer' Curbeam (NASA), MSP2
	Arne Christer Fuglesang (ESA), MSP3
	Joan Elizabeth Miller Higginbotham (NASA), MSP4
	Sunita Lyn 'Suni' Williams (NASA), MSP5

J. O'Sullivan, *In the Footsteps of Columbus*, Springer Praxis Books,
DOI 10.1007/978-3-319-27562-8_12

Docking

STS-116

Docking Date/Time:	11 December 2006, 22:12 UTC
Undocking Date/Time:	19 December 2006, 22:10 UTC
Docking Port:	PMA-2, Destiny Forward

Landing

Landing Date/Time:	22 December 2006, 22:32 UTC
Landing Site:	Runway 15, Shuttle Landing Facility, Kennedy Space Center
Landing Vehicle:	Space Shuttle Discovery
Landing Mission:	STS-116
Landing Vehicle Crew:	Mark Lewis 'Roman' Polansky (NASA), CDR William Anthony 'Bill' Oefelein (NASA), PLT Nicholas James MacDonald Patrick (NASA), MSP1 Robert Lee, Jr. 'Beamer' Curbeam (NASA), MSP2 Arne Christer Fuglesang (ESA), MSP3 Joan Elizabeth Miller Higginbotham (NASA), MSP4 Thomas Arthur Reiter (ESA), MSP5

ISS Expeditions

ISS Expedition:	Expedition 14
ISS Crew:	Michael Eladio 'LA' Lopez-Alegria (NASA), ISS-CDR Mikhail Vladislavovich Tyurin (RKA), ISS-Flight Engineer 1 Thomas Arthur Reiter (ESA), ISS-Flight Engineer 2 replaced by Sunita Lyn 'Suni' Williams (NASA), ISS-Flight Engineer 2

12.1 THE ISS STORY SO FAR

A Space Shuttle and a Soyuz visited the station between Thomas Reiter's launch on STS-121 and the visit by Christer Fuglesang.

STS-115 launched on 9 September, after delays due to both tropical storms and fuel cell problems. The crew installed the combined P3/P4 truss segments during three spacewalks, to give the ISS a new set of solar panels.

Soyuz TMA-9 was launched on 18 September 2006 with Anousheh Ansari, the first female 'space tourist'. She replaced Japanese Daisuke Enomoto, who had failed his pre-flight medical. She arrived at the ISS with the Expedition 14 crew on 20 September and landed with Expedition 13 in Soyuz TMA-8 on 29 September. See the Astrolab mission for STS-115 and Soyuz TMA-9.

The ISS after installation of the P3/P4 truss segments and solar arrays by STS-115. (NASA)

12.2 CHRISTER FUGLESANG

Early Career

Christer Fuglesang was born on 18 March 1957 in Stockholm, Sweden. After graduating from the Bromma Gymnasium in Stockholm in 1975 he attended the Royal Institute of Technology (KTH) in that city, from which he gained a master's in engineering physics in 1981. He received a doctorate in experimental particle physics from the University of Stockholm in 1987, and was made an associate professor in particle physics there in 1991.

While a graduate student, he worked at the European Organisation for Nuclear Research (CERN) near Geneva on the UA5 experiment investigating proton-antiproton collisions. In 1988 he became a Fellow of CERN, where he worked on the CPLEAR experiment studying the subtle CP-violation of kaon particles. A year later, Christer became a Senior Fellow and head of the particle identification subdetector, and then worked on the early development of the Large Hadron Collider. In November 1990 he obtained a position at the Manne Siegbahn Institute of Physics in Stockholm but continued to work at CERN into 1991.

Fuglesang became a member of the ESA astronaut corps in May 1992 and started a varied career that saw him travel between Houston, Cologne and Moscow. He followed the introductory training at the European Astronaut Centre (EAC) with a 4 week training programme at the Gagarin Cosmonaut Training Centre in Star City, Moscow, to prepare for the EuroMir missions. In May 1993 Fuglesang and Thomas Reiter were selected for the

Christer Fuglesang. (NASA)

EuroMir95 mission and started training to prepare for their flight engineer tasks, spacewalks, and Soyuz spacecraft operations. Reiter was selected as the prime crew in March 1995 with Fuglesang as his backup. In this role, Fuglesang served as the prime Crew Interface Coordinator in the Russian Mission Control Centre in Kaliningrad, near Moscow.

Between March and June 1996, Christer was given specialised training on Soyuz operations for undocking, re-entry and landing of the Russian spacecraft. From August 1996 to April 1998 NASA provided Mission Specialist training at the Johnson Space Center in Houston. From May to October 1998 he returned to the Russian training regime, working on the Soyuz TM spacecraft operations for undocking, re-entry and landing. With his Soyuz Return Commander certificate he would be able to command a three-person Soyuz capsule on its return from space.

In October 1998 he returned to Houston on assignment to the Astronaut Office. He served as the Crew Support Astronaut for ISS Expedition 2, then worked on equipment for the station and for use during spacewalks.

His scientific research and development work during this period included the SilEye experiment that investigated flashes perceived in astronauts' eyes. This was conducted on Mir between 1995 and 1999, and continued on board the ISS in the form of the Alteino detector and the ALTEA facility. He also initiated the DESIRE Project which simulated and estimated radiation inside the ISS. In 2006 he was appointed Affiliated Professor at KTH.

Previous Missions

N/A

12.3 THE CELSIUS MISSION

Celsius Mission Patches

The Celsius mission honoured the Swedish scientist and astronomer Anders Celsius, who was famous for inventing the standard international temperature scale that bears his name.

The principal graphic element of the logo was a map of Scandinavia in light blue stripes and a dark blue stripe representing the flight path of the Space Shuttle Discovery.

As with other patches of this era, the STS-116 patch design showed the orbiter trailing a plume. This time the plume represented the flags of America and Sweden in honour of Mission Specialist Christer Fuglesang. The labels '116' and '12A.1' on the wings of the orbiter indicated the mission number and station assembly designation. The constellation Ursa Major on the left-hand side of the patch provided pointers to the North Star that was prominently superimposed on the P5 truss of the ISS. These astronomical elements were borrowed from the state flag of Alaska to honour pilot Bill Oefelein. In addition the seven stars of the constellation were said to depict the Mercury astronauts. Suni Williams was to be delivered to the station to join the permanent crew. The tab attached to the bottom of the patch bearing the name 'Williams' reflected delays in the launch date and uncertainty concerning the passenger's identity.

The STS-116 crew with Fuglesang on the right. (NASA)

Celsius Mission Patch. (ESA)

STS-116 Mission Patch. (NASA)

Celsius Mission Objectives

ESA astronaut Christer Fuglesang became the first Swedish astronaut to venture into space when he flew as a Mission Specialist on Space Shuttle Discovery for the STS-116 mission that undertook ISS Assembly Flight 12A.1.

The STS-116 mission had many objectives:

- The P5 truss segment was a 'spacer' to be installed between the P4 and P6 segments, both of which carried solar array assemblies. The P6 segment had been installed early in the assembly process by being temporarily placed on the

Z1 truss atop the Unity node. STS-120 in October 2007 was to transfer P6 to its permanent position at the end of P5, thereby completing the port side of the integrated truss.

- NASA astronaut Sunita Williams was to replace ESA astronaut Thomas Reiter as the second Flight Engineer of the Expedition 14 crew.
- Although primarily an ISS assembly flight including spacewalks for Fuglesang, a number of experiments in human physiology and radiation dosimetry were to be conducted for the ESA Celsius mission.
- Delivery of 2.5 tonnes of supplies, equipment and research payloads in the SpaceHab Single Module.

Science

ALTEA

ALTEA was a study of the effects of cosmic radiation on the brain. Since the Apollo lunar missions, many astronauts had reported seeing flashes of light when their eyes were closed. It was thought that these flashes were caused by cosmic ray particles.

ALTEA was developed by the Italian Space Agency to follow up on another Italian project called Altino that was performed by Italian ESA astronaut Roberto Vittori on his Marco Polo mission to the ISS.

"The ALTEA study fits well with Christer's research, as he initiated a similar project called SilEye that ran on the Russian space station Mir between 1995 and 1999," explained Mark Pearce, a research associate of Fuglesang who was also an associate professor at KTH.

ALTEA was a helmet containing six particle detectors which provided the data needed to make a 3D reconstruction of how cosmic ray particles travelled through the brain. During the experiment the test subject would press a button when he or she saw a flash and an EEG was taken of the astronaut's brain activity.

DESIRE

Dose Estimation by Simulation of the ISS Radiation Environment (DESIRE) was a computer model of the ISS.

"The first step is to calibrate the software," said Mark Pearce. "This entails measuring the radiation levels inside the ISS and comparing them to the measurements made in the lab."

To achieve this, the predicted levels of radiation on the ISS were to be compared to readings from dosimeters – devices used to measure exposure to hazardous environments – that were installed in the Destiny module and the Unity node.

The second part of the project was to simulate the environment inside the European Columbus laboratory. "An accurate system for assessing these kinds of hazards will be extremely important for long-duration voyages such as, for example, a journey to Mars. Of course to achieve this it is vital to have dependable software," Pearce explained.

The DESIRE project was funded by ESA and the Swedish National Space Board.

Timeline

During his training for the Celsius mission, Christer Fuglesang wrote a diary that was hosted by the ESA website. This is a summary of his experiences prior to launching to the ISS.

Training Monday, 13 November to Friday, 17 November 2006

On Monday the STS-116 crew flew in NASA T-38s from Ellington Air Force Base in Houston to the Kennedy Space Center in Florida.

The following day, they practised driving an armoured personnel carrier from the launch pad to a safe distance, as they would during an emergency evacuation of the pad. In addition, they received a security briefing and the final fitting of their spacesuits. Fuglesang, Curbeam and Patrick reviewed photography and filming of the ET that was normally undertaken immediately after achieving orbit. The launch was expected to be at night, so it wasn't likely that they would be able to photograph the tank but they prepared in case the schedule changed. They ended the day with a dinner with some of the KSC management in the Beach House that was normally available to Shuttle prime crews while they were at the Cape.

Wednesday saw the crew visit the launch pad again. After a press conference, they ascended the launch tower in order to inspect the 'zip line' which would rapidly deliver them to the ground in the event of an evacuation, although they did not rehearse this activity. Then they reviewed Discovery's cargo bay, where almost everything was in place: the docking system, the SpaceHab module, the P5 truss segment, the cargo pallet bearing apparatus for transfer to the exterior of the station, the robotic arm and its OBSS extension.

On Thursday, there was a dress rehearsal of launch day, including wearing full spacesuits, getting fastened into their seats and simulating an entire countdown. This tested all communications with the control room and between the flight deck and the mid-deck.

On Friday Fuglesang and Curbeam practised EVA 2 in the Neutral Buoyancy Laboratory (NBL) in Houston.

Training Thursday, 23 November to Thursday, 30 November 2006

After the last training session in the NBL it was Thanksgiving on 23 November and the crew had two days off. Fuglesang received a visit from his daughter, who was attending a college in northern Texas.

In reviewing this time, he wrote, "I went through my notes and counted the number of times I've been in the spacewalk-suits in the pool to train for STS-116: 48 times since the summer of 2002. It is almost 250 hours in total; approximately twice as many hours as were planned before the Columbia accident. But it isn't a record, because STS-114, the first flight after Columbia, had over 70 practice sessions in the pool. When we came out of the pool yesterday there were a couple of rounds of cake, speeches of thanks and photographs with many of those who have been working with us there for several years."

Also this week he trained for amateur radio on the ISS (ARISS), reviewed procedure manuals, and rehearsed loading tasks.

On 30 November the crew of STS-116 entered quarantine, which meant that they slept in the JSC crew quarters. They were free to go to the gym and to their offices but only after the end of the normal working day. The limited group of people that were allowed to meet them were required to undergo a medical examination beforehand; this special list included their instructors, spouses, and children over ten years of age. They also started to change their body clocks to suit the mission time, so that once in Florida they would sleep between 03:45 and 11:45. They would still be able to entertain visitors while at the Cape and Fuglesang would see his family.

The week consisted of medical examinations, including the 'basal measurement data' that would be compared to data during and after the flight. Fuglesang wrote, "They took nine tubes of blood from me, a lot of which are for the experiments. One of these [experiments] will be carried out for ESA to study how much the radiation in space damages chromosomes. Another experiment will study latent herpes virus, which can be a risk during a long space journey. Apart from blood, urine over a 24 hour period was also required, so during one day I had to carry a refrigeration bag around with me, containing the plastic bottles that I was to fill."

He also went through the 'tilt-table' test to see how well the body regulates blood pressure when transitioning from a horizontal to a vertical position. He was a test subject for a new treatment called midodrine. He wrote, "I will be the first to test it during a flight. The only drawback is that if I feel ill during landing for other reasons (which is quite usual) then I cannot take the most common medicine that does help (phenergan), as mixing them can result in unpleasant side effects."

On Tuesday he had television interviews with Swedish TV4 morning news and spoke with Maud Olofsson, the Swedish space minister who was in the studio. On Wednesday he received a telephone call from the Swedish Prime Minister Fredrik Reinfeldt, who wished him luck.

Training Saturday, 2 December to Friday, 8 December 2006

On Saturday the crew had briefings with the flight directors and discussed the current difficulties on board the ISS, in particular the recent inability of the docked Progress freighter to boost the station's orbit because of its asymmetric configuration. They also practised with the cameras that they were to use to photograph the aurora borealis.

On Monday, Fuglesang and Shuttle pilot Oefelein went up in a T-38 to execute loops in order to simulate g-forces and parabolic arcs to simulate weightlessness. NASA Administrator Mike Griffin visited the crew quarters for several hours on Tuesday, and Fuglesang had a session in the gym.

The launch attempt on Thursday, 7 December was scrubbed due to weather so the astronauts left the Shuttle and headed back to crew quarters for another day.

Fuglesang wrote in his diary, "Three hours before take-off time I was strapped to my seat on my back and then I only had to wait, except for a few tests of the radio communication now and then. I browsed through my notes to refresh my recollection of what would happen, and was half asleep for long stretches of time. Funnily, I never even felt butterflies in my stomach – maybe because I was dubious of our getting away all the time, because of the weather. When the break ('scrub' in NASA lingo) was called, I had my own small task to perform: to secure the pyro-box, which can blow the entrance hatch away, with a securing pin."

Flight Day 1 – Saturday, 9 December 2006

After another day of enforced rest due to the weather, Discovery launched at 7:47 p.m. CST, the first night launch in more than four years. Once in orbit, the payload bay doors were swung open to allow the radiators to operate. The early tasks included setting up computers and powering up and checking out the robotic arm.

Flight Day 2 – Sunday, 10 December 2006

As was now routine, the first full day in space was spent examining the Shuttle's heat shield, wing leading edges and nose. Also as usual, the crew prepared for docking by installing the centreline camera and extending the outer ring of the docking system. Curbeam and Fuglesang tested their spacesuits to be ready for their first spacewalk.

Flight Day 3 – Monday, 11 December 2006

After the backflip manoeuvre and photographic inspection by the station crew, Discovery docked with the ISS at 4:12 p.m. CST and the hatches were opened at 5:54 p.m. After the obligatory safety briefing and a welcoming ceremony which marked the first time two ESA astronauts were on board the ISS at the same time, Williams installed her seat liner in the Soyuz TMA-9 spacecraft to make her a member of Expedition 14. Reiter transferred his seat liner to the Shuttle and joined the STS-116 crew.

After a vibration sensor alarm, the Shuttle crew were asked to inspect the port wing of Discovery and they sent images to Mission Control for review. Once the inspection was complete, the Shuttle's robotic arm hoisted the P5 truss segment out of the payload bay and handed it over to Canadarm2, in readiness for the following day's spacewalk.

Spacewalkers Curbeam and Fuglesang spent the night 'camping out' in the Quest airlock with the pressure reduced to 10.2 psi. This procedure purged their bloodstreams of nitrogen bubbles to protect against decompression sickness when they went to the even lower pressure of their spacesuits.

Flight Day 4 – Tuesday, 12 December 2006

During the first spacewalk of the mission, which lasted 6 hours 36 minutes, Curbeam and Fuglesang attached the P5 segment of the truss and replaced a failed camera that was to support future assembly tasks.

Afterwards, Curbeam congratulated Dr. John C. Mather of NASA's Goddard Space Flight Center, who had just been honoured with the Nobel Prize for his research on the Big Bang theory of how the universe originated, using data obtained by an instrument flown on the Cosmic Background Explorer Satellite.

Having reviewed photographs, video and sensor data, Mission managers passed the Shuttle's heat shield for a safe return to Earth.

Flight Day 5 – Wednesday, 13 December 2006

Before the solar arrays on the P4 truss could be activated to rotate and follow the Sun, a pair of solar arrays of the P6 truss needed to be retracted out of the way. Flight controllers and astronauts issued a succession of commands to retract and redeploy, step by step fashion, the accordion-style arrays. The guidewires which had been deployed for the previous 6 years snagged, so it became a laborious task. Only 17 of the 31 bays were able to be retracted, but this was sufficient to allow clearance for the P4 solar array to be brought online. The first step in this process was to open valves to allow 300 pounds of ammonia coolant to flow into the truss and its radiators. Eventually, these radiators would provide permanent cooling for the avionics and electronics on the station. When the P4 truss was activated, it successfully tracked the Sun. Mission Control had yet to decide whether to add a fourth spacewalk in order to finally retract the P6 array.

Meanwhile, other astronauts made a start on transferring a total of 4,107 pounds in the SpaceHab module and 1,107 pounds in the Shuttle mid-deck to the station.

Later on, Oefelein, Patrick, Curbeam and Fuglesang conducted interviews with CBS Radio, Fox Radio and Space.com.

The kink that developed in the port-side solar array of the P6 truss segment during the first attempt to retract that array on 13 December 2006. (NASA)

Flight Day 6 – Thursday, 14 December 2006

Before the second spacewalk got underway, some of the station's systems were powered down and attitude control of the joint ISS/Discovery complex was assigned to the Shuttle, which would use its thrusters. After sleeping in the Quest airlock at 10.2 psi, Curbeam and Fuglesang headed out. Flight controllers powered down about half of the station's systems to enable the spacewalkers to connect power channels 2 and 3 of the P4 truss; channels 1 and 4 were assigned to the third EVA. Once these connections were complete, the systems were powered up again. One of the external thermal control system loops was now shedding excess heat into space and the DC-to-DC converters were correctly regulating solar array voltages. Curbeam and Fuglesang also relocated the two Crew Equipment and Translation Aid (CETA) carts in order to make room for the third EVA.

Later Fuglesang wrote of his spacewalk, "To float along the outside of the Station, to watch Earth down below glide past, to see the curved horizon with a thin, blue layer of atmosphere bordering on black space. I need much more time to describe all that is great and all the wonderful emotions that I felt. Sometimes it was difficult; for example, to fit in the foot-supports out on the corners where my handle didn't really fit so well, or to find the right way when darkness fell during a night pass. It was annoying when the extension to the pistol grip tool got loose and disappeared into the darkness, but apart from that I was really pleased with the first spacewalk. I had the experience of my life! Ground control has done a fantastic job. The only difficulty for my part was that the connections in that rats' nest were tougher and required more strength than when we practised in the pool. It is complicated to reach them and I wanted to be as careful as possible so that none of the other cables or anything else was damaged. When I stood on the end of the robotic arm during the relocation of the CETA carts it was incredibly beautiful. We came in over Europe during a night pass. I could see lights from several cities, and up north, towards Sweden, the whole horizon was covered by the aurora! Soon afterwards I was met by a fantastically splendid sunrise."

An attempt to transfer attitude control back to the ISS and its gyroscopes wasn't successful. Flight controllers believed that the station was experiencing a higher than usual amount of atmospheric drag due to recent solar activity and this was applying a torque which interfered with the control algorithm.

Flight Day 7 – Friday, 15 December 2006

Media tasks continued with Lopez-Alegria and Tyurin conducting interviews with KNX Radio in Los Angeles and with National Public Radio. Fuglesang and Reiter participated in a VIP call with Swedish dignitaries.

Flight controllers continued to troubleshoot the stuck solar array of the P6 truss, believing that the guide wire was snagged in a swivelling grommet. They swivelled the array repeatedly but to no avail. Expedition 14's Williams deployed the P6 solar array blanket slightly and then retracted it through the same distance, again in vain. Mission managers were assessing the need for an extra fourth spacewalk, but would wait until the result of some investigative tasks that had been added to the third spacewalk.

The entire joint crew also continued to transfer food, supplies and equipment from SpaceHab and the Shuttle to the station.

Curbeam and Williams 'camped out' in the Quest airlock in preparation for the third spacewalk of the mission, which they were to perform the following day.

Flight Day 8 – Saturday, 16 December 2006

Curbeam and Williams made a joint Shuttle/ISS spacewalk to connect the remaining two channels of the P4 truss. As previously, some systems were powered down in order to accommodate this activity. The spacewalkers also relocated debris shield panels from inside the station to a storage point outside for installation on the Zvezda module at some future time. They also installed another grapple fixture for Canadarm2. Then they made their way to the P6 truss solar array to manually attempt to free the stuck panels. They positioned themselves on either side of the array and, using what were called the "Beamer Shake" and the "Suni Shake" they took turns shaking the array blanket box while the crew inside the station reeled in the array one bay at a time. In total, Curbeam shook the blanket 19 times and Williams shook it 13 times. Their colleagues, coordinating with flight controllers on the ground, initiated eight retraction cycles. With the array now 65% retracted and only 11 bays still deployed, it was decided to add an extra day and a fourth spacewalk to the mission in order to further remedy the problem.

After further revisions to the software to accommodate the station's asymmetrical configuration, attitude control was successfully transferred back to its gyroscopes.

Meanwhile, the crews were ahead of schedule in transferring equipment and supplies, and Lopez-Alegria and Tyurin carried out maintenance by replacing a component on the carbon dioxide removal system.

An excerpt from Fuglesang's diary:

> So much is happening that I don't have time to write e-mails. Yesterday (16 December) we talked to the media and it was great. The interest is phenomenal.
>
> It was a bit embarrassing during the press event together with the whole crew, when almost half of the event went to Sweden and all the questions were for me. What you didn't hear was all the humorous remarks around me! (Smiley Face emoticon)
>
> But the atmosphere amongst the crew is really good, and we all get along and have fun together. Later, I had a videoconference with my family. It was really nice to see them and talk to them for 20 minutes.
>
> We have had quiet evenings. The lights are turned off inside the Shuttle, so we tend to go to bed on time. That means I can't write any e-mails. I like to go to bed a bit later, and have stayed up with Thomas and Micha in the Russian module.
>
> I have slept in the Russian airlock the last two nights. There are two windows there and I could, in peace and quiet, watch Earth quickly move by beneath us. Unfortunately though, I have no computer there to look up our exact position.
>
> I look forward to another spacewalk in two days! It seems a certainty now. And yet one day more in space!

Flight Day 9 – Sunday, 17 December 2006

Canadarm2 was repositioned in readiness for Monday's spacewalk by Curbeam and Fuglesang. The internal transfer of apparatus between vehicles was estimated to be 70% complete at this stage.

Flight Day 10 – Monday, 18 December 2006

After another 'camp out' by Curbeam and Fuglesang in the Quest airlock, the crew awoke to *Good Vibrations* by the Beach Boys, played to encourage successful vibrations when the crew made a final attempt to free the stuck solar panels.

During the 6 hour 38 minute spacewalk, Curbeam was transported to the P6 array on Canadarm2. While Fuglesang was stationed on the truss itself, shaking the containment 'blanket boxes', Curbeam manually freed the guide wires which had jammed in the guiding grommets and then folded the stuck hinges between the solar arrays. Williams and Higginbotham operated the station's robotic arm while Oefelein acted as spacewalk coordinator. Applause broke out in Mission Control when the arrays slid into the retention box and the latches were closed at 6:34 p.m. CST.

The P6 truss and its solar arrays had been temporarily installed on the Z1 truss atop the Unity node by STS-97 in November 2000 to power the station during the early stages of its assembly. The folding up of the arrays was in preparation for relocating P6 to its permanent place on the end of the port-side truss.

While the spacewalk was underway, the crews inside continued to transfer equipment and supplies to and fro between the Shuttle and station.

Flight Day 11 – Tuesday, 19 December 2006

During the farewells, ISS commander Lopez-Alegria and flight engineer Tyurin rang their ship's bell and saluted Discovery's crew as they exited the station for their own vehicle. Once undocked, Shuttle commander Polansky radioed, "From the crew of Discovery, we wish you smooth sailing. Thank you for the hard work, and we hope you enjoy the new electrical system on the station."

Having delivered Sunita Williams to the Expedition 14 crew, Discovery now carried Thomas Reiter home after spending half a year on the ISS.

Flight Day 12 – Wednesday, 20 December 2006

A 6 hour inspection of the orbiter's heat shield was conducted using the 50-foot-long Orbiter Boom Sensor System. Crew members not involved in this task spent their time packing apparatus ready for landing.

Due to the extension of the mission and the Shuttle's limited resources, the landing could only be postponed by one day if the Kennedy Space Center was not available, so the backup sites at Edwards Air Force Base, California and White Sands, New Mexico were activated.[14]

[14] White Sands Space Harbour was only used as a landing strip once in the Shuttle's history, by STS-3 on 30 March 1982; the intended site, Edwards Air Force Base was flooded and the crew had trained at White Sands, so chose it over Kennedy Space Centre because at that stage of the Shuttle Program the wide open desert strips were considered safer than the narrow Shuttle Landing Facility.

The crew released a number of small satellites as part of various experimental programmes:

- Micro-Electromechanical System-Based PICOSAT Inspector (MEPSI) might one day use on-board imagery to assess spacecraft damage.
- Radar Fence Transponder (RAFT) was a US Naval Academy experiment to test technology for new spacecraft design.
- Atmospheric Neutral Density Experiment (ANDE) consisted of two spherical microsatellites to measure the density and composition of the atmosphere at the altitude of low Earth orbit. They were to be tracked from the ground.

From Fuglesang's ESA diary:

We have just undocked and made a half turn around the ISS. It was amazing to see the ISS from above, flying over Earth. My main task was to press the undock button and then to take pictures: technical images for different uses, as well as good pictures for media. In total we took close to 300 photos.

It has been an incredible time on board the ISS with the last spectacular spacewalk crowning the experience. What a feeling when we finally succeeded to pack the solar arrays in their boxes. Fantastic teamwork involving everyone on board and the ground control.

The people in Houston have worked around the clock to come up with plans which laid a good foundation for the spacewalk, but in the end it was real-time solutions that got the work done. You can guess that we are all very pleased!

However, it became a long day. After the spacewalk, we spent many hours to get everything in order for today's undocking. I think I worked almost solidly for 15 hours.

We ended the day with a social gathering in the Service Module and we had a nice chat mixed with other things – such as watching Earth and playing in weightlessness – well past our scheduled bedtime.

On another note, our sleeping rhythm is slowly changing (due to orbital dynamics) so we get to bed half an hour earlier every two days – which doesn't suit me.

It has been overwhelming with new impressions and experiences the last ten days, but something very special happened yesterday. It was when I was hanging at the top of the Space Station: We were mending the solar panel with improvised tools and methods, it was night, Beamer [Robert Curbeam] tried to pull loose a jammed wire and I took a glance at Earth. I think we were over Europe, and beneath me I saw large parts of the ISS and far below I saw lights from many cities. To the left was yet another spectacular aurora, shimmering and flowing in light-green. I put out the lamps on my helmet (the light reflected on the Station) and just enjoyed the unbelievable experience!

Flight Day 13 – Thursday, 21 December 2006

Routine preparations for the landing got underway, including stowing gear, test firing the Shuttle's thrusters and verifying flight control surfaces. Fuglesang and Higginbotham deployed the last of the three ANDE microsatellites, then the crew spoke with CNN and ABC. Anchorage resident Oefelein also answered questions from students at the Challenger Learning Center of Alaska in Kenai.

Fuglesang continued to update his ESA website diary as he prepared to return home:

> I saw Sweden! A couple of hours ago we flew over Germany and I watched towards the North. It was night, but there was a clear sky over northern Europe and I could clearly see Denmark and the south of Sweden up to Stockholm. There was an aurora over it all.
>
> It felt like another fulfilment of the journey ('mission success!') I was ready with the camera to take pictures of the aurora and I got what was visible of Sweden, Denmark, the Baltic Sea, northern Germany and Poland.
>
> Unfortunately, it is really difficult to obtain sharp pictures in the night when you need aperture times of a few seconds and the Shuttle is moving about 15 kilometres during those seconds.
>
> You also have to be fast: it only takes a couple of minutes and then what you wanted to capture has disappeared below the horizon. I will attach the least bad picture. It is the first photograph of Sweden from space taken by a Swede (Smiley Face emoticon).

Flight Day 14 – Friday, 22 December 2006

The last day in space for the STS-116 crew started with the festive *Home for the Holidays* by Perry Como, which was played for the entire crew at the requested of the flight control team in Houston.

After high winds postponed landing opportunities in both Florida and California, Discovery was cleared for Runway 15 of the Shuttle Landing Facility, where it landed at 4:32 p.m. CST.

Reflections on the mission

Once settled back on terra firma, Christer Fuglesang gave an interview to ESA on 25 January 2007 relating his personal experiences of spaceflight.

ESA: *What was it like to be inside the Shuttle for the launch?*

CF: The launch was a real highlight! I was never really nervous, which did kind of surprise me. I didn't really dare to believe that we were really going, because of the weather. The launch can also be scrubbed just one second before take-off due to some technical problem. Until the big solid boosters are lit and you start to move, you never know. Once we left, it was of course a wonderful feeling: 'Yes! We are really going!' When we got into space, everyone was shouting and laughing.

E: *What was it like to see the ISS for the first time?*

CF: First it was just a really big bright star. When it came closer and you can see the details, it was big and beautiful. When you get close enough to dock, it's really big!

E: What was it like when you first entered the Station?

CF: You enter into the [Destiny] lab. It is such a big space that you can be in the middle and not be able to touch anything. You feel a bit dizzy for the first minutes because you are used to the Shuttle where there is less space.

E: Can you describe what it was like when you stepped out of the airlock for the first spacewalk?

CF: It was different than planned! We had a very well-choreographed plan for the way everything was supposed to be done. When Beamer egressed, he managed somehow to get a door open where the hand control of the SAFER [spacewalk emergency backpack] is. So suddenly we had a new problem that needed to be solved. I actually had to go out the airlock to try to fix this. It wasn't something we had trained in the pool, so I wasn't sure that I would be able to. But it worked out, and that gave me self-confidence. It is a wonderful view when you are out there, and you can see Earth and see the big Station. Translating along the truss, I enjoyed that a lot. You can just give yourself a little push and you float a few metres without touching anything.

E: How did you feel at the end of the second spacewalk, when you had to return inside?

CF: I felt a little bit sad. Particularly because we had resources to stay out for another hour and I was hoping they'd come up with something else for us to do. But no one said anything. I tried to hang outside there for as long as I could before we had to go inside. I was very pleased when I got to do the third one!

E: You trained extensively for this mission. Was there still anything that surprised you?

CF: Something that was complicated was to go to the toilet... particularly what the Americans call 'number twos'. Due to weightlessness the intestines are not the normal way, so you have to work really hard to get things going. It can get quite uncomfortable.

E: Did you get much of an opportunity to look down at Earth, and what where the most impressive things that you saw?

CF: I didn't have as much opportunity as I would have liked to. We were very, very busy. After undocking we started to get a bit more time. The orientation of the Shuttle meant we also got a better view. I was particularly pleased the first time I saw Sweden. We also saw an aurora over Sweden, and that was beautiful. But one of the best passes was the very last day. It was night-time over Europe. We came in over Ireland and over England; I could see London. You could clearly see the Netherlands because there was so much light. Then I saw all of the Scandinavian countries, even the southern coast of Norway; I could see clouds covering Oslo which were lit up. And I could see up to the middle of Sweden and Finland;

Helsinki. On the opposite side of the gulf, Tallinn and St. Petersburg. It is just like flying over a map. The light tells you where the cities are, and then just the complete darkness over the water. It was a beautiful pass.

E: *What have you been doing since the landing?*

CF: I had to go into NASA on Christmas Eve and Christmas Day to provide data for some of the experiments, and then on the 26th we had a big medical exam. On the 27th we started with meetings to prepare for the debriefings and the presentations which we have to start giving very soon. I had just four days off over the New Year. Since then we've been in daily debriefings. I have just come back from EAC [ESA's European Astronaut Centre in Cologne, Germany] where we had debriefings on Monday and Tuesday this week.

E: *What kind of things do you report back on during the debriefings?*

CF: Anything from minor technical details that didn't work very well – for example, a [helmet] camera that was lost during one of the EVAs because a screw hadn't worked properly – through to the overall message for us. Excellent teamwork helped to make this mission such a success. There was a really good connection between us and the ground crew; they trusted us and we trusted them. It was like we were not only their prolonged arm, but also kind of a prolonged brain to help to give inputs.

E: *How was your re-adaptation to gravity when you came back?*

CF: My balance was very affected. It felt a bit like you had been drinking heavily. But it came back fairly quickly. On the second day it was barely noticeable, and by the third it was completely back. The first time I went jogging, five days after we returned, I got a lot more muscle soreness than I would usually get for such a short run.

E: *When does your mission completely come to an end?*

CF: I will be on the road for at least half of the time through to April. We are going to visit all the NASA centres with the crew. We are making a crew trip to Europe with the highlights being Scandinavia, the EAC and ESTEC. There are a few things still for the experiments (taking post-flight data) that I will be finished with them in another couple of months.

E: *Did you expect that Sweden would be enthusiastic about your flight?*

CF: Two months before the launch I saw how it was building up, so I did expect some interest, but I never imagined that it would be to that extent. When we had an in-flight call and it was both the Crown Princess and the Deputy Prime Minister that was really nice!

E: *Do you have any longer-term plans as an astronaut?*

CF: I will spend a couple of months in Europe this summer, partly working at the EAC. Then I will be back here in Houston in the autumn and I hope to get another assignment with the Shuttle. I would like to do a long-duration mission.

12.4 POSTSCRIPT

Subsequent Missions

See the Alisse mission.

Christer Fuglesang Today

In May 2010 Christer Fuglesang took over as Head of Science and Application Division within the Directorate of Human Spaceflight and Operations at the European Space Research and Technology Centre (ESTEC) in Noorwijk, the Netherlands.

He is currently seconded to the Swedish KTH Royal Institute of Technology in the Department of Physics and the Department of Aeronautical and Vehicle Engineering, teaching both particle physics and human spaceflight. As a prominent member of Vetenskap och Folkbildning (the Swedish Sceptics Association) he identifies strongly with sceptics and atheists.

Christer Fuglesang with Stephen Hawking at the Hawking Radiation Conference at KTH Royal Institute of Technology, Stockholm, August 2015. (www.kth.se)

In 2012 Fuglesang received the Royal Institute of Technology 2012 Alumni of the Year award. He has also received the following honours:

- Honorary Doctorate from Umeå University, Sweden, 1999.
- Honorary Doctorate from the University of Nova Gorica, Slovenia, 2007.
- NASA Space Flight Medal, 2007 and 2009.
- His Majesty The King's Medal in Sweden, 2007.
- NASA Exceptional Service Medal, 2010.
- Royal Institute of Technology Alumni of the Year award, 2012.

13

Esperia

Mission

ESA Mission Name:	Esperia
Astronaut:	Paolo Angelo Nespoli
Mission Duration:	15 days, 2 hours, 24 minutes
Mission Sponsors:	ESA/ASI
ISS Milestones:	ISS 10A, 38th crewed mission to the ISS

Launch

Launch Date/Time:	23 October 2007, 15:38 UTC
Launch Site:	Pad 39-A, Kennedy Space Center
Launch Vehicle:	Space Shuttle Discovery (OV-103)
Launch Mission:	STS-120
Launch Vehicle Crew:	Pamela Ann Melroy (NASA), CDR
	George David 'Zambo' Zamka (NASA), PLT
	Scott Edward Parazynski (NASA), MSP1
	Stephanie Diana Wilson (NASA), MSP2
	Douglas Harry Wheelock (NASA), MSP3
	Paolo Angelo Nespoli (ESA), MSP4
	Daniel Michio Tani (NASA), MSP5

Docking

STS-120	
Docking Date/Time:	25 October 2007, 12:40 UTC
Undocking Date/Time:	5 November 2007, 10:32 UTC
Docking Port:	PMA-2, Destiny Forward

J. O'Sullivan, *In the Footsteps of Columbus*, Springer Praxis Books,
DOI 10.1007/978-3-319-27562-8_13

Landing

Landing Date/Time: 7 November 2007, 18:01 UTC
Landing Site: Runway 33, Shuttle Landing Facility, Kennedy Space Center
Landing Vehicle: Space Shuttle Discovery
Landing Mission: STS-120
Landing Vehicle Crew: Pamela Ann Melroy (NASA), CDR
George David 'Zambo' Zamka (NASA), PLT
Scott Edward Parazynski (NASA), MSP1
Stephanie Diana Wilson (NASA), MSP2
Douglas Harry Wheelock (NASA), MSP3
Paolo Angelo Nespoli (ESA), MSP4
Clayton Conrad Anderson (NASA), MSP5

ISS Expeditions

ISS Expedition: Expedition 16
ISS Crew: Peggy Annette Whitson (NASA), ISS-CDR
Yuri Ivanovich Malenchenko (RKA), ISS-Flight Engineer 1
Clayton Conrad Anderson (NASA), ISS-Flight Engineer 2
replaced by
Daniel Michio Tani (NASA), ISS-Flight Engineer 2

13.1 THE ISS STORY SO FAR

There had been two Space Shuttle flights and two Soyuz flights to the ISS since Christer Fuglesang's Celsius mission in December 2006.

Soyuz TMA-10 in April 2007 had delivered the Expedition 15 crew and the billionaire Charles Simonyi.[15] The Hungarian born Simonyi had led the group in Microsoft that developed the Word and Excel applications. He returned to Earth on Soyuz TMA-9 after a mission lasting a total of 14 days, along with the Expedition 14 crew.

STS-117, Assembly Flight 13A, delivered the S3/S4 truss segments and their solar arrays to the station in June 2007, along with additional Expedition 15 crewmember Clayton Anderson to replace Sunita Williams. Over four EVAs, S3/S4 was installed and the remaining solar arrays of the P6 truss segment sitting on Z1 were stowed in preparation for relocating P6 to its final destination. Problems during the joint Shuttle/ISS operations included the failure of the Russian attitude control computer system and the temporary loss of altitude control. The former was solved by replacing power cables and the latter was no great surprise, given the addition of the combined S3/S4 truss segments which caused the station gyros to go offline.

[15] During a second visit to the ISS on board Soyuz TMA-14 in March 2009 Simonyi became the first (and so far only) 'space tourist' to make a second flight.

The ISS after the STS-118 mission, showing the S3/S4 truss segments and solar arrays delivered by STS-117 matching the P3/P4 truss segments delivered by STS-115. (NASA)

In August STS-118, ISS Assembly Flight 13A.1, delivered the S5 truss and carried the SpaceHab Logistics Single Module, the latter containing food, clothing and scientific equipment for the crew. Of note, the crew included Barbara Morgan as a Mission Specialist. She had served as backup to Christa McAuliffe, who died on board Challenger in 1986. After returning to teaching, Morgan continued to work with NASA's Education Division in the Office of Human Resources and Education. In 1998 she was selected as part of NASA Astronaut Group 17 and qualified as a fully-fledged Mission Specialist.

In October 2007 Soyuz TMA-11 delivered the Expedition 16 crew to the ISS along with Malaysian spaceflight participant Shukor Al Masrie Muszaphar. The orthopaedic surgeon conducted life science and biotechnology experiments as well as publicity activities.

13.2 PAOLO NESPOLI

Early Career

Paolo Nespoli was born on 6 April 1957 in Milan, Italy. He was drafted by the Italian army in 1977 and became a non-commissioned officer and parachute instructor at the Scuola Militare di Paracadutismo of Pisa. In 1980 he joined the 9° Btg d'Assalto 'Col Moschin' of Livorno, where he was a member of the Special Forces. From 1982 to 1984, he was assigned to the Italian contingent of the Multinational Peacekeeping Force in Beirut, Lebanon. On his return to Italy he was commissioned as an officer and remained with the Special Forces.

Paolo Nespoli. (NASA)

Paolo resumed his education in 1985, graduating from the Polytechnic University of New York in 1988 with a bachelor's degree in aerospace engineering and following this up in 1989 with a master's in aeronautics and astronautics. Having left the army in 1987, he returned to Italy to work as a design engineer for Proel Tecnologie in Florence, which manufactured ion propulsion units for satellites and spacecraft. There he conducted mechanical analysis and supported the qualification of the flight units of the Electron Gun Assembly, one of the main parts of the Italian Space Agency's Tethered Satellite System. He was awarded the Laurea in Ingegneria Meccanica by the Università degli Studi di Firenze, Italy, in 1990.

In 1991 Paolo joined the European Astronaut Centre (EAC) in Cologne, Germany, as an astronaut training engineer. He contributed to basic training for European astronauts and was responsible for the preparation and management of astronaut proficiency maintenance, as well as the Astronaut Training Database – one of the systems used in the training process. In 1995 he worked on the EuroMir project at ESTEC in Noordwijk, the Netherlands, where he headed the team which prepared, integrated, and supported the Payload and Crew Support Computer that was used on the Mir space station. In 1996 he went to the NASA Johnson Space Center in Houston, Texas, and worked in the Spaceflight Training Division, training crews for the ISS.

In July 1998 Paolo joined the ASI astronaut corps and one month later he joined ESA's European astronaut corps. He was promptly assigned to Houston as a member of NASA's Astronaut Group 17, known as the Penguins, together with ESA astronauts Léopold Eyharts, Hans Schlegel, and Roberto Vittori. In 2000 Paolo qualified to fly on the Space Shuttle and to work on board the ISS. In July 2001 he completed the Space Shuttle robotics arm course and in September 2003 completed advanced skills training for spacewalks. In August 2004 he continued his training at the Gagarin Cosmonaut Training Centre in Star City, Moscow, learning to operate the Soyuz spacecraft.

After returning to the Johnson Space Center he was assigned to Space Shuttle mission STS-120 in June 2006.

Previous Missions

N/A

13.3 THE ESPERIA MISSION

Esperia Mission Patches

As STS-120 was to deliver the Italian-built Node 2 to the ISS, the mission name Esperia was derived from the ancient Greek name for the Italian peninsula.

The patch was designed by Giorgetto Giugiaro's ItalDesign. It depicted a comet travelling from a blue Earth to a red Mars via the Moon, suggesting the likely trajectory for future space exploration. A stylistic ISS was shown in grey, with its grey orbit around Earth depicting the first steps to Mars. The Italian connection was confirmed by placing the Italian flag adjacent to the 'Esperia' banner.

Once again, the STS-120 patch followed the routine of depicting the orbiter rising above Earth's horizon and into space. Node-2 was visible in the payload bay. A large star represented the ISS. The crew were to move the P6 truss segment from its initial position, indicated by red points, to its final position, indicated by gold points. On the other side of the orbiter, the Moon and Mars reflected the logo for the Vision for Space Exploration (VSE) of President George W. Bush. The constellation of Orion was also a reference to the Orion spacecraft that was proposed as part of the VSE architecture.

Esperia Mission Objectives

Italian ESA astronaut Paolo Nespoli was to be launched to the ISS on the European Esperia mission aboard Space Shuttle Discovery as STS-120. Its main payload for ISS Assembly Flight 10A was the Node 2 module called 'Harmony'.

The Esperia mission was a flight opportunity for the Italian Space Agency that stemmed from an agreement with NASA for the provision of three Multi-Purpose Logistics Modules. The cooperation with ESA led to the assignment of Paolo Nespoli to this flight, and an additional agreement between ESA and ASI was signed to this effect. A principal focus was his role in coordinating spacewalks on this mission.

The STS-120 crew with Paolo Nespoli on the right. (NASA)

The mission objectives included:

- The Node 2 module was developed for NASA by ASI, and its structural design had a lot in common with the Multi-Purpose Logistics Modules (MPLM) and the European Columbus laboratory. As the first European-built module to be permanently attached to the station, it would add 74 cubic metres of living and working volume, and would permanently host the Columbus (ESA) and Kibo (JAXA) laboratories.

Esperia Mission Patch. (www.spacefacts.de)

- The solar arrays of the P6 truss segment that had been temporarily installed on the Unity node by STS-97 in November 2000 had provided power to the station during the early stages of its assembly. The port-side P6 arrays were retracted by STS-116 and the starboard-side arrays by STS-117 in readiness for STS-120 moving the P6 segment to its permanent place at the end of the port side of the integrated truss structure.
- During his mission, Paolo Nespoli was to carry out a number of experiments for the European scientific community in the area of human physiology and biology.
- NASA astronaut Clayton Anderson was to be replaced as Expedition 16 Flight Engineer by his colleague Dan Tani.[16]

[16] Dan Tani's wife, Jane Egan, is a native of Cork, Ireland, as is the author.

STS-120 Mission Patch. (NASA)

Timeline

Flight Day 1 – Tuesday, 23 October 2007

The Space Shuttle Discovery launched on time at 10:38 CDT carrying Node 2 'Harmony' to the ISS. The crew included Mission Specialist Paolo Nespoli, a European astronaut who was making his first spaceflight.

Once in space, the payload bay doors were opened to expose the radiator panels and the Ku-Band antenna was deployed as part of the routine of configuring the vehicle for orbital flight.

Flight Day 2 – Wednesday, 24 October 2007

The first full day in space began with a 5 hour inspection of Discovery's heat shield using the Orbiter Boom Sensor System. Meanwhile the centreline camera was set up and the outer ring of the docking system was extended.

On board the ISS, the Expedition 16 crew conducted a leak check of PMA-2 in preparation for the arrival of the Shuttle and inspected the EVA suits in the Quest airlock. Anderson readied the 400 and 800 millimetre lenses on the cameras that were to be used during the forthcoming backflip manoeuvre by the orbiter, as part of the routine to verify the integrity of the thermal protection system. Anderson and Malenchenko also rehearsed this photo session.

Paolo Nespoli on board Discovery during STS-120. (NASA)

Flight Day 3 – Thursday, 25 October 2007
History was made on this day, when two female spacecraft commanders greeted one other in space. Expedition 16 commander Peggy Whitson welcomed the STS-120 commander Pam Melroy and her crew aboard the ISS when the hatches were opened.

The first task for Anderson and Tani was to swap their Soyuz seat liners, thereby making Tani a member of the ISS crew.

Flight Day 4 – Friday, 26 October 2007
On the day that the Harmony node was to be installed on the station, the Discovery crew woke up to *Rocket Man* by Elton John, which was played for Wheelock in advance of his first spacewalk.

On the first spacewalk of the mission, Parazynski and Wheelock removed and stowed the S-band Antenna Structural Assembly that was to be returned to Earth. Then they secured the power and data grapple fixture to Harmony, removed contamination covers, and disconnected the power cables that linked Harmony to Discovery. Wilson, Anderson and Tani then used Canadarm2 to transfer Harmony from the payload bay to its temporary berth on the port side of Unity. Nespoli coordinated spacewalk activities. After the Shuttle had gone, the Expedition 16 crew were to move PMA-2 from the front of Destiny onto Harmony and then place Harmony on the front of Destiny.

Flight Day 5 – Saturday, 27 October 2007

The day on which the Italian-built Harmony node was to be opened kicked off with an Italian wakeup call *Bellissime Stelle* (Beautiful Stars) by Andréa Bocelli, which was chosen for ESA astronaut Paolo Nespoli.

The hatch was opened at 7:24 a.m. CDT, and Whitson and Nespoli were the first to enter the new module. They wore surgical masks and goggles to prevent injury from any loose debris.

Whitson commented upon the name of the node, which had been chosen by schoolchildren, "We think Harmony is a very good name for this module because it represents the culmination of a lot of international partner work and will allow international partner modules to be added on."

Nespoli added, "It's a pleasure to be here in this very beautiful piece of hardware. I would like to thank everybody who worked hard in making this possible and allowing the Space Station to be built even further."

In the afternoon, Anderson, Whitson, Malenchenko, Melroy, Zamka, Wilson, and Wheelock all gathered in Harmony to participate in interviews with CBS News, FOX News, and WHAM-TV of Rochester, New York; the latter being Melroy's hometown.

Tani and his spacewalking partner, Scott Parazynski, spent the night 'camped out' in the Quest airlock at reduced pressure to prepare for their spacewalk the next day.

Flight Day 6 – Sunday, 28 October 2007

During their 6 hour 33 minute spacewalk, Tani and Parazynski disconnected cables from the P6 truss segment so that Canadarm2, operated by Wilson and Wheelock, could lift it off Unity. It was parked overnight 'in mid-air' on the end of the arm.

Tani also visually inspected the station's starboard Solar Alpha Rotary Joint and gathered samples of shavings which he found beneath the joint's multi-layer insulation covers. This task had been added to assist engineers investigate possible causes of the increased friction observed during the past month and a half as the joint rotated for solar array positioning. Mission managers decided to limit the use of the rotary joint while engineers continued to assess the anomaly. The two spacewalkers also mated the power and data grapple fixture and reconfigured connectors to allow the radiator on the S1 truss to be deployed at a later date.

Having analysed all the evidence, the management team certified Discovery's thermal protection system for re-entry.

Flight Day 7 – Monday, 29 October 2007

Mission management informed Discovery of changes to the intended schedule. There would be an extra docked day. After analysing photos of debris found inside the station's starboard Solar Alpha Rotary Joint, managers had decided to change the objectives and tasks of the fourth spacewalk. The testing of a gun for repairing thermal protection tiles had been deferred to a later mission in order to enable the spacewalk to further inspect the joint. On the flight plan, this additional docked-day had been inserted between the fourth and fifth spacewalks and it provided for some crew off-duty time, along with ample time to prepare apparatus for the fifth spacewalk. And the seemingly healthy port-side SARJ was to be inspected during the third spacewalk for comparison with its ailing partner.

On board the ISS, astronauts continued to activate and check out the avionics and systems racks inside the Harmony node. In parallel, with Anderson and Tani operating Canadarm2, the P6 that had spent the night on the station's arm was handed to the Shuttle's arm operated by Wilson and Zamka. Next the Mobile Transporter on which Canadarm2 was riding was driven to the end of the port truss, whereupon Canadarm2 retrieved the P6 and again held it 'in mid-air' overnight. Meanwhile, Nespoli helped spacewalkers Parazynski and Wheelock prepare the airlock for their next spacewalk.

There was another press conference in the Harmony node, during which the crew spoke to ABC News, NBC News and CNN News. And Nespoli talked to Italian students from the IIS Deambrosis-Natta School, near Genoa, and the Engineering Faculty of the University of L'Aquila using amateur radio as part of the ongoing ESA ARISS programme.

Flight Day 8 – Tuesday, 30 October 2007

During their third spacewalk of the mission, Parazynski and Wheelock successfully installed the P6 truss segment at its permanent position. They also put a spare main bus switching unit onto a stowage platform.

As an added task, Parazynski investigated the port Solar Alpha Rotary Joint for comparison to the starboard joint, from which suspicious shavings had been gathered on a previous EVA. He described the port SARJ as being in "pristine" condition.

As the P6 solar arrays were unfurling, one panel was torn as the array reached the 80% deployed state. The deployment of that array was halted at that point. Meanwhile, its counterpart was unfurled 100% without difficulty.

Space Station Program Manager Mike Suffredini reported that the damaged array was supplying just 3% less power than it would if it had unfurled fully. "This will take time and needs to be worked," he said. "But my personal opinion is we've got the time to work this issue, so we can be methodical about it. And we will."

After re-entering the station, Wheelock noticed a small hole in the outer layer of the thumb of his right glove, which would require investigation before the fourth spacewalk.

Flight Day 9 – Wednesday, 31 October 2007

The plan for the mission and in particular the fourth spacewalk objectives, were changed again as the torn solar array took priority. The spacewalk which had been planned for Thursday was postponed to Friday, or possibly Saturday if extra time was needed for preparation.

Station Program Manager Mike Suffredini said the priority for the remainder of the mission was to complete the deployment of the array. "We give this team a little time to start thinking about creative solutions, and it doesn't take them long to blow you away with what they come up with."

The inspection of the starboard Solar Alpha Rotary Joint planned for Thursday would now be put off until a later date. The mission's fifth spacewalk (which had been planned for Saturday) would be performed by the ISS crew after the Shuttle had departed.

Whitson and Tani worked inside the Harmony module, deploying the Zero Gravity Stowage Rack and removing the anti-vibration mount launch bracket from the common cabin air assembly.

Nespoli received a call from Giorgio Napolitano, President of the Italian Republic. "Good morning Mr. Nespoli, and nice to see you too Commander Melroy."

Speaking from his official residence at Palazzo del Quirinale in Rome, the President expressed his pride and satisfaction at the Italian contribution to the international challenge that was the ISS. He also stressed the need to continue this effort to stimulate the young generation to take an interest and play a leading role in the future. "The name of this mission, Esperia, underlines very well the Italian role," he said. "The mission bridges Italy with Europe, with the United States, indeed with the whole world. It is very important that this mission is a great success."

In response, Nespoli said, "It is crucial that our country continues to work in the space sector, and participate in international projects. And we must also stimulate younger generations, supporting and keeping our dreams alive."

The Italian President said, "This link with the young generation and the schools is really superb. I believe the seeds you are planting will soon blossom. To have faith in the future, our youngsters need hope, motivation and dreams."

In his second amateur radio link-up to Italian students, Nespoli spoke to the Liceo Scientifico 'G. Galilei', Civitavecchia, and ITI-LST 'Mottura', Caltanissetta.

Flight Day 10 – Thursday, 1 November 2007

Using strips of aluminium, a hole punch, a bolt connector, and 20 metres of wire, astronauts created solar array hinge stabilisers that should allow them to take the pressure off of the damaged hinges on the solar array. The contraption worked in the same manner as a cufflink, with the wire being passed through a hole on the solar array and the strip of aluminium supporting it from below.

In parallel with preparations for the unique spacewalk, the crew completed the initial outfitting of the Harmony node.

The two crews also spoke with former President Bush during a visit to the Johnson Space Center.

Flight Day 11 – Friday, 2 November 2007

The astronauts spent the day repositioning the station's Mobile Transporter and Canadarm2 from the end of the port truss (where it had installed the P6 segment) back to S0 in the centre. Once in place, it removed the Orbiter Boom Sensor System from the payload bay and handed it to the Shuttle's robot arm. Canadarm2 then returned to the end of the port truss in readiness for the forthcoming spacewalk. Meanwhile, Parazynski and Wheelock studied detailed plans for the repair task, and their colleagues assisted them by insulating tools with Kapton tape for protection against any electrical currents.

Meanwhile, Whitson and Anderson installed a computer router to complete the initial outfitting of the Harmony node.

Flight Day 12 – Saturday, 3 November 2007

To initiate the day's tasks, Canadarm2 retrieved the Orbiter Boom Sensor System from the Shuttle's arm and Parazynski mounted a platform on the end of the boom. He travelled for 45 minutes to reach the working position, some 50 metres along the P6 truss and 27.5 metres out along the solar array. He then severed the snagged wire and installed home-made 'cufflink' stabilisers in order to strengthen the array's structure in the vicinity of the damage. Wheelock assisted from a position close to the truss by monitoring the separation between Parazynski and the array. Then they watched out for complications as engineers on the ground completed the deployment of the array.

"One of the most satisfying days that I've ever had in Mission Control," said Derek Hassman, lead ISS flight director.

Flight Day 13 – Sunday, 4 November 2007

Prior to closing the hatches at 12:43 p.m. CST, the last of 2,020 pounds of equipment and scientific samples were loaded aboard the Shuttle, including the metal filings that had been collected from the starboard Solar Array Rotary Joint.

Flight Day 14 – Monday, 5 November 2007

Zamka backed Discovery away from the ISS at 4:32 a.m., leaving behind the Harmony node and the relocated and repaired solar array. When 122 metres in front of the station, he began a fly-around so that his crewmates could video and photograph the reconfigured station.

Flight Day 15 – Tuesday, 6 November 2007

In preparation for returning home, Discovery's crew tested flight control systems and thrusters, stowed equipment, and set up a special reclining seat on the mid-deck for Clay Anderson because he had spent more than five months on the ISS.

Aboard the station, Whitson, Malenchenko and Tani had some time off in advance of the EVAs to relocate PMA-2 from the front of Destiny to the end of Harmony as a preliminary to placing Harmony onto the end of Destiny ready for the next Shuttle.

Flight Day 16 – Wednesday, 7 November 2007

The morning call for the Shuttle was *Chitty Chitty Bang Bang* by Sherman and Sherman, which was played for Pamela Melroy as a prelude to her commanding the touch down at 12:01 p.m. on Runway 33 of the Shuttle Landing Facility in Florida.

Simonetta di Pippo, Director of ASI's space science and exploration (Osservazione dell'Universo) programme, said, "The Esperia Mission has been a full success. First of all, I want to stress the strong spirit of collaboration which animates ASI and ESA in its management. And of course NASA gave us a sense of strong collaboration too. The launch was perfect. A complex mission. An impressive crew. Paolo Nespoli did a great job. The landing brings this mission to a close, but it must be considered as an important step in the exploitation of the ISS and beyond, for the long-term exploration endeavour which will see again Italy, together with ESA, in close cooperation with NASA."

13.4 POSTSCRIPT

Subsequent Missions

See the Magisstra mission.

Paolo Nespoli Today

Paolo Nespoli is still an active astronaut at ESA and is scheduled to return to the ISS in 2017 as part of Expedition 52/53.

Paolo Nespoli at ESA Science Week, Dublin, 17 November 2012. (www.science.ie, courtesy of Cian O'Regan)

He has received the following honours:

- NASA Spaceflight Medal, 2007.
- Commendatore Ordine al Merito della Repubblica Italiana, 2007.
- Cavaliere dell'Ordine della Stella della solidarietà Italiana, 2009.

14

Columbus, Phase I

Mission

ESA Mission Name:	Columbus
Astronaut:	Hans Wilhelm Schlegel
Mission Duration:	12 days, 18 hours, 21 minutes
Mission Sponsors:	ESA
ISS Milestones:	ISS 1E, 39th crewed mission to the ISS

Launch

Launch Date/Time:	7 February 2008, 19:45 UTC
Launch Site:	Pad 39-A, Kennedy Space Center
Launch Vehicle:	Space Shuttle Atlantis (OV-104)
Launch Mission:	STS-122
Launch Vehicle Crew:	Stephen Nathaniel Frick (NASA), CDR
	Alan Goodwin 'Dex' Poindexter (NASA), PLT
	Stanley Glen Love (NASA), MSP1
	Leland Devon 'Lee' Melvin (NASA), MSP2
	Rex Joseph Walheim (NASA), MSP3
	Hans Wilhelm Schlegel (ESA), MSP4
	Léopold Paul Pierre Eyharts (ESA), MSP5

J. O'Sullivan, *In the Footsteps of Columbus*, Springer Praxis Books,
DOI 10.1007/978-3-319-27562-8_14

Docking

STS-122
Docking Date/Time: 18 February 2008, 09:24 UTC
Undocking Date/Time: 19 December 2006, 22:10 UTC
Docking Port: PMA-2, Harmony Forward

Landing

Landing Date/Time:	20 February 2008, 14:07 UTC
Landing Site:	Runway 15, Shuttle Landing Facility, Kennedy Space Center
Landing Vehicle:	Space Shuttle Atlantis (OV-104)
Landing Mission:	STS-122
Landing Vehicle Crew:	Stephen Nathaniel Frick (NASA), CDR Alan Goodwin 'Dex' Poindexter (NASA), PLT Stanley Glen Love (NASA), MSP1 Leland Devon 'Lee' Melvin (NASA), MSP2 Rex Joseph Walheim (NASA), MSP3 Hans Wilhelm Schlegel (ESA), MSP4 Daniel Michio Tani (NASA), MSP5

ISS Expeditions

ISS Expedition:	Expedition 13
ISS Crew:	Peggy Annette Whitson (NASA), ISS-CDR Yuri Ivanovich Malenchenko (RKA), ISS-Flight Engineer 1 Daniel Michio Tani (NASA), ISS-Flight Engineer 2 replaced by Léopold Paul Pierre Eyharts (ESA), ISS-Flight Engineer 2

14.1 THE ISS STORY SO FAR

There were no missions between Paolo Nespoli's flight on STS-120 and the arrival of Hans Schlegel and Léopold Eyharts on STS-122.

14.2 HANS SCHLEGEL

Early Career

Hans Schlegel was born on 3 August 1951 in Überlingen, Germany. He studied at Hansa Gymnasium in Cologne and was an exchange student in Lewis Central High School in Council Bluffs, Iowa, USA.

Between 1970 and 1972 Schlegel served as a paratrooper with the Federal Armed Forces, reaching the rank of second lieutenant. He went on to study physics at the Rheinisch Westfälische Technische Hochschule at the University of Aachen in Germany. He remained there between 1979 and 1986 as a solid state physicist carrying out research into the electronic transport properties and optical properties of semiconductors, then worked as a specialist in non-destructive testing methodology at the Institut Dr. Förster Gmbh & Co. KG in Reutlingen.

Schlegel joined the astronaut group of the German Aerospace Centre (Deutches Zentrum fur Luft- und Raumfarht; DLR) in 1988 and in 1993 he flew on the STS-55 mission as a Payload Specialist for the German-sponsored Spacelab D2 research module.

In August 1995 he went to the Gagarin Cosmonaut Training Centre in Moscow and trained as backup for Reinhold Ewald's EuroMir97 mission. During that mission in February 1997, he was the Crew Interface Coordinator responsible for ground-to-air communications. Between June 1997 and January 1998 he received additional training and certification as a flight engineer for the Mir space station. In 1998 he joined ESA's astronaut corps, together with Frank De Winne, Léopold Eyharts, André Kuipers, Paolo Nespoli and Roberto Vittori. In 1998 Hans joined NASA Astronaut Group 17, known as the Penguins; as indeed did Eyharts, Nespoli and Vittori. He was assigned to the Capcom Branch of the Astronaut Office, speaking to astronauts on the ISS. He became a lead Capcom and a Space Station Capcom Instructor. From 2002 to 2004 he worked in the Robotics Branch and as ISS Capcom and was Lead ISS Capcom for Expedition 10. In May 2005 Hans was appointed ESA Lead Astronaut at JSC. In July 2006 he was assigned to the STS-122 mission which was to deliver ESA's Columbus Laboratory to the ISS. Meanwhile, during the preparations for that mission, he worked as Shuttle Capcom, as ISS Capcom Instructor, and in the ISS Branch as lead for systems and crew interfaces, heading up a team of twelve.

Previous Mission

STS-55

In April 1993 the second German Spacelab D2 mission was launched on STS-55. The Spacelab D1 mission had been performed by STS-61A in October 1985. Both missions were largely funded by the German Research Institute for Aviation and Space Flight (Deutsche Forschungsanstalt für Luft- und Raumfahrt; DLR) with that funding paying for the science programme on the flight and two German astronauts; in the case of D2, the physicists Ulrich Walter and Hans Schlegel. A total of 88 scientific experiments were conducted during the 10 day mission for a varied programme which included life sciences, technology applications, Earth observation, astronomy, and atmospheric physics.

The STS-55 crew with Hans Schlegel second from the left in the back row. (www.spacefacts.de)

14.3 THE COLUMBUS MISSION

Columbus Mission Patches

The Columbus laboratory was named in honour of Christopher Columbus, the Italian explorer who set sail westward across the Atlantic Ocean in 1492 and claimed the island of Hispaniola for the Spanish crown.

The ESA logo highlighted the fact that the Columbus laboratory was the main ESA component of the ISS. The light blue circle symbolised Earth and the dark blue ellipse symbolised the orbit of STS-122 carrying Columbus to space. The white stripe across Earth symbolised both the path from east to west that Christopher Columbus had taken on his voyage of discovery and the path of the Columbus laboratory from west to east from the launch pad in Florida into orbit and to the ISS. The stars were to represent the eleven ESA Member States which contributed to ESA's human spaceflight programme.

The STS-122 crew with Hans Schlegel on the right. (NASA)

The STS-122 patch showed a sailing ship transforming into an orbiter. It representing the east to west journey of Christopher Columbus transforming into the west to east voyage of the Shuttle as it carried the module that bore his name to the ISS.

Columbus Mission Objectives

On 7 February 2008 Space Shuttle Atlantis launched as STS-122, Assembly Flight 1E, carrying the Columbus laboratory. When the first European laboratory dedicated to long-term experimentation in weightlessness was installed on the starboard side of the Harmony node on 11 February it brought to fruition many years of organisation and hard work.

The crew of the Columbus assembly and commissioning mission included two ESA astronauts, Léopold Eyharts from France and Hans Schlegel from Germany.

The Columbus mission had two phases:

1. The STS-122 mission was to attach the Columbus module to the ISS, activate it and begin the commissioning process which included the installation of external apparatus. This first phase will be described in this chapter.

Columbus Mission Patch. (ESA)

2. When STS-122 left, Léopold Eyharts was to remain on the ISS as a member of Expedition 16 and continue the commissioning of Columbus by activating its internal experiment facilities as well as undertaking European scientific, public relations and educational tasks, and additional activities in his role as the station's second flight engineer. He would then return to Earth with STS-123 in March 2008. This phase will be discussed in the next chapter.

   The main objectives of the STS-122 mission were:

   - The Columbus laboratory module was to be transported to the ISS inside the Shuttle's payload bay and be installed on the starboard side of the Harmony node.
   - ESA astronauts Schlegel and Eyharts were to make a start on commissioning the systems and experiment facilities of the new laboratory.

STS-122 Mission Patch. (NASA)

- Two European external experiment facilities, EuTEF and SOLAR, were to be installed on the exterior of the Columbus laboratory during a spacewalk.
- Eyharts was to replace NASA's Daniel Tani as the second flight engineer of Expedition 16, thereby becoming the second ESA astronaut to join a long-duration crew. (Albeit for only a few weeks.)
- Schlegel and Eyharts were to undertake a number of experiments for the European scientific community, including runs of the first experiments in the experiment facilities in Columbus.
- The first Columbus experiments involving the weightless environment inside the ISS were in the areas of human physiology and biology, fluid science, and radiation dosimetry. Those that required exposure to open space addressed a variety of scientific topics including exobiology, solar science and material science, in addition to various monitoring and sensor technologies.
- As well as the standard logistics for the Shuttle and Expedition crews, the mission was also to deliver equipment to be installed inside and outside the Columbus laboratory.

Columbus Laboratory

As ESA's largest single contribution to the ISS, the Columbus module was Europe's first permanent research facility in space. It offered European scientists full access to a weightless environment that cannot possibly be duplicated on Earth. The state-of-the-art facility had a volume of 75 cubic metres and carried a broad suite of research equipment. In addition there were external platforms to support experiments and applications in space science, Earth observation and advanced technologies.

The laboratory had room for ten internationally standardised racks to accommodate experiment equipment: eight payload racks in the sidewalls and two in the 'ceiling'. Each rack was the size of a telephone booth and could host autonomous and independent laboratories, complete with power and cooling systems. Video and data links would send results to researchers on Earth. These racks were tailored to obtain the maximum amount of research from a minimum of volume, and they were to be shared by ESA and NASA.

Five of the payload racks were assigned to ESA experiments:

- Biolab could carry out experiments on micro-organisms, cells and tissue cultures, as well as small plants and insects.
- The European Physiology Modules Facility would investigate the effects of long-duration spaceflight on the human body.
- The Fluid Science Laboratory would study the strange behaviour of liquids in microgravity. These experiments can deliver far-reaching benefits, such as improved ways to manufacture metals and to clean up oil spills.
- The European Drawer Rack was a modular and flexible experiment carrier system for a large variety of scientific disciplines.
- The European Transport Carrier would accommodate items for transport and stowage, and it would also serve as a workbench.

Two external facilities were to be attached to Columbus in order to benefit from exposure to the vacuum and radiation environment of space:

- The European Technology Exposure Facility (EuTEF) was to expose samples to the space environment.
- SOLAR was a platform to study solar-related phenomena.

Columbus Control Centre

A worldwide network of control centres support the crew of the ISS around the clock. In Europe the Columbus Control Centre located at the German Aerospace Centre DLR in Oberpfaffenhofen, near Munich, Germany, known by the call-sign 'Col-CC', is the direct link to European astronauts on the ISS. Its primary tasks are to command and control the European laboratory's systems, to coordinate operations of European payloads on the ISS, and to operate the European ground network providing communication services such as voice, video and data to partner facilities. It has two control rooms: one for continuous real-time operations and the other for activities such as training controllers and carrying out simulations. The control team ensures that astronauts in the Columbus laboratory work safely and that the payload facilities function properly. They monitor astronauts operating inside the laboratory, and they configure the systems that maintain the air quality, power the experiments, and remove excess heat from experiments.

The Columbus Control Centre is also responsible for safety under the overall authority of Space Station Mission Control in Houston. In this role, it reacts to changes during a mission, coordinating decisions and establishing priorities.

An astronaut's involvement with a research project in the Columbus laboratory can range from continuously monitoring the experiment to simply installing it, letting it run automatically, and then removing it. All autonomous systems and experiments are monitored and coordinated by the control team on the ground. Researchers on Earth can control and monitor experiments in the laboratory by relaying commands and experiment data directly from their workplaces. Dedicated connections with eight User Support and Operation Centres across Europe make this possible. All downlinked data is routed through the Columbus Control Centre. Although engineering data is archived there, scientific and facility data is distributed to User Support and Operations Centres for processing and archiving.

The User Support and Operations Centres are based at national centres all across Europe and are responsible for specific operations of ESA experiments in the Columbus laboratory. The flow of data between the Columbus Control Centre and the User Support and Operations Centres is considered in generating mission plans and timelines for both the flight controllers and the astronauts.

The Columbus Control Centre is also connected to the European Astronaut Centre in Cologne, Germany, which is responsible for ESA astronaut medical support, monitoring, and safety during missions.

Because the Columbus laboratory also hosts non-European experiments, decisions taken by the Columbus Control Centre are coordinated with the NASA Space Station Mission Control Center in Houston and the Huntsville Operations Support Center in Alabama, as well as the Mission Control Centre in Moscow.

Columbus Facilities

Biolab

Biological experiment facility in Columbus

Biolab can support biological experiments on micro-organisms, cells, tissue cultures, small plants, and small invertebrates. Performing life science experiments in space identifies the role that weightlessness plays at all levels of an organism, ranging from the effects upon a single cell up to complex organisms, including humans.

Accommodation and transport

Biolab was preinstalled inside the Columbus laboratory. Standard experiment containers and vials are transported separately by cargo ferries such as the European Automated Transfer Vehicle, the Russian Progress vehicles, or commercial vehicles.

Biolab was designed to support biological experiments. (ESA)

Operational concept

The biological samples, together with their ancillary items, are transported from Earth in experiment containers or in small vials. The latter case applies if the samples require storage in the Minus Eighty Laboratory Freezer (MELFI) aboard the station prior to use.

On-orbit, experiment containers are inserted into Biolab for processing, whereas frozen samples are first thawed-out in the Experiment Preparation Unit that is installed inside the

BioGlovebox. Once this manual loading is complete, the automatic processing of the experiment is initiated by crewmembers.

In addition to weightlessness, experiments can also be performed under simulated Earth gravity in a centrifuge in order to compare results (and a reference experiment is performed on the ground). During processing of the experiment, the facility handling mechanism transports the samples into the facility's diagnostic instrumentation so that a scientist on the ground can employ teleoperations to participate in the preliminary analyses. A typical experiment can run from one day to three months.

Utilisation scenario

The Facility Responsible Centre for Biolab has the overall responsibility to operate it according to the needs of individual Experiment Container providers, who can monitor the processing of experiments from own User Home Bases.

European Drawer Rack

Multi-discipline flexible experiment carrier in Columbus

Perceiving a need in the scientific community for medium-sized, dedicated experiment equipment for space research in order to reduce research costs and development times, ESA developed the European Drawer Rack as a flexible experiment carrier for a large variety of scientific disciplines. It provides the accommodation and resources for experiment modules in two types of standard ISS housings known as International Subrack Interface Standard drawer units and ISS Lockers. It can accommodate up to three of these drawer units, each having a volume of 72 litres, as well as four lockers having a volume of 57 litres. This strategy gives a rapid turn-around capability and provides increased flight opportunities for users wishing to undertake experiments which don't require a complete rack. The overall design of the facility is optimised to accommodate three to four payloads in parallel.

Resource management

The resource management includes the monitoring of resource allocations to individual payloads, but the operating concept of the European Drawer Rack assumes payloads to be largely autonomous. The facility computer distributes ISS data to payloads and sends payload data to Earth and to the laptop of the European Drawer Rack, whose data management system supports all modes of payload operation ranging from fully automatic to step-by-step execution by an astronaut. In addition to distributing the resources of the laboratory to the experiment modules, the European Drawer Rack provides services such as air cooling and conversion of the 120 volt Columbus power supply to 28 volts for payloads.

Initial configuration

The initial configuration of the European Drawer Rack included one experiment module, the Protein Crystallisation Diagnostics Facility. This multi-user material science instrument is designed to study protein crystallisation in space, in particular the conditions under which good zeolite crystals can be grown. This can only be determined in

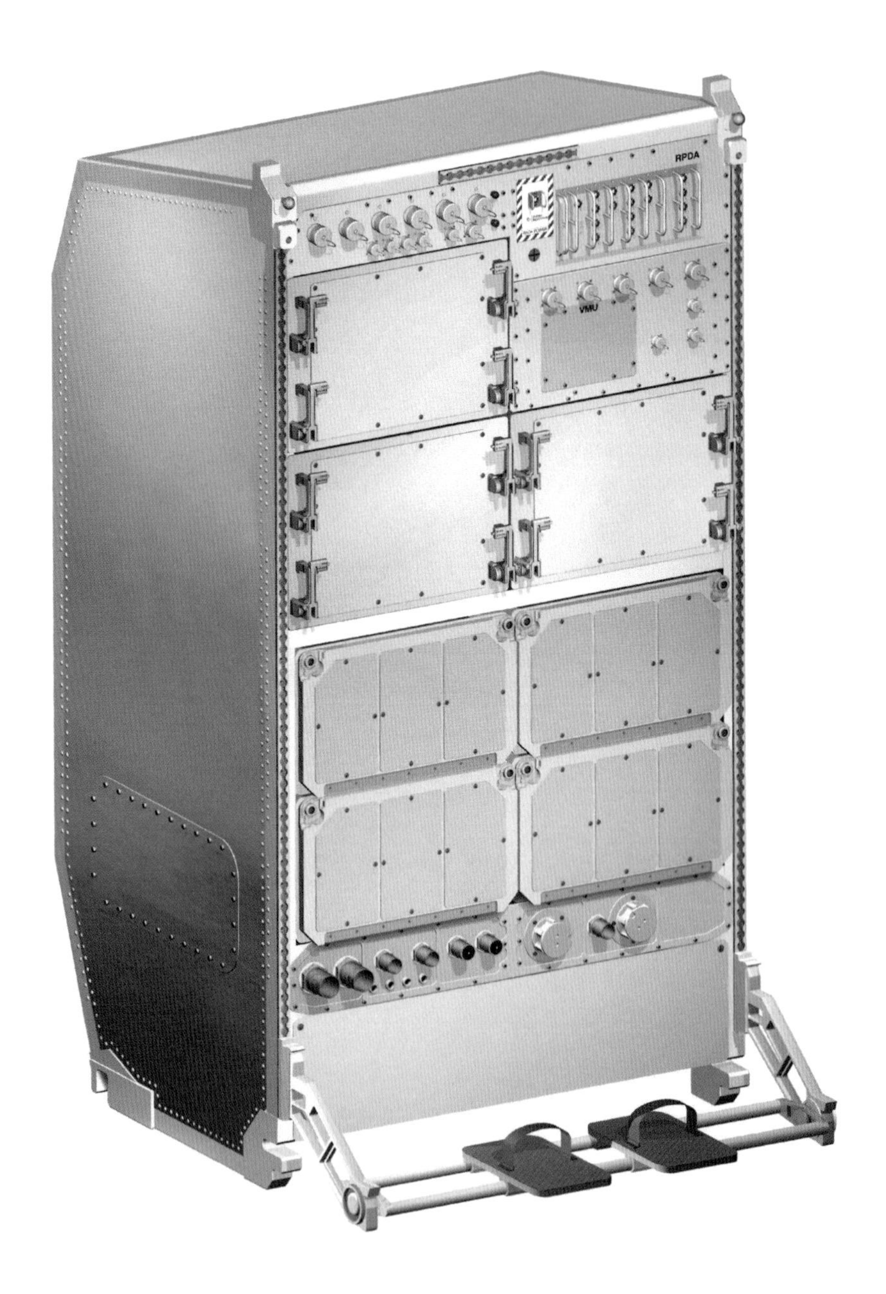

European Drawer Rack. (ESA)

weightlessness, because gravity influences the outcome. The results generated will have benefits in various industrial applications.

A second module was to be launched on a later mission. This Facility for Adsorption and Surface Tension would establish a link between emulsion stability and characteristics of droplet interfaces, a research project that has industrial applications and is linked to investigations such as foam stability and drainage.

European Physiology Modules

Research facility for human physiology experiments in Columbus

A European Physiology Module is an International Standard Payload Rack which is equipped with Science Modules containing apparatus to investigate the effects of long-duration spaceflight on the human body. The results ought to increase our understanding of terrestrial problems such as ageing, osteoporosis, balance disorders, and loss of muscle.

Accommodation and Transport

The European Physiology Modules facility was preinstalled in the Columbus laboratory for launch. New Science Modules and other apparatus will be transported initially by Space Shuttles using the Multi-Purpose Logistics Module, but later by other craft such as the European Automated Transfer Vehicle, the Russian Progress and Soyuz vehicles, and commercial vehicles.

Operational concept

In order to be able to evaluate the on-board data, it is necessary to collect reference data on the crew who serve as the test subjects for the experiments both prior to and after the mission. For this reason, the facility provides Baseline Data Collection Models which are functional copies of the instruments used on board. These models are readily transportable to ensure availability of the equipment for the crewmember's pre-launch and post-flight activities.

Utilisation scenario

The Facility Responsible Centre for the European Physiology Module has the overall responsibility to operate the facility according to the needs of individual Science Modules. The Principal Investigators can monitor the execution of their experiments from local User Home Bases.

European Transport Carrier

Multi-user logistics carrier in Columbus

The European Transport Carrier transports payloads that are unable to be launched with ESA facilities because of stowage or transport limitations. In orbit, it functions as a work-bench and a stowage facility to support experiments with Biolab, the Fluid Science Lab, the European Physiology Modules, and the European Drawer Rack.

European Physiology Modules. (ESA)

European Transport Carrier. (ESA)

Accommodation and transport
Prior to being preinstalled inside the Columbus laboratory for launch, the European Transport Carrier was used in the Multi-Purpose Logistics Modules which flew to the ISS on board the Space Shuttle on many occasions, being reconfigured during ground turn-around to accommodate specific transport and stowage needs for each flight.

Utilisation scenario
On board the ISS, the doors of the European Transport Carrier are unlocked for access to the facility and then the Zero-g Stowage Pockets in the upper and lower portions (which obviously can only be used in weightlessness) are set up. The stowage containers can be used for various kinds of payload stowage activities. When the Carrier is prepared for download, all of the Zero-g Stowage Pockets are first emptied, and then the pockets are folded and stowed away. The doors are then locked again and the whole rack is transferred for return to Earth. It is designed to serve up to 15 flights.

Fluid Science Laboratory
Fluid physics research facility in Columbus
The Fluid Science Laboratory is a multi-user facility to study the dynamics of fluids in a weightless state. This allows investigations on fluid dynamic effects; phenomena which are normally masked by gravity-driven convection, sedimentation, stratification, and fluid static pressure. This research will include diffusion-controlled heat and mass transfer in crystallisation processes, interfacial mass exchange, simulation of geophysical fluid flows, emulsion stability, and more.

Accommodation and Transport
The Fluid Science Laboratory was preinstalled in the Columbus laboratory. Prepared Experiment Containers are transported separately by space ferries such as the European Automated Transfer Vehicle or Russian Progress vehicles.

Operational concept
An Experiment Container is used for each experiment, or category of experiments. An astronaut will removed it from storage and insert it into the Central Experiment Modules drawer, where it is cycled through an experiment and diagnostics calibration processing prior to the experiment being initiated.

A Experiment Container has the fixed dimensions of 400 x 270 x 280 mm with a typical mass of about 25 kg (a maximum of 40 kg). The fluid cell assembly (including the process stimuli and control electronics) are accommodated within this volume. An Experiment Container may also be equipped with dedicated experiment diagnostics to complement the standard diagnostics provided by the Fluid Science Laboratory itself.

The control concept allows operating modes consisting of fully automatic, semi-automatic and fully interactive experiment processing (i.e. step-by-step command keying

Fluid Science Laboratory. (ESA)

by an astronaut). All these modes may be initiated either by an astronaut or from the ground in almost real-time.

Utilisation scenario

The Facility Responsible Centre for the Fluid Science Laboratory has overall responsibility to operate the facility to meet the needs of individual Experiment Container Providers. The individual Principal Investigators can monitor the processing of their experiments from a User Home Base. A facility like Fluid Science Laboratory, which can be used repeatedly with different Experiment Containers, allows shorter individual mission preparation times and thereby contributes to a faster scientific development in a specific field.

Microgravity Science Glovebox

ESA's Microgravity Science Glovebox allows astronauts on the ISS to perform a wide variety of experiments in a controlled environment that is completely isolated. The gloves are the access points through which an astronaut manipulates experiments in the fields of material science, biotechnology, fluid science, combustion science, and crystal growth research.

Scientific gloveboxes are common on Earth. To build a glovebox that will last at least 10 years in weightlessness was a much tougher proposition, however. The Microgravity Science Glovebox had to fit into a standard ISS equipment rack and be sufficiently versatile to accommodate a very large range of experiments and materials, including several which no one had thought of during the design stage.

After being delivered in the Multi-Purpose Logistics Module Leonardo, carried in the cargo bay of Space Shuttle Endeavour in June 2002, the Microgravity Science Glovebox was promptly set up in the Destiny module but it was to be transferred as necessary between the laboratory modules as they were added to the station. It has a working volume of 255 litres, and its broadest usable volume is 40 cm in diameter. It incorporates an airlock for transferring payload and equipment (maximum 40 litres), and can provide negative pressure with air circulation and filtration. It can provide a nitrogen atmosphere and vent its atmosphere to vacuum. In addition to 200 watts of air cooling it can provide 800 watts by using a cold plate. It supplies power at 120 volts and 28 volts DC. It provides an analogue video link, four video cameras, and has a trio of video recorders and a hard disk.

Muscle Atrophy Research and Exercise System

The Muscle Atrophy Research and Exercise System (MARES) is a three-in-one muscle-measurement machine on the ISS which monitors an astronaut's muscles during an exercise work out. Researchers wanted to understand why muscle strength decreases during spaceflight, in order to prepare for long-duration missions and safe space tourism. MARES is an exercise bench that offers detailed data about how muscles behave. Looking at muscle contraction at any single moment gives little information but MARES provides a full overview of muscle speed and force whilst bending an elbow or a knee joint.

Our bodies are amazing machines that perform wonderful feats daily without us thinking of it. For example, hold a glass in your hand and fill it with water and your arm muscles

Microgravity Science Glovebox. (ESA)

will automatically hold the glass stable despite the changing weight as it fills. MARES can chart this fine motor control as well as giving a precise overview of muscle torque and speed. An astronaut is required to move their joints in such a manner as to follow a graph or a dot on the display screen as a motor in the system generates counterforce.

As ESA project manager Joaquim Castellsaguer i Petit, said, "There is no equipment on Earth other than MARES that can offer this type of fine control and measure the user's reaction." The machine is too complex for an everyday exercise regime, but is perfect for testing and analysing exercise routines in order to counteract the loss of muscle mass seen in astronauts. The next generation of space exercise machines will probably include variable counterforce and MARES can be used for preliminary testing. Special motors were created for the apparatus: two high-power and a third for low power. "When we started designing the machine, no motor in the world could power MARES," Joaquim explained. The equipment is controlled and run from the CADMOS User Support and Operations Centre in Toulouse, France.

In addition to the MARES used in the Columbus laboratory, there is a ground model in Toulouse for reference and another at the European Astronaut Centre in Cologne, Germany for training.

Technical specifications

Torque and angular position/velocity measurements and training on joint movements, both left and right:

- Knee flexion/extension
- Ankle flexion/extension
- Trunk flexion/extension
- Hip flexion/extension
- Shoulder flexion/extension
- Elbow flexion/extension
- Wrist flexion/extension; supination/pronation; radial/ulnar deviation.

Force and linear position and velocity measurements and training are supported on the following multi-joint movements:

- Whole arm linear press (front, overhead and intermediate trajectories)
- Whole leg linear press (front, down and intermediate trajectories)

and MARES supports exercise motions in these modes:

- Isometric
- Isokinetic (concentric and eccentric)
- Isotonic (concentric and eccentric)
- Simulation of ideal elements: spring, friction and inertia
- Parameter control following a predefined pattern: position control, velocity control, torque/force control, power control
- Quick release of free motion

including complex combinations of these modes.

The maximum ratings were:

- Torque (Nm): ±450 continuous, ±900 peak (200 ms)
- Force (N): ± 250
- Angular velocity (rad/s): ±9 (515°/s) concentric and eccentric
- Linear velocity (m/s): ± 0.5
- Mechanical power (W): 2,700 continuous, 4500 peak.

Accuracy

- Torque: ±0.3 Nm for low torques, ±0.5% for high torques; 500 Hz
- Force: ±0.125 N; 500 Hz
- Angular velocity: ±0.2°/s for low velocities, ±0.5% for high velocities; 200 Hz
- Linear velocity: ±1 mm/s for low velocities, ±0.1% for high velocities; 200 Hz
- Angular position: ±0.5°; 200 Hz
- Linear position: ±0.5 mm; 200 Hz.

Size and power consumption:

- Mass: 200 kg
- Main box size: 95 x 45 x 45 cm
- Power: 300 watts average, 1,000 watts peak.

This apparatus therefore gives ESA an exceptional opportunity to investigate how muscles are affected by prolonged exposure to weightlessness.

External Payloads

SOLAR

The SOLAR facility was to monitor the Sun with unprecedented accuracy across most of its spectral range. Apart from contributing to solar and stellar physics this information is of great importance for modelling the atmosphere of our own planet, including its chemistry, in support of climatology.

It used three instruments that complemented each other to measure the solar irradiance across the electromagnetic spectrum spanning 17 nanometres (nm) to 100 micrometres (μm), where 99% of the solar energy is emitted:

- SOVIM (SOlar Variable and Irradiance Monitor) covers near-ultraviolet, visible and thermal regions of the spectrum (200 nm to 100 μm).
- SOLSPEC (SOLar SPECtral Irradiance measurements) covers the 180 to 3,000 nm range.
- SOL-ACES (SOLar Auto-Calibrating Extreme UV/UV Spectrophotometers) measures the extreme-ultraviolet and ultraviolet spectral ranges.

SOLAR's observations would improve our understanding of the Sun and allow scientists to create accurate computer models in order to predict its behaviour. The more accurate the data that we obtain, the better we will understand how our star influences Earth.

To jump ahead in the story of this study, SOLAR was launched with Columbus in February 2008 and then installed on the exterior of the module during a spacewalk. It was designed to work for only 18 months but its operation exceeded expectations. In 2012 the ISS turned itself to enable SOLAR to track the Sun for a full solar day (approximately one month). This was the first occasion on which the station changed its attitude solely in order to optimise a scientific experiment.

Vessel-ID

As the ISS circles Earth, the Vessel-ID experiment on Columbus monitors ships crossing the oceans beneath. The Automatic Identification System (AIS) is the marine equivalent of air traffic control. All international vessels, cargo ships above certain weights, as well as all passenger ships, are required to carry Class-A AIS transponders that broadcast data such as the ship's identification, position, course, speed, cargo, and voyage information. AIS allows port authorities and coast guards to track ships but the system uses basic VHF radio signals with a range of just 74 km. It is useful near coasts and when communicating with ships, but is not practicable in the open ocean. AIS signals travel much farther upwards though, and the ISS is ideal for a space-based AIS receiver because it travels in a relatively low orbit.

A Vessel Identification System operating on a global basis would benefit law enforcement, fishing control, border control, and maritime safety. It would be easier to combat piracy, illegal dumping, and fishing in protected areas. However, AIS was never designed to be a global monitoring system and its signals were not transmitted with satellites in mind. A demonstration of the system's capabilities from space was needed.

During a spacewalk in November 2009 the Columbus module was fitted with a VHF antenna and other hardware to capture the AIS signals. In May 2010 a control computer called ERNO-Box and its associated hardware was installed. The Norwegian User Support and Operation Centre in Trondheim, Norway, receives data almost continuously from ESA's Columbus Control Centre in Germany. On a good day, some 400,000 ship reports are received from more than 22,000 different ships. In October 2011 the total number of reports exceeded 110 million from over 82,000 different ships. These were analysed in order to develop new algorithms for next-generation receivers. An upgrade of the ground system in November 2011 has allowed for data transfer in almost real time, a crucial requirement for practical use. And the work to improve receiver algorithms continues.

European Technology Exposure Facility

The European Technology Exposure Facility (EuTEF) was launched with Columbus and installed on the module's External Payload Adaptor during a spacewalk. The Adaptor comprised an adapter plate, the Active Flight Releasable Attachment Mechanism, and various connectors and harness. The aim of this facility was to expose samples of materials to the space environment. As a programmable, fully automated, multi-user facility, EuTEF provided modular and flexible accommodations for a range of technology payloads. These could be mounted directly on the adapter plate or a support structure that would elevate them for optimum exposure either to the direction of flight or aimed away from Earth.

The facility was specifically designed to enable the rapid turnaround of experiments, and for this first flight it had nine different instruments:

- MEDET, the Material Exposure and Degradation Experiment (CNES, ONERA, University of Southampton, ESA).
- DOSTEL, radiation measurements (DLR Institute of Flight Medicine).
- TRIBOLAB, a testbed for the tribology properties of materials in space (INTA, INASMET).
- EXPOSE, photobiology and exobiology (Kayser-Threde, under ESA contract).
- DEBIE-2, a micrometeoroid and orbital debris detector (Patria Finavitec, under ESA contract). This instrument was very similar its predecessor on the ESA Project for On-Board Autonomy (Proba) technology demonstration micro-satellite launched in 2001.
- FIPEX, an atomic oxygen detector (University of Dresden), sharing a standard berth with DEBIE-2.
- PLEGPAY, plasma electron gun payload for plasma discharge in orbit (Thales Alenia Space, under ASI contract).
- EuTEMP, to measure EuTEF's thermal environment during its unpowered transport from the Shuttle to the Columbus External Payload Facility (EFACEC, under ESA contract).
- EVC: an Earth Viewing Camera developed by ESA/Carlo Gavazzi Space for public outreach activities.

The facility was retrieved and returned to Earth by STS-128 in September 2009 to be refurbished and flown again with different payloads.

Timeline

Flight Day 1 – Thursday, 7 February 2008
Space Shuttle Atlantis lifted off from the Kennedy Space Center at 1:45 p.m. CST with Commander Steve Frick, Pilot Alan Poindexter, and Mission Specialists Leland Melvin, Rex Walheim, Stanley Love, Hans Schlegel and Léopold Eyharts.

"The launch of Columbus marks the onset of a new era. We have long waited for this moment in European human spaceflight and space-related sciences," announced Daniel Sacotte, ESA's Director for Human Spaceflight, Microgravity and Exploration Programmes. "The first decision to study the Columbus facility was taken back in 1985. At that time, it was planned as a contribution to NASA's Space Station Freedom project. As the world changed, the station was redesigned and became a truly international programme. We were able to initiate full-scale development of Columbus some twelve years ago. Today Columbus is a reality, a laboratory in space far more capable than what we proposed in 1985, and even than what we planned in 1995, as we benefited from the delays in ISS assembly to improve our design and our equipment. Columbus is now a world-class space laboratory, ready for a decade of exciting science experiments."

ESA Director General, Jean-Jacques Dordain, said, "When the hatch is opened and the astronauts enter Columbus to switch on and commission its science payloads, it will be a great day for Europe, and I see this day coming very soon now. It's been a long road since the early days of our Spacelab laboratory on board the Space Shuttle. With Columbus, and the upcoming ATVs, we've grown from the status of passenger to that of fully-fledged partner. For the first time, a European-manned facility will be permanently operated in orbit, under the control of a European centre. The know-how that we acquired in order to reach this point will be of tremendous importance in preparing for the future of human spaceflight – in Earth orbit and beyond – with our international partners. I wish to pay tribute to all the teams at ESA and the industrial contractors in Europe for this success, as well as to the ESA Member States which, in spite of all the changes in the configuration of the ISS throughout the years, have lent their support to allow ESA to become a highly trustworthy partner in this endeavour. I take the opportunity also to acknowledge NASA, its Administrator [Michael D. Griffin], and all his teams involved in this extraordinary achievement, for their dedication in maintaining a robust schedule in their difficult task of assembling the International Space Station."

Also on this day, the unmanned Progress M-63 freighter which had been launched on 5 February docked at the Pirs module with welcome supplies.

Flight Day 2 – Friday, 8 February 2008

The crew carried out the standard 5 hour inspection of the orbiter's heat shield using the robotic arm and the OBSS, extended the outer ring of the docking system and installed the centreline camera. In addition Walheim, Schlegel and Love checked out the spacesuits that they were to wear for the three spacewalks assigned to the mission.

On board the ISS, Whitson, Malenchenko and Tani conducted a leak check of PMA-2 on the front of the Harmony node, where the Shuttle was to dock.

Flight Day 3 – Saturday, 9 February 2008

Prior to ISS docking, Frick performed the obligatory backflip so that Whitson and Malenchenko could photograph the Shuttle for analysis of its thermal protection system. Ground specialists were concerned about minor damage to a thermal blanket on the starboard Orbital Manoeuvring System pod, similar to damage on STS-117, although the location of this particular blanket meant that it wouldn't experience as much heat during re-entry.

During the docking, one of the five general purpose computers on the Shuttle malfunctioned, but only two, a prime and a backup were required and therefore the operation continued without problem.

Expedition 16 welcomed the Atlantis crew into the ISS at 12:40 p.m. Immediately, ESA astronaut Léopold Eyharts and NASA's Dan Tani exchanged Soyuz seat liners, thus making Eyharts a member of Expedition 16 and Tani a member of the STS-122 crew.

A "crew medical issue" meant the mission's first spacewalk had to be postponed from Sunday to Monday. It is thought that Schlegel temporarily lost his voice. Walheim would therefore be joined by Love rather than Schlegel on that EVA.

Flight Day 4 – Sunday, 10 February 2008

Mission specialists Walheim, Love, Poindexter and Schlegel devoted most of the day to finalising the checklists for the spacewalk.

Flight Day 5 – Monday, 11 February 2008

To herald the first spacewalk of the mission, the wakeup song *Fly Like an Eagle* by Steve Miller was played for Leland Melvin.

During the EVA, Love and Walheim installed a grapple fixture on Columbus for the robot arm to grip and connected electrical and data links to the module. Then Canadarm2, operated by Melvin, Tani and Eyharts, lifted Columbus from the payload bay and transferred it to the CBM on the starboard side of the Harmony module.

"The European Columbus module is now part of the ISS," Expedition 16 astronaut Eyharts radioed to Mission Control in Houston at 3:44 p.m. CST.

"Another great day for the European Space Agency. A great day for our European industry. A great day for Europe in general," said Alan Thirkettle, ESA's ISS Programme Manager. "Now we have four of the international partners with their elements in space. It is really becoming the International Space Station." Alluding to the Japanese laboratory, he added, "We are very much looking forward to having the fifth partner join us next month."

As Columbus was moving into place, Walheim and Love began to replace a large nitrogen tank that had been used in pressurising the station's ammonia cooling system.

The Columbus module being moved to the Harmony module. (NASA)

Flight Day 6 – Tuesday, 12 February 2008

Expedition 16's Eyharts and Atlantis's Schlegel opened the hatches to the Columbus laboratory at 8:08 a.m. CST.

"This is a great moment, and Hans and I are very proud to be here and to ingress for the first time the Columbus module," Eyharts said.

"I think it starts a new era now, the volume of the European scientific module, Columbus, and the ISS are connected for many, many years of research in space in cooperation internationally," Schlegel added.

They brought the computers, ventilation and cooling systems online. Although the cooling system went into partial shutdown when connected to the station's system, once temperature fluctuations had been eliminated it was brought back on line. And when the communications system became available, the Columbus Control Centre in Oberpfaffenhofen began supporting operations.

Later, Walheim and Schlegel 'camped out' in the Quest airlock in preparation for their spacewalk the following day.

Frick, Melvin, Love and Walheim talked to the FOX News programme *Fox and Friends*, KGO-TV in San Francisco, and *The Tavis Smiley Show* on PBS. Later Frick, Poindexter and Schlegel spoke with CBS News and the Pittsburgh television stations KDKA-TV and WPXI-TV.

"We are very pleased indeed to see the crew inside the laboratory," said Alan Thirkettle, the ESA ISS Programme Manager. "We are equally happy with the fact that the Columbus Control Centre in Oberpfaffenhofen is controlling all of the activities now."

Kirk Shireman, NASA's ISS Deputy Programme Manager, added, "It's really exciting to see the crew moving in to this new module. It's really neat to see how the Space Station is coming to its full potential."

Flight Day 7 – Wednesday, 13 February 2008

The primary objective of the mission's second EVA, which was performed by Walheim and Schlegel, was to complete the replacement of a Nitrogen Tank Assembly on the P1 truss segment that had been used to pressurise the station's ammonia cooling system. In addition, they were to make some minor repairs to the debris shield that protected the Destiny module against impacts. They also carried out a number of 'get ahead' tasks such as installing thermal covers on the trunnion pins that had supported the Columbus laboratory while it was in the Shuttle's payload bay.

Mission Control extended Atlantis's mission by an extra day in order to continue the activation of the new laboratory, and cleared the vehicle's thermal protection system for re-entry.

Flight Day 8 – Thursday, 14 February 2008

In the morning, Frick, Schlegel, Tani, Whitson, Malenchenko and Eyharts spoke with the German Chancellor, Angela Merkel, at the Federal Ministry of Economics and Technology in Berlin. Also present were ESA Director General Jean-Jacques Dordain, former astronaut Thomas Reiter of the German Space Agency, Professor Jan Wörner, Chairman of the

Board of DLR, Federal Minister Michael Glos, and Evert Dudok, CEO of Astrium Satellites.

"We are proud as Germans and Europeans that we could contribute to the ISS with Columbus," declared Angela Merkel, "Europe now has a permanent base for research in space."

"We as a crew are feeling very good," said Schlegel. The view of Earth was fantastic. "I realised that our Earth is nothing other than a big mother ship. The colours are so soft and I really hope that the images that we send down help to raise everybody's conscience on how important it is to protect our Earth."

Later on, Tani, Whitson and Malenchenko spoke with NBC News, WOI-TV and WBBM radio in the ongoing effort to maintain public awareness of activities aboard the ISS.

Flight Day 9 – Friday, 15 February 2008

During the third and final spacewalk of the STS-122 mission, Walheim and Love transferred the SOLAR experiment to the Columbus module. Melvin operated Canadarm2, to guide them and their apparatus to the proper locations. They then transferred a failed Control Moment Gyroscope to the payload bay for return to Earth. Finally, they installed the European Technology Exposure Facility (EuTEF) on Columbus. On their way back, they inspected a damaged handrail on the outside of the Quest airlock to determine whether it was likely to have impaired the gloves of some astronauts on previous EVAs.

SOLAR was installed on the zenith position on the external platform. EuTEF was placed in the upper forward-facing position. The plan was to retrieve both packages after several years and return them to Earth.

Flight Day 10 – Saturday, 16 February 2008

To allow the Shuttle crew to recover from the previous day's spacewalk they were allowed to sleep an extra 30 minutes.

Atlantis fired its thrusters for 36 minutes in order to boost the station's altitude by about 1.4 miles in preparation for the STS-123 mission. Meanwhile, the Columbus outfitting and transfer operations continued. All members of the joint crew participated in a news conference involving participants at the Johnson Space Center, the Kennedy Space Center, ESA's European Astronaut Centre in Cologne, Germany, and the Headquarters of CNES (the French space agency) in Paris.

Flight Day 11 – Sunday, 17 February 2008

Prior to closing the hatches, the seven-member crew of Atlantis (including Tani but not Eyharts) completed the final cargo transfers between the two vehicles. Atlantis had launched with one of the heaviest mid-decks in the history of the Shuttle. The return payload of 2,040 pounds would make it the heaviest mid-deck at landing. In addition, a final leak check of the vestibule between the Harmony and Columbus modules was done. Overnight, Atlantis transferred 92 pounds of oxygen to a tank on the Quest airlock.

Flight Day 12 – Monday, 18 February 2008

The wakeup call *Over the Rainbow/What a Wonderful World* by Hawaiian-Japanese vocalist Israel Kamakawiwo'ole with his ukulele was played for Tani.

After undocking, Atlantis moved 122 metres in front of the ISS, then started a loop around it to obtain video and photography of Columbus. Following the separation manoeuvre, the Shuttle crew used the OBSS to carry out a final inspection of their thermal protection system. Meanwhile, Tani continued to exercise to prepare his cardiovascular system for a return to Earth gravity after having spent 120 days in weightlessness.

Flight Day 13 – Tuesday, 19 February 2008

After stowing all the loose items, Atlantis's crew worked through the routine de-orbit preparations by test firing each of the RCS thrusters and activating one of the auxiliary power units for the hydraulic power needed to verify the orbiter's flight control surfaces. In the afternoon the heaters of the four aft vernier engines failed but this was of little consequence because these thrusters would not be required for de-orbit and re-entry.

The crew talked to ABC News, CNN, and WRIC-TV in Richmond, Virginia, which was near the home of Melvin. Because the g-loads of re-entry would be worse for Tani, returning from a long-term mission, he was to use a recumbent seat and this was installed on the mid-deck. Melvin and Schlegel stowed the Ku-band communications antenna in preparation for closing the payload bay doors on the following day.

Flight Day 14 – Wednesday, 20 February 2008

At 8:07 a.m. CST, Atlantis and its seven-member crew landed on Runway 15 at the Kennedy Space Center.

14.4 POSTSCRIPT

Subsequent Missions

At the time of writing, the Columbus mission was Hans Schlegel's last spaceflight.

Hans Schlegel Today

Despite not having flown in space for 8 years, at the time of writing (late 2015) Schlegel is still listed as the oldest active astronaut in the ESA astronaut corps.

He continues to promote space and ESA by conducting public talks such as 'Working and Living in Space – A practical report of a twice flown German', which he presented in October 2015 at the Dallas Goethe Centre in conjunction with the American Council on Germany and the German American Club of Dallas.

Hans Schlegel, September 2015, Cologne, Germany (www.CollectSpace.com, courtesy of Christoph Kaspari)

He is a member of the following organisations:

- Deutsche Physikalische Gesellschaft (German Physical Society).
- AFS – Interkulturelle Begegnungen (American Field Service Germany).

He has also been awarded the following honours:

- Verdienstkreuz 1. Klasse des Verdienstordens der Bundesrepublik Deutschland (Federal Service Cross 1st Class, Federal Republic of Germany).
- Medal of Friendship of Russia.
- NASA Exceptional Achievement Medal.

15

Columbus, Phase II

Mission

ESA Mission Name:	Columbus
Astronaut:	Léopold Paul Pierre Eyharts
Mission Duration:	48 days, 4 hours, 53 minutes
Mission Sponsors:	ESA
ISS Milestones:	ISS 1E, 39th crewed mission to the ISS

Launch

Launch Date/Time:	7 February 2008, 19:45 UTC
Launch Site:	Pad 39-A, Kennedy Space Center
Launch Vehicle:	Space Shuttle Atlantis (OV-104)
Launch Mission:	STS-122
Launch Vehicle Crew:	Stephen Nathaniel Frick (NASA), CDR
	Alan Goodwin 'Dex' Poindexter (NASA), PLT
	Stanley Glen Love (NASA), MSP1
	Leland Devon 'Lee' Melvin (NASA), MSP2
	Rex Joseph Walheim (NASA), MSP3
	Hans Wilhelm Schlegel (ESA), MSP4
	Léopold Paul Pierre Eyharts (ESA), MSP5

Docking

STS-122

Docking Date/Time:	18 February 2008, 09:24 UTC
Undocking Date/Time:	25 March 2008, 00:25 UTC
Docking Port:	PMA-2, Harmony Forward

J. O'Sullivan, *In the Footsteps of Columbus*, Springer Praxis Books,
DOI 10.1007/978-3-319-27562-8_15

Landing

Landing Date/Time:	27 March 2008, 00:39 UTC
Landing Site:	Runway 15, Shuttle Landing Facility, Kennedy Space Center
Landing Vehicle:	Space Shuttle Endeavour (OV-105)
Landing Mission:	STS-123
Landing Vehicle Crew:	Dominic Lee Pudwill Gorie (NASA), CDR Gregory Harold 'Box' Johnson (NASA), PLT Robert Louis Behnken (NASA), MSP1 Michael James Foreman (NASA), MSP2 Takao Doi (JAXA), MSP3 Richard Michael Linnehan (NASA), MSP4 Léopold Paul Pierre Eyharts (ESA), MSP5

ISS Expeditions

ISS Expedition:	Expedition 16
ISS Crew:	Peggy Annette Whitson (NASA), ISS-CDR Yuri Ivanovich Malenchenko (RKA), ISS-Flight Engineer 1 Léopold Paul Pierre Eyharts (ESA), ISS-Flight Engineer 2 replaced by Garrett Erin 'Big G' Reisman (NASA), ISS-Flight Engineer 2

15.1 THE ISS STORY SO FAR

See the previous chapter.

15.2 LÉOPOLD EYHARTS

Early Career

Léopold Eyharts was born on 28 April 1957 in Biarritz, France. After graduating as an aeronautical engineer from the French Air Force Academy of Salon-de-Provence in 1979 he qualified as a fighter pilot in Tours in 1980 and graduated from the EPNER French test pilot school in Istres in 1988. He flew Jaguars from Istres Air Force Base starting in 1980 and in 1985 became a flight commander at Saint-Dizier Air Force Base. In 1990 he was promoted to Chief Test Pilot at the Brétigny-sur-Orge Flight Test Centre near Paris.

In 1990 Léopold joined the astronaut corps of the Centre National d'Études Spatiale (CNES) and worked on the programme in Toulouse to develop the Hermes spaceplane. He was also one of the test pilots and engineers in charge of the CNES programme to provide brief periods of weightlessness by flying parabolic arcs using Caravelle aircraft. He had two training sessions at the Gagarin Cosmonaut Training Centre near Moscow, in 1991 and 1993. He took part in evaluating the Buran Space Shuttle training in Moscow, flying the Tupolev-154 Buran simulator. In 1994 he made qualification flights of the Airbus A300 as a replacement for the Caravelle on parabolic flights; the A300 became operational in this role in 1995.

In July 1994 Léopold was assigned as backup for Claudie André-Deshays (later Haigneré) on the Franco-Russian Cassiopée mission which was flown in August 1996. In January 1998 he launched on Soyuz TM-27 for on his own CNES mission to Mir, Pégase, during which he carried out experiments in medical research, neuroscience, biology, fluid physics and technology. In 1998, Eyharts joined the ESA astronaut corps along with Frank De Winne, André Kuipers, Paolo Nespoli, Hans Schlegel and Roberto Vittori. He was immediately sent to NASA's Johnson Space Center in Houston as a member of the 1998 Astronaut Group 17, known as the Penguins, together with Nespoli, Schlegel and Vittori. He received technical assignments from the Astronaut Office there and worked in the ISS Operations Branch as a section chief for Space Station systems, software, and on-board information technology.

Léopold was backup to Thomas Reiter for Expedition 13/14, ESA's first long-duration mission to the ISS. From October 2004 he trained with American and Russian backup crewmembers at both the Gagarin Cosmonaut Training Centre near Moscow and the Johnson Space Center.

Previous Mission

Soyuz TM-27 Pégase

In January 1998 Léopold Eyharts launched on Soyuz TM-27, accompanied by Talgat Musabayev and Nikolai Budarin who were to become Mir Expedition 25. Eyharts stayed on board for almost 21 days and returned home with the retiring EO-24 crew. His experiment programme was essentially a repeat of the 1996 Cassiopée mission by Claudie André-Deshays, but there is merit in repeating experiments.

The Soyuz TM-27 crew with Léopold Eyharts on the left. (www.spacefacts.de)

15.3 THE COLUMBUS MISSION CONTINUED

Columbus Mission Patches

See the previous chapter for the Columbus and STS-122 mission patches. The patch for Expedition 16 portrayed the ISS at the tip of the Astronaut Badge, with the ISS transiting the full Moon and a nearly complete annular eclipse of the Sun. The ISS was depicted in its final configuration, symbolising the role of this expedition in preparing for the arrival and commissioning of the two international partner modules and their related components. As was usual for this period, due to the addition of short-term members to the expedition the individual names were omitted in order to avoid confusion.

The Expedition 16 crew with Eyharts third from the right. (NASA)

Timeline

Expedition 16 Week ending Friday, 29 February 2008
After Atlantis left, Léopold Eyharts settled in as a member of the Expedition 16 crew. During his first week he activated the first experiment inside the Columbus laboratory.

Expedition 16 Mission Patch. (NASA)

This Waving and Coiling of Arabidopsis Roots at Different g-levels (WAICO) experiment was to study the growth of two types of Arabidopsis seed. A wild variety and a genetically modified version were grown on-board in both weightlessness and Earth-normal gravity environments, the latter being achieved by means of a spinning centrifuge. The seeds were monitored by real-time video link to the lead scientist, Professor Guenther Scherer of the Leibniz Universität Hannover, Germany. The long-term objective of such experiments is to be able to grow nutritional fresh produce in space during a highly extended mission, such as a voyage to Mars lasting several years.

At the end of the experimental run, Biolab would automatically flush the cultivation box using a fixative to preserve the seeds in their final state of growth. Eyharts was to photograph the plants and return them to Earth with him on STS-123 for analysis by the investigators.

Expedition 16 Week ending 11 March 2008

On 6 March 2005 the Earth Viewing Camera (EVC) finally transmitted images to the European Space Technology and Research Centre (ESTEC) in the Netherlands. A second image, this one commanded from the ground, was obtained the next day. The camera was 0.4 x 0.28 x 0.16 metres and weighed 7.8 kg. It used a commercial off-the-shelf sensor provided by Kodak whose 2 x 2K detector could capture colour images of Earth's surface spanning an area of 200 x 200 km from the altitude of the ISS.

"It was really exciting to see the first image arriving from space after the long period of developing the camera and testing it in orbit," said Massimo Sabbatini, ESA's Principal Investigator for the EVC. "This success would not have been possible without the major contribution of Carlo Gavazzi Space and the hard work of the integration and ESTEC operations teams. We are just starting to experiment with the various camera parameters to adjust for the vast range of lighting conditions we encounter. That's why the second picture is slightly blurred. The ISS is travelling at about 7 km per second, so we have to adjust the exposure time to compensate for this rapid motion. At that speed the camera moves over hundreds of metres on the ground in a matter of milliseconds. The camera is intended to be a valuable resource for public outreach and education. We hope to encourage teachers and students to use it as a tool for studying all aspects of Earth observation from space; imaging, telemetry, telecommunications links and orbit predictions. We're also hoping to receive requests for images of particular regions over which the ISS is passing."

On 11 March 2008 STS-123 Endeavour lifted off for a 16 day mission to the ISS. Its payload was the first segment of the Japanese permanent module, specifically the Experiment Logistics Module-Pressurised Section, and also the Special Purpose Dextrous Manipulator (SPDM) known as 'Dextre' that was a sophisticated 'hand' for Canadarm2.

Expedition 16 Week ending 14 March 2008

Soon after the arrival of Endeavour on 13 March, Eyharts stowed his Soyuz seat liner on the Shuttle and Garrett Reisman installed his in Soyuz TMA-11 to replace Eyharts as an Expedition 16 member.

The pallet containing the Dextre unit was moved to the P1 truss segment, and both of the related Orbital Replacement Unit/Tool Changeout Mechanisms (OTCM) were installed.

Reisman and Linnehan performed the first spacewalk of the mission and installed the Japanese ELM-PS in a temporary position on the zenith CBM of the Harmony node. The intention was that when Japan's Kibo Pressurised Laboratory Module (PLM) was delivered by STS-124 and installed opposite the Columbus module, the ELM-PS would be removed from Harmony and placed atop the main Japanese module.

Expedition 16 Week ending 21 March 2008

Astronauts entered the ELM-PS for the first time on 15 March. Meanwhile Canadarm2 transferred the Shuttle OBSS to the exterior of the ISS, where it was to remain until STS-124 had delivered the Kibo PLM because that module was so bulky that there would be no room in the payload bay for the OBSS. That orbiter would therefore not be able to inspect its thermal protection system until after it retrieved the OBSS from the station.

During the second spacewalk, on 16 March, Linnehan and Foreman continued the assembly of the Dextre 'hand'. Then Reisman and Behnken tested it on 17 March, in particular verifying that all of its brakes were working properly. The third EVA on 18 March saw Linnehan and Behnken complete the assembly of Dextre and install various spare parts outside the Quest airlock. However, their attempt to place the MISSE-6 experiment on the exterior of the Columbus module was foiled when latching pins failed to engage. On the fourth spacewalk of the mission, Behnken and Foreman evaluated a new way of repairing Space Shuttle thermal tiles, this task having been deferred from a previous mission.

Expedition 16 Week ending 27 March 2008

On 22 March the final spacewalk of the STS-123 mission saw Behnken and Foreman store the OBSS, successfully install the MISSE-6 package, and carry out a follow-up inspection of the starboard Solar Alpha Rotary Joint whose operation had been impaired for some time.

Endeavour touched down at 7:39:08 p.m. CDT on 26 March, taking Léopold Eyharts home after nearly 49 days in space. This ended ESA's second residency on the ISS. The next mission would be historic because it would mark the first time that an ESA astronaut would be in command of the ISS.

Léopold Eyharts at SpaceUp Paris in 2014. (www.spaceup.fr)

15.4 POSTSCRIPT

Subsequent Missions

At the time of writing, the Columbus mission was Léopold Eyharts' last spaceflight.

Léopold Eyharts Today

Eyharts has received the following honours:

- Chevalier of the Légion d'honneur.
- Chevalier of the Ordre National du Mérite.
- Médaille d'Outre-Mer.
- Silver National Defence Medal.
- Cavalier of the Order of Courage (Russia).
- Cavalier of the Order of Friendship (Russia).

16

Oasiss

Mission

ESA Mission Name:	Oasiss
Astronaut:	Frank De Winne
Mission Duration:	187 days, 20 hours, 41 minutes
Mission Sponsors:	ESA
ISS Milestones:	ISS 19S, 47th crewed mission to the ISS

Launch

Launch Date/Time:	27 May 2009, 10:34 UTC
Launch Site:	Pad 1, Baikonur Cosmodrome, Kazakhstan
Launch Vehicle:	Soyuz TMA
Launch Mission:	Soyuz TMA-15
Launch Vehicle Crew:	Roman Yuriyevich Romanenko (RKA), CDR Frank De Winne (ESA), Flight Engineer Robert Brent Thirsk (CSA) Flight Engineer

Docking

Soyuz TMA-15	
Docking Date/Time:	29 May 2009, 12:34 UTC
Undocking Date/Time:	1 December 2009, 03:56 UTC
Docking Port:	Zarya Nadir

J. O'Sullivan, *In the Footsteps of Columbus*, Springer Praxis Books,
DOI 10.1007/978-3-319-27562-8_16

Landing

Landing Date/Time:	1 December 2009, 07:16 UTC
Landing Site:	TBA
Landing Vehicle:	Soyuz TMA
Landing Mission:	Soyuz TMA-15
Landing Vehicle Crew:	Roman Yuriyevich Romanenko (RKA), CDR Frank De Winne (ESA), Flight Engineer Robert Brent Thirsk (CSA) Flight Engineer

ISS Expeditions

ISS Expedition:	Expedition 20
ISS Crew:	Gennadi Ivanovich Padalka (RKA), ISS-CDR Michael Reed Barratt (NASA), ISS-Flight Engineer 1 Koichi Wakata (JAXA), ISS-Flight Engineer 2 Replaced by Timothy Lennart Kopra (NASA), ISS-Flight Engineer 2 Replaced by Nicole Marie Passonno Stott (NASA), ISS-Flight Engineer 2 Frank De Winne (ESA), ISS-Flight Engineer 3 Roman Yuriyevich Romanenko (RKA), ISS-Flight Engineer 4 Robert Brent Thirsk (CSA), ISS-Flight Engineer 5
ISS Expedition:	Expedition 21
ISS Crew:	Frank De Winne (ESA), ISS-CDR Roman Yuriyevich Romanenko (RKA), ISS-Flight Engineer 1 Robert Brent Thirsk (CSA), ISS-Flight Engineer 2 Jeffery Nels Williams (NASA), ISS-Flight Engineer 3 Maksim Vikorovitch Surayev (RKA), ISS-Flight Engineer 4 Nicole Marie Passonno Stott (NASA), ISS-Flight Engineer 5

16.1 THE ISS STORY SO FAR

There was considerable assembly activity and crew traffic to and from the station between the delivery of the Columbus module by STS-122 and the arrival of Frank De Winne on Soyuz TMA-15.

In March 2008 Endeavour flew the STS-123 mission, Assembly Flight 1J/A, which delivered the Japanese Experiment Logistics Module-Pressurised Section (ELM-PS). This was the first part of the Kibo system. The final component of Canadarm2 was also delivered – the Special Purpose Dextrous Manipulator (SPDM) or Dextre. The Shuttle's Orbiter Boom Sensor System which enables an orbiter to inspect its thermal protection system is usually returned to Earth and installed on the next vehicle. However, because Discovery would carry the Pressurised Laboratory Module (PLM) of Kibo, there would not be room in the payload bay for the OBSS and so it was left with the ISS for Discovery to collect later.

The ISS as seen by departing STS-119 showing the Kibo complex on the bottom left, the Columbus module on bottom right, and the S6 truss segment with its solar arrays on the extreme right. (NASA)

In April 2008 Soyuz TMA-12 delivered the Expedition 17 crew and the South Korean spaceflight participant So Yeon Yi, who returned on Soyuz TMA-11 with the Expedition 16 crew. Their return to Earth was an eventful one, because a bolt between the descent module and the service module did not separate, causing the vehicle to enter the atmosphere sideways and imposing a load of 8.5g for a short period. After the bolt sheared, the capsule suffered a heavy landing, coming down 450 km short of its planned target as a result of performing a purely ballistic re-entry.

And then STS-124 launched in May 2008 with the largest module ever taken to the ISS. Assembly Flight 1J carried the 14,787 kg Kibo Pressurised Laboratory Module (PLM) that incorporated a robot arm called the Japanese Experiment Module Remote Manipulator System (JEMRMS). There was also a swap of Expedition 17 flight engineers. The OBSS was retrieved to enable Discovery to inspect its thermal protection system prior to returning to Earth.

Unrelated to the ISS but a milestone for human spaceflight, Shenzhou 7, China's third manned mission, was the first to carry a three-person crew and included the first EVA to be performed by a taikonaut.

Soyuz TMA-13 launched in October 2008 carrying Expedition 18 members along with 'space tourist' Richard Garriott, a computer game entrepreneur and son of the Skylab and Shuttle astronaut Owen Garriott. He narrowly lost out on the record of being the first son of a space traveller to fly in space, because when he returned to Earth on Soyuz TMA-12 the spacecraft commander was Sergei Volkov, son of Salyut 7 and Mir cosmonaut Alexandr Volkov.

November's STS-126 was a Utilisation and Logistics Flight, ULF2. The Leonardo MPLM carried 6.5 tons of stores and equipment. Repair and maintenance work was performed on the starboard Solar Alpha Rotary Joint that had not been operating correctly for some time, and once again two ISS flight engineers swapped places.

STS-119 delivered the S6 truss segment and its solar arrays to the ISS in March 2009 to complete the integrated truss structure and bring the power generation capacity of the ISS to 120kW, paving the way for operating a permanent six-person crew. In addition, Koichi Wakata became the first Japanese member of an Expedition crew.

After two members of Expedition 19 arrived in Soyuz TMA-14 to join Wakata, the arrival of Soyuz TMA-15 delivered Expedition 20 to form the first six-person crew. It was on this latter flight that Charles Simonyi made his second visit to the ISS, becoming the first 'space tourist' to fly to the ISS more than once. He returned to Earth in Soyuz TMA-13.

STS-125 was the last Hubble servicing mission in May 2009. Because it would be unable to reach the ISS in the event that damage to its thermal protection system prevented it from returning to Earth, Endeavour was on the launch pad ready to undertake a rescue mission (flying as STS-400), and when this proved unnecessary it was assigned to fly to the ISS in July as STS-127.

16.2 FRANK DE WINNE

Early Career

See the Odissea mission.

Previous Missions

See the Odissea mission.

16.3 THE OASISS MISSION

Oasiss Mission Patches

The mission name, Oasiss, was chosen from 520 suggestions in response to an ESA competition. The winner was Jan Puylaert from Ghent in Belgium. It referred both to the ISS as an oasis in space and to Earth as an oasis for humankind.

The name also complemented De Winne's role as a goodwill ambassador for UNICEF Belgium. In support of the UNICEF 2009 Water, Sanitation and Hygiene (WASH) campaign, several events were timed to coincide with his flight in order to draw public attention to the availability and cleanliness of water, which is critically important for human life.

As De Winne stated, "Water is a scarce resource on board the ISS; responsible use and recycling in space can assist in developing efficient water processing applications for Earth, which are particularly important for the developing countries."

In the patch, Earth was depicted as a drop of water with a man acting as the trunk of a tree feeding water to the branches that were growing out of his arms. The rocket illustrated the Soyuz which was to carry De Winne to the ISS. A single white star symbolised future humans reaching other planets.

The angel for the Soyuz TMA-15 patch came from a design by 15-year-old Yura Menkevich of the Kemerovo region of western Siberia and was chosen by the Soyuz

The Soyuz TMA-15 crew with Frank De Winne on the right. (www.spacefacts.de)

The Expedition 20 crew with De Winne on the left in the front row. (NASA)

The Expedition 21 crew with De Winne third from the right. (NASA)

commander Roman Romanenko. In the original design the angel was holding what seemed to be Sputnik, and one of the angel's wings had the colours of the American flag whilst the other had the colours of the Russian flag. Dutch artist Erik van der Hoorn incorporated the angel into his design, which represented the flags of Russia, Belgium and Canada. The angel graciously sailed towards the ISS, symbolised by a red circle composed of the outer bands of the flags of the crew's home countries. The orbit extended into one of the blue bands of Earth to emphasise the strong link between the space programme and our home planet. Two groups of three stars symbolised a safe launch and a safe landing. Together the six stars also commemorated the fact that, beginning with this mission, the station was to have a permanent six-person crew.

The Expedition 20 patch symbolised the first six-person ISS crew with six gold stars. The astronaut symbol extended from the base of the patch to the star at the top in order to represent the international team, both on the ground and in space, which worked together to advance our knowledge of living and working in space. The ISS in the foreground represented where we are now and the important role the station plays in meeting our exploration goals, with the blue, grey and red arcs symbolising Earth, the Moon and Mars.

Oasiss Mission Patch. (ESA)

The Expedition 21 patch was designed by Frank De Winne's wife Lena. It was inspired by a fractal of six, symbolising the teamwork of the six-person crew. This theme was repeated in the six stars in the centre and the six stars in the upper segment. The top segment showed the Moon and Mars, indicating the role of the ISS in furthering the process of exploration. The left segment showed children on Earth in the sunshine, representing our home. The right segment depicted the ISS, the Soyuz and the Shuttle orbiter as the current state of affairs in space exploration.

Soyuz TMA-15 Mission Patch. (www.spacefacts.de)

Oasiss Mission Objectives

On 27 May 2009, Soyuz TMA-15 lifted off from the Baikonur Cosmodrome carrying ESA astronaut Frank De Winne. Two days later it docked with the ISS and he began his 6 month Oasiss mission as a member of the first six-person Space Station Expedition crew. Later in the flight, he became the first European to command the weightless research centre. This was his second mission to the ISS because he had performed the 11 day Odissea mission in 2002. It marked many important milestones for ESA, the European astronaut corps, European science, and the European control centres, as well as being of significance to ESA in terms of cooperation with international partners.

The main objectives of the Oasiss mission were:

- The arrival of the Soyuz TMA-15 crew brought Expedition 20 to six members, as De Winne, Romanenko and Thirsk joined Gennadi Padalka, Michael Barratt and Koichi Wakata. For the first time an ISS crew represented the five main partners in the project: USA, Russia, Europe, Japan and Canada.

Expedition 20 Mission Patch. (NASA)

- From May to October 2009, De Winne would be a flight engineer on Expedition 20, following in the footsteps of Thomas Reiter in 2006 and Léopold Eyharts in 2008. But in October he was to become the first ESA astronaut to command the ISS when Expedition 20 was succeeded by Expedition 21.
- During the Oasiss mission the European programme of scientific experiments and technology demonstrations would continue, particularly involving the Columbus laboratory. These science experiments and technology demonstrations came predominantly from scientific institutions all across Europe and were specifically tailored to a long-duration mission. This research focused upon human physiology, biology, radiation dosimetry, exobiology, fluid physics and materials sciences.
- In terms of educational and promotional activities, De Winne was to conduct five live lessons from space, concentrating on water as a resource.
- This mission would see the arrival of the first Japanese H-II Transfer Vehicle with supplies for the Japanese laboratory. De Winne would be one of two operators when Canadarm2 'snatched' the vehicle from its station-keeping position close alongside

Expedition 21 Mission Patch. (NASA)

the ISS and berthed it temporarily on the Harmony module. He was also to be the prime operator of the Japanese robotic arm for transferring scientific payloads to the Japanese external payload facility positioned outside the Kibo Laboratory.

- During the Oasiss mission the Russians were to launch Mini Research Module 2, called Poisk, on an unmanned Soyuz-U rocket and dock it with the Zvezda module.
- ESA's Columbus Control Centre at Oberpfaffenhofen, near Munich, Germany, would be the hub of European activity for this mission; monitoring and coordinating De Winne's activities, coordinating with Mission Control in Houston and Mission Control in Moscow, and with the European Astronaut Centre in Cologne, Germany as well as the User Support and Operations Centres throughout Europe.
- De Winne's mission was authorised under the International Space Station agreement with the international programme partners by which ESA was entitled to an 8.3% share in the station's resources following the attachment of ESA's Columbus

laboratory. This would allow ESA to send one European astronaut to the station for a long-duration six month mission every year.
- During his mission De Winne was to be joined by Swedish ESA astronaut Christer Fuglesang, flying as a Mission Specialist on board STS-128.

Timeline

During his Oasiss mission, Frank De Winne wrote a diary for the ESA website. Here is a summary of his experiences on the ISS. Note that the entries were neither regular or comprehensive.

Expedition 20 Week ending 31 May 2009

Launch on board Soyuz TMA-15 was on 27 May 2009, and the docking with the ISS came two days later. ESA astronaut Frank De Winne, Russian cosmonaut Roman Romanenko, and Canadian Space Agency astronaut Robert Thirsk were welcomed on the ISS by the Expedition 20 residents: Russia's Gennadi Padalka, NASA's Michael Barratt and Japan's Koichi Wakata. They comprised the first six-person ISS crew and the first to involve all five main partners in the station's assembly and operation.

"This is a very exciting moment for the ISS partners and a major milestone for human spaceflight and exploration. With Frank, Roman and Bob having joined the other three ISS crewmembers, we've reached the full six-person crew capability. We have had a very intense two weeks in Europe with the roll out of the Node 3 module, the selection of six new members of the European astronaut corps, and now this milestone," said Simonetta Di Pippo, ESA's Director of Human Spaceflight. "This opens up new and exciting opportunities on the utilisation of the Station for scientific and research activities but also in preparation for future exploration missions to more distant destinations."

Expedition 20 Week ending 14 June 2009

Padalka and Barrett made a 4 hour 54 minutes spacewalk on 9 June to prepare Zvezda for the arrival of the Russian Mini Research Module 2 called Poisk. They installed a new antenna and photographed the Strela-2 crane. A day later, they depressurised the Zvezda transfer compartment whilst inside wearing spacesuits in order to install a docking cone. This was officially a spacewalk even though they did not exit the station.

Expedition 20 Week ending 21 July 2009

On 30 June Progress M-02M was undocked from Pirs and was de-orbited with its cargo of trash. The last of the older type, Progress M-67, would dock at the aft port of Zvezda on 29 July to continue the Russian supply effort.

Having acclimatised to his new environment, De Winne was busy at his station maintenance and scientific experiment tasks. This week he worked on the off-gassing of the Internal Thermal Control System (ITCS). This used water to maintain the temperature of the air at about 22.5 degrees, and the bubbles of gas that formed in the water had to be removed on a routine basis.

As De Winne wrote in his diary, "I have also been doing some voluntary science. These are tasks that we can more or less choose to do, but that aren't vitally important. I have been demonstrating the Bernoulli effect. The Dutch-Swedish scientist Daniel Bernoulli showed that an increase in the speed of a liquid or gas is accompanied by a fall in the pressure in the liquid or gas. Amongst other things, the Bernoulli effect causes the 'lift' of an airplane. As a pilot, it particularly interested me to demonstrate this effect in weightlessness. I've also been taking some samples of all the water supplies here on the ISS. Just like on Earth, here in space water is very important to us. This links nicely to water quality and access to water on our planet. Being goodwill ambassador for UNICEF Belgium, I am especially interested in promoting the WASH campaign that highlights the problems of access to drinking water, sanitation and hygiene. The name of my mission, Oasiss, also refers to the importance of water for life on our planet. Every day, five thousand children die either because they don't have access to water or because they live in bad hygienic circumstance. All the water samples I collected were fine, so there's nothing for us to worry about. We can carry on using our drinking water."

A delay in launching STS-127 for Assembly Flight 2J/A because of a hydrogen leak meant that De Winne had to wait until 17 July for Endeavour to bring the Japanese Kibo laboratory to the station. Its arrival with a crew of seven astronauts placed the station's resources under pressure. De Winne wrote, "For the first time there were 13 of us around the dining table in Unity. It was a special feeling of life, noise and fun!"

On 20 July, which was the 40th anniversary of the Apollo 11 lunar landing, De Winne reflected on the progress made, and yet to come. "A very special anniversary! Ten years from now there will once again be Earth-dwellers walking on our celestial neighbour. No doubt our work on the ISS will help to make a return to the Moon a little bit easier."

Expedition 20 Week ending 9 August 2009

In his diary De Winne wrote, "Meanwhile there's plenty of work to be done on the Station. Not only are we carrying out experiments, but also all kinds of other tasks, such as the repair of the Advanced Resistive Exercise Device (ARED) that we use to perform our daily exercise. We use it to perform some of about two and a half hours' of exercise per day. The advantage of ARED, is that we can do many different exercises with just one piece of apparatus. It helps us counteract bone loss and it also stimulates our muscles. This power training is important when you remain in space for a long time. ARED works just like a piece of gym equipment on Earth, but up here the 'weights' are replaced by vacuum cylinders that work more or less like a bicycle pump, but in reverse. I tested ARED together with my Canadian and Japanese crewmates, Wakata and Thirsk."

Japan's Koichi Wakata returned to Earth with Endeavour on 31 July, landing in Florida. His place on the Expedition 20 crew was taken by NASA astronaut Timothy Kopra. Meanwhile De Winne was preparing for the visit by his Swedish ESA colleague Christer Fuglesang, scheduled to launch later in the month as part of the STS-128 Discovery Assembly Flight 17A crew.

Expedition 20 Week ending 16 August 2009

In August, De Winne wrote of the mundane tasks that face astronauts, in particular the blocked toilet during the visit of STS-127 which obliged the record-breaking 13 crew-members to share three toilets. "The toilet in the Destiny module, officially known as the Waste and Hygiene Compartment (WHC), became 'blocked' during the recent visit of seven astronauts to the International Space Station with Space Shuttle Endeavour. Together with Mike Barratt and our Russian commander Gennadi Padalka, we managed to fix it. Fortunately I had some spare time to work on the repair. We replaced a pump, a filter, a control panel and a container that catches fluids. There is another toilet in the Russian Zvezda module and there was also one in the Space Shuttle. Of course, a toilet works slightly differently here in space compared to down on Earth. In weightlessness the toilet uses an air flow. The WHC separates liquid and solid waste and is part of a unique recycling system here on the Station, the Water Recovery System (WRS). Urine and other waste water is recycled into drinking water. This kind of system will be very important in the future when we send human missions to Mars. They won't be able to receive new supplies from Earth, as we can here on the ISS."

On 7 August, De Winne, who was trained for Canadarm2 and the robotic arm on the Kibo module, relocated PMA-3 from one side of the Unity node to the other, to make room for the arrival of the new Tranquility node and Cupola. He and Thirsk operated Canadarm2 for this 6 hour 30 minute operation. Afterwards De Winne said, "Finally I got to do some of the real robotics operations. The task was not extremely challenging, but still with robotics you always need to be very concentrated and work very correctly, and take care not to make the smallest mistake. It went very well, actually almost as in the training; even better. The arm flies very stable. The hand controllers are better than in the simulators, because they aren't used 8 hours per day! It was a real great pleasure to work with Bob. We were both very concentrated and very calm, but still worked at a good pace in order to stay ahead of the timeline."

Expedition 20 Week ending 23 August 2009

Much of this week was spent preparing for the arrival of STS-128 which was to launch on 29 August with the Leonardo MPLM, and the first Japanese H-II Transfer Vehicle (HTV) scheduled a fortnight later (see the Alissé mission below for details of these arrivals).

De Winne wrote, "With the arrival of Discovery, for the second time during my mission there will be thirteen of us on the Station. And together with my colleague Christer Fuglesang there will be two ESA astronauts in space at the same time. This is symbolic of the important European contribution to the ISS programme."

De Winne also used the Microgravity Science Glovebox (MSG), now located inside the European Columbus laboratory. With a volume of 255 litres, this glovebox provided a safe and totally isolated environment in which to undertake all manner of experiments. He utilised it for an experiment called Investigating the Structure of Paramagnetic Aggregates from Colloidal Emulsions (InSPACE2), that was to investigate very special fluids that change their properties in the presence of a magnetic field. Specifically, these 'smart' materials become solid in the presence of a magnetic field and fluid when the field is no longer present. They could be put to many possible uses, including the development of new brake systems, robotics, landing gear for aircraft, and systems to dampen vibrations.

Expedition 20 Week ending 31 August 2009

On 21 August De Winne repaired the Data Management System (DMS) at the heart of the Columbus software. It comprised a number of computers and other electronic equipment. In replacing the faulty Command & Monitoring Unit 1, one of four such units in the DMS, he worked with colleagues in the Columbus Control Centre in Germany. The repair was a success.

On 29 August Discovery made a night launch from the Kennedy Space Center in Florida, with its crew including ESA's Christer Fuglesang.

In his diary, De Winne explained why the ISS needed to raise its orbit regularly, "Yesterday the Station was flying at an average altitude of 348 km. Its closest approach [the perigee] was at 342 km and the furthest [the apogee] was at 354 km. The ISS rotates around Earth roughly every 90 minutes. That is nearly 16 times a day! The angle of this orbit to the equator is just less than 52 degrees. As a result of atmospheric drag, the Station gradually gets closer to Earth. In the past 24 hours it will have dropped in altitude by 44 metres. The amount varies each day because of fluctuations in solar activity which alters the density of the upper layers of the atmosphere. But there is no danger that we can get too close to Earth because the ISS regularly gets a 'push' upwards; for example by the spacecraft that bring supplies to the ISS, such as ESA's Automated Transfer Vehicle [ATV-1, named 'Jules Verne', visited in 2008]. Another interesting fact is the first ISS module, Zarya, launched in November 1998, has now made more than 61,750 orbits around Earth."

Earlier in August he helped to repair the Fluids & Combustion Facility (FCF). It was located in the Destiny module and included the Combustion Integrated Rack (CIR) that contained a 100-litre furnace in which combustion effects in weightlessness could be studied. It was the only such facility on board the ISS.

In his August diary entry, De Winne promoted a quiz organised by the ESA's Human Spaceflight Education Team and UNICEF that was aimed at European children aged 12–14 on the theme 'Water on Earth and in space'. He said, "As goodwill ambassador for UNICEF Belgium, this is a theme very close to my heart. The name of my mission, Oasiss, also refers to water. And UNICEF is currently running the WASH campaign around projects for water, sanitation and hygiene."

On 31 August 2009, STS-128 Discovery docked with the ISS and De Winne welcomed his ESA colleague Christer Fuglesang on board the station; he left when the Shuttle departed on 8 September. See the next chapter for Fuglesang's Alissé mission.

Expedition 20 Week ending 13 September 2009

At 19:01 CEST (17:01 UT) on 10 September 2009 an H-IIB heavy launch vehicle lifted off from the Tanegashima Space Centre in Japan with the first unmanned H-II Transfer Vehicle (HTV), carrying about 4.5 tonnes of supplies for the ISS.

Expedition 20 Week ending 20 September 2009

Awaiting the arrival of HTV-1, De Winne wrote, "This mission is very important to me. Unlike other spacecraft that visit [such as the Russian Progress and the ESA ATV freighters], HTV-1 will not dock directly with the ISS, but will first approach to within about 10 metres. Nicole Stott, Bob Thirsk and I will then use the robotic arm, Canadarm2, to grab hold of the Japanese spacecraft and attach it to the Harmony module. So the training in Japan before my mission will be put to good use. I have also had the opportunity to practise the procedures here in space."

When the spacecraft arrived on 17 September, it approached to a point just below the ISS and then halted in place. After Stott and De Winne had grappled it using Canadarm2 it was berthed to the nadir of the Harmony node. Once the HTV subsystems had been activated, Thirsk and De Winne performed vestibule outfitting procedures in preparation for ingress and unloading.

Expedition 20 Week ending 27 September 2009

During a 20 minute video link-up on 21 September 2009 De Winne demonstrated a live experiment to children at science museums in Barcelona (Spain), Thessaloniki (Greece), Milan (Italy) and Mechelen (Belgium). This 'Do objects have weight in space?' experiment was proposed by teachers in response to a call for experiments which could be carried out aboard the ISS in order to demonstrate the effects of freefall. It involved calculating the mass of an object by measuring the time that it took to oscillate whilst hooked onto a spring. This illustrated the difference between the concepts of weight and mass, where weight depends on the force of gravity and mass is intrinsic to the object regardless of where it is. The live call by De Winne was the highlight of a 3 hour programme connecting the four museums, which also saw the schoolchildren perform the experiment themselves. The audience was also treated to a space show, a quiz, and a lecture from a space expert.

ESA's Director of Human Spaceflight, Simonetta Di Pippo, said, "One of our missions is to bring the fascination of human spaceflight and exploration as close as possible to our fellow citizens and in particular to our youth to inspire them and motivate them to include engineering and scientific studies in their educational path. Our astronauts are our best ambassadors. They're very effective in providing demonstrations of physical phenomena. They are very good in supporting the teachers, adding also the inspirational element that by studying hard you can reach the sky, and beyond!"

On 24 September De Winne used the Japanese robotic arm to transfer experiment payloads to the External Facility that projected from the far end of the Japanese Kibo laboratory.

This month also saw him testing the WEarable Augmented Reality technology (WEAR) designed by Space Applications in Zaventem, Belgium. This was a technology that predicted the Google Glass. Already tested by De Winne on the ground, it was a heads-up display system which was worn on the user's head and projected data onto a partially see-through video screen located in front of the eyes. It could be used to read text, refer to schematics, or watch video, and the ISS astronauts were to test its utility during normal activities.

De Winne also worked on the Sodium Loading in Microgravity (SOLO) experiment, which was a human physiology investigation that compared the level of salt intake and bone loss in microgravity. As he wrote, "I'm eating specially prepared meals, some of which are very low in salt and others that have a very high salt content [compared to the normal diet aboard the ISS]. For this experiment I also have to regularly measure my mass and take urine and blood samples that are frozen and will later be taken down to Earth for analysis in the lab."

Expedition 20 Week ending 4 October 2009

In a 29 September press release, ESA confirmed a 1 week extension to De Winne's Oasiss mission, with his landing in the Soyuz TMA-15 spacecraft rescheduled for 1 December. Meanwhile, Soyuz TMA-16 docked at the aft port of Zvezda on 2 October to deliver Jeffrey Williams, Maxim Surayev and the Cirque de Soleil billionaire Guy Laliberté as a Canadian spaceflight participant.

Expedition 20/21 Week ending 11 October 2009

On 6 October 2009, De Winne announced the winner of the ESA/UNICEF Water Quiz during a live transmission from space. Benedetto Lui, age 14 from Italy, had correctly answered all the questions and given the closest answer to a tie-break question. By participating in this quiz, the children could find out more about water on Earth and in space, and also about their relationship. The 10 questions were posted on the ESA/UNICEF Water Quiz website between 9 and 30 September. More than 500 children registered to participate, with many of the entrants coming from the Czech Republic, which was one of ESA's newest Member States, and from De Winne's own Belgium.

De Winne gave another lesson from space on 8 October, when he linked up via video with the Free University in Brussels. At the end of a day of learning about living and working in space, 300 children saw him give three demonstrations of how water behaves in space. To wrap up, he answered questions from five children in the audience in Brussels.

In an auspicious day for ESA, on 11 October 2009 De Winne became the first European astronaut to command the International Space Station. There had been a change of command ceremony earlier in the week, but the official start of Expedition 21 with De Winne in command was when the Expedition 20 commander, Gennadi Padalka, departed in Soyuz TMA-14 together with NASA astronaut Michael Barratt and Guy Laliberté, the Canadian spaceflight participant who had visited for 8 days.

De Winne's Expedition 21 crewmembers were Jeff Williams and Max Suraev, Nicole Stott (who arrived on STS-128 in August), and Roman Romenenko and Robert Thirsk (both of whom had flown up with De Winne).

Expedition 21 Week ending 18 October 2009

In his diary entry of 12 October 2009, De Winne wrote of his new duties, "As ISS Commander, I am responsible for the daily operations on the Station and for ensuring the crew works as one integrated team. This is by far the most important task, to ensure that the atmosphere amongst the crew is such that everyone can perform their tasks to the best of their abilities."

On 15 October he linked up with the Columbus Control Centre in Oberpfaffenhofen, where ESA Director of Human Spaceflight Simonetta Di Pippo and Chairman of the German Aerospace Centre Johann-Dietrich Wörner were present. Di Pippo began by announcing, "One more achievement, one more milestone for Europe, ESA, and human spaceflight and exploration: the commandership of the International Space Station is today the responsibility of a European astronaut, Frank De Winne from Belgium. We are confirming our ever increasing momentum. Is there a better way to start a month in which we shall see the first ESA/EU International Conference on Human Space Exploration?" There was even more to celebrate, as Di Poppo explained. "This commandership could not occur at a more appropriate moment, as we are enjoying the flawless working of the European laboratory Columbus, harvesting a wealth of data, we're working relentlessly to prepare ATV 'Johannes Kepler' for launch, and we have a new generation of European astronauts in training. It is a well-deserved confirmation of the quality of our effort. It bodes well for the future of Europe in human spaceflight and exploration."

In response, De Winne made it clear that his serving as commander of the ISS was the result of the outstanding cooperation of all international partners involved. He thanked all of the space agencies for their support and especially the Columbus Control Centre and its staff.

Next the international media representatives at the Columbus Control Centre were invited to pose questions to De Winne and his crew. Then Di Pippo concluded the event by once again emphasising the importance of ESA astronauts. "The astronauts are our best ambassadors, especially when they are recognised by their peers. Through them, ESA is entrusted with such a great responsibility as the ISS commandership. Frank De Winne is performing a great mission, Roberto Vittori, Paolo Nespoli, and André Kuipers are training for future missions, and we have just brought forward flight opportunities for ESA to early 2014. Our new recruits are progressing in their basic training. We will be important players in global human spaceflight and exploration for many more years to come."

Meanwhile, the Russians launched the Progress M-03M freighter with supplies. On arrival at the ISS it docked at Pirs.

Expedition 21 Week ending 1 November 2009

It is worth reproducing De Winne's diary entry for 27 October in full because it nicely expresses his personal feelings as well as his responsibilities as an astronaut:

> It is exactly 5 months ago today that I launched to the International Space Station from Baikonur Cosmodrome. In just over a month I will return to Earth again. But there's a lot to do before then!
>
> In the past week I have participated in several public relations events to mark my becoming ISS Commander. It is a big honour for Europe and the European astronaut corps, but of course, also for me personally. Not a lot has changed as far as everyday work goes. I do have to make sure to keep an eye on the clock though, because I lead the morning and evening planning conferences.
>
> In space, we conduct research in a wide variety of areas. One important area of research looks at how the human body reacts to weightlessness. Together with Bob Thirsk and Jeffrey Williams, I have just participated in the SLEEP experiment. That stands for Sleep-Wake Actigraphy & Light Exposure during Spaceflight. We have a special device that, amongst other things, measures our sleep patterns.
>
> In our Space Station crammed full with all of its technologies, it isn't so surprising that something breaks down now and then. Recently it was MELFI-2 [a minus-eighty freezer for laboratory samples] that was playing up. We had to replace an electronics box. In doing this a few litres of coolant escaped, so we had to clean up. Free-floating water can pose a big problem in the ISS, with all the electrical cables

and connectors everywhere. Fortunately Jeff and Nicole [Stott] were immediately on hand to help me fix the leak and clear up the mess. Meanwhile I have managed to fix the MELFI and it is working fine now.

On 17 October we received a visitor. Not astronauts this time, but an unmanned Russian Progress M-03M [designated 35P by NASA] that docked with the ISS. These Progress spacecraft are regularly used to resupply the ISS. This one has 2.5 tonnes of supplies on board, including nearly 800 kg of spare parts, life support equipment and other hardware, 870 kg of fuel, and 420 kg of water.

A few days from now, we will unberth the Japanese H-II Transfer Vehicle from the ISS.[17] Filled with unwanted waste from the Station, the cargo spacecraft will then burn up on re-entry into Earth's atmosphere. Nicole and I have already spent quite some time practising procedures to detach HTV-1. I have also spent many hours transferring cargo between it and the ISS, in particular to the Japanese Kibo laboratory.

30 October was also a very special anniversary for me. It was exactly 7 years ago on that day that I launched to the ISS for the first time, on my Odissea mission. That was a relatively short spaceflight of just under 2 weeks; very different to my half-year stay this time around! The ISS has also grown considerably in size during that time.

The Expedition 21 crew with Frank De Winne at the centre of the back row. (NASA)

[17] The HTV-1 spacecraft was unberthed on 30 October and burned up as intended in Earth's atmosphere high above the Pacific Ocean on 2 November.

Expedition 21 Week ending 22 November 2009

The Russian Mini-Research Module 2 (MRM-2) or Poisk was launched on 10 November from the Baikonur Cosmodrome and 2 days later docked with Zvezda, opposite Pirs. It provided the Russian segment of the ISS with an additional airlock for spacewalks and (like Pirs) had a port at the end for Soyuz and Progress vehicles. Although at 4.6 metres long and 2.6 metres in diameter these modules were small, they provided some additional space to accommodate scientific equipment.

Expedition 21 Week ending 29 November 2009

Atlantis arrived at the ISS on 18 November as STS-129, Assembly Flight ULF3, with the first ever ExPRESS Logistics Carrier in its payload bay. This was loaded with two additional gyroscopes, two nitrogen tank assemblies, two pump modules, an ammonia tank assembly, a spare latching end effector for Canadarm2, a spare trailing umbilical system for the Mobile Transporter, and a high-pressure gas tank – all of which were to be unloaded and transferred to storage sites on the ISS during spacewalks.

As De Winne wrote in his diary, "I have seen all three [surviving] Space Shuttles. After a visit of Endeavour in July and Discovery in August, this time it was Atlantis that came to dock with the Space Station and bring some cargo. For the Belgians, the Shuttle Atlantis is especially dear. It was Atlantis that flew the first Belgian astronaut, Dirk Frimount, around the Earth for 9 days in 1992 on a mission that conducted atmospheric observations."

When Atlantis undocked on 25 November it took with it Expedition 21's Nicole Stott, who had not been replaced by anyone from the Shuttle crew. As De Winne explained, "Thus for the last moments of my mission, the ISS had only five crew members instead of the usual six. On my return with Bob and Roman, the rest of the crew, American Jeffrey Williams and Russian Maxim Surayev, will be alone for some weeks before a new trio arrives. We will have to wait until next April until the ISS once again has a full crew of six."

De Winne wrapped up, "I hope that my space mission has been an inspiration for young people in particular, encouraging them towards working in science or technology. They will go even further into space, to the Moon and towards the planet Mars. It is this deep desire to explore that makes us human beings."

Expedition 21 1 December 2009

Russian cosmonaut Roman Romanenko, ESA astronaut Frank De Winne and Canadian Space Agency astronaut Robert Thirsk departed the ISS in Soyuz TMA-15 and landed safely in Kazakhstan at 13.15 local time (08.15 CET) on 1 December.

Simonetta Di Pippo, ESA Director of Human Spaceflight, was in the Mission Control Centre near Moscow at that time and she happily announced, "ESA's Oasiss mission was a total success. We were able to achieve all the goals that we set ourselves in terms of science, space operations, education and outreach. With one of our astronauts aboard we can conduct considerably more scientific experiments and increase our ability to exploit the ISS. Europe should have at least one astronaut on board the ISS every year. This year's two successful missions [Oasiss and Alissé] demonstrate the soundness of our exploitation programme and they prepare the ground for confirmation of Europe's participation in the ISS lifetime extension. I am already looking forward to the next two European missions to the ISS in 2010 and the subsequent missions."

16.4 POSTSCRIPT

Subsequent Missions

At the time of writing, the Oasiss mission was Frank de Winne's last spaceflight.

Frank De Winne Today

See the Odissea mission.

17

Alissé

Mission

ESA Mission Name:	Alissé
Astronaut:	Arne Christer Fuglesang
Mission Duration:	13 days, 20 hours, 53 minutes
Mission Sponsors:	ESA
ISS Milestones:	ISS 17A, 49th crewed mission to the ISS

Launch

Launch Date/Time:	29 August 2009, 03:59 UTC
Launch Site:	Pad 39-A, Kennedy Space Center
Launch Vehicle:	Space Shuttle Discovery (OV-103)
Launch Mission:	STS-128
Launch Vehicle Crew:	Frederick Wilford 'Rick' Sturckow (NASA), CDR
	Kevin Anthony Ford (NASA), PLT
	Patrick Graham Forrester (NASA), MSP1
	Jose Moreno Hernandez (NASA), MSP2
	Arne Christer Fuglesang (ESA), MSP3
	John Daniel 'Danny' Olivas (NASA), MSP4
	Nicole Marie Passonno Stott (NASA), MSP5

Docking

STS-128

Docking Date/Time:	31 August 2009, 00:54 UTC
Undocking Date/Time:	8 September 2009, 19:26 UTC
Docking Port:	PMA-2, Harmony Forward

J. O'Sullivan, *In the Footsteps of Columbus*, Springer Praxis Books,
DOI 10.1007/978-3-319-27562-8_17

Landing

Landing Date/Time:	12 September 2009, 00:53 UTC
Landing Site:	Runway 22, Edwards Airforce Base
Landing Vehicle:	Space Shuttle Discovery (OV-103)
Landing Mission:	STS-128
Landing Vehicle Crew:	Frederick Wilford 'Rick' Sturckow (NASA), CDR Kevin Anthony Ford (NASA), PLT Patrick Graham Forrester (NASA), MSP1 Jose Moreno Hernandez (NASA), MSP2 Arne Christer Fuglesang (ESA), MSP3 John Daniel 'Danny' Olivas (NASA), MSP4 Timothy Lennart Kopra (NASA), MSP5

ISS Expeditions

ISS Expedition:	Expedition 20
ISS Crew:	Gennadi Ivanovich Padalka (RKA), ISS-CDR Michael Reed Barratt (NASA), ISS-Flight Engineer 1 Timothy Lennart Kopra (NASA), ISS-Flight Engineer 2 Replaced by Nicole Marie Passonno Stott (NASA), ISS-Flight Engineer 2 Frank De Winne (ESA), ISS-Flight Engineer 3 Roman Yuriyevich Romanenko (RKA), ISS-Flight Engineer 4 Robert Brent Thirsk (CSA), ISS-Flight Engineer 5

17.1 THE ISS STORY SO FAR

There was only one visiting mission since Frank De Winne arrived on the ISS on Soyuz TMA-15 in May 2009. STS-127, Assembly Flight 2JA, installed the Exposed Facility (EF) and the Experimental Logistics Module-Exposed Section (ELM-ES) to complete the Japanese Kibo laboratory. There was also an Expedition 20 crew exchange in which Timothy Kopra replaced Koichi Wakata.

17.2 CHRISTER FUGLESANG

Early Career

See the Celsius mission.

Previous Mission

See the Celsius mission.

17.3 THE ALISSÉ MISSION

Alissé Mission Patches

The name Alissé was picked out from 190 competition entries, the winner being Jurgen Modlich from Baierbrunn in Germany. So-called 'trade winds' are the prevailing pattern of easterly surface winds in the tropics. On setting sail in 1492, Christopher Columbus let the Alizé wind blow his flotilla of three ships westward across the Atlantic Ocean.[18]

The logo for the Alissé mission featured the wing of a bird enclosing images of the ISS and Space Shuttle, either side of two sets of horizontal lines. The horizontal lines symbolised different aspects of the mission: the Shuttle closing on the ISS during its orbital rendezvous, and also the two spacewalks to be undertaken by Christer Fuglesang. They also represent the two ESA astronauts on board the ISS during the mission. The four individual lines also acknowledged the four participating space agencies. Once again the mission name incorporated the name 'ISS'.

The STS-128 crew with Christer Fuglesang second from the right in the back row. (NASA)

[18] The ships were named *Niña*, *Pinta* and *Santa Maria*.

Alissé Mission Patch. (ESA)

The STS-128 patch showed the Space Shuttle Discovery with its payload bay doors open and the Leonardo Multi-Purpose Logistics Module in its payload bay. Earth and the ISS wrapped around the astronaut symbol to symbolise a continuous human presence in space. Strangely the Station seems to have originated from Florida, despite many of its elements having been launched from the Baikonur Cosmodrome. The names of the Shuttle crewmembers bordered the patch in an unfurled manner and were linked at either side by the US and Swedish flags.

Alissé Mission Objectives

This was to be ESA astronaut Christer Fuglesang's second spaceflight, and the principal focus of the mission was to be his spacewalk activities as an STS-128 Mission Specialist. He was also to perform experiments and various educational and public relations activities for the Alissé mission. He was to participate in two of the three spacewalks scheduled for when Discovery was docked with the ISS to carry out maintenance activities, prepare for the arrival in February 2010 of the European-developed Tranquility node, and recover

STS-128 Mission Patch. (NASA)

external experiments for return to Earth after spending one and a half years on the exterior of the Columbus module. He would also oversee cargo transfers to and from the Multi-Purpose Logistics Module.

The specific objectives were as follows:

- Three EVAs were planned, with Fuglesang participating in the second and third. Overall, the main tasks would be:
  - Run external cabling in preparation for the eventual installation of the Node 3 module named Tranquility.
  - Install a new Ammonia Tank Assembly on the P1 truss segment.
  - Retrieve the European Technology Exposure Facility (EuTEF) and the NASA Materials on the ISS Experiments (MISSE) experiment and stow them in Discovery's payload bay for return to Earth.
  - Install a Rate Gyro Assembly.
  - Install the Payload Attachment System on the S3 truss segment.

- The Leonardo MPLM was to deliver:
  - The European-built Minus Eighty Laboratory Freezer (MELFI-2) to be installed in the Japanese Kibo laboratory.
  - ESA's Materials Science Laboratory facility to be installed in the Destiny laboratory.
  - The Atmospheric Revitalization System for the Tranquility node.
  - A new treadmill exercise device (COLBERT) to be installed in the Harmony node.
  - Miscellaneous experiment facilities and consumables.
  - A new crew quarters (sleeping compartment) to assist in accommodating a permanent six-person ISS crew.
- NASA astronaut Nicole Stott was to replace NASA astronaut Timothy Kopra (who had been delivered by STS-127) as an ISS Expedition 20 flight engineer.
- During his mission Christer Fuglesang would be a subject for a number of ESA and non-ESA experiments in the field of human physiology, and would conduct an educational experiment based on cosmic radiation.

Timeline

Flight Day 1 – Friday, 28 August 2009

Space Shuttle Discovery lifted off from the Kennedy Space Center at 11:59 p.m. EDT (early the next day, Universal Time) for Assembly Flight 17A to deliver supplies and research facilities to the ISS.

After reaching orbit the crew began the usual on-orbit procedures, including opening the payload bay doors, checking out of the robotic arm and surveying the payload bay. The primary payload was the Leonardo MPLM that was to be temporarily attached to the Harmony node for moving items both to and from the ISS, with Fuglesang supervising the transfers.

Speaking after personally watching the launch, Simonetta Di Pippo, ESA's Director for Human Spaceflight explained, "When Christer Fuglesang last flew to the ISS he was welcomed aboard by an ESA astronaut who was completing a 6 month mission there. This time, again, he will join Frank De Winne, who is conducting a long-duration stay in the ISS (and is soon to become its commander). So how better to illustrate the strength of Europe's presence in human spaceflight? We currently have astronauts training for two more ISS increments and one Shuttle flight on an ASI opportunity. At the same time, our next generation of newly selected astronauts is about to start basic training. European astronauts are in space to stay, and we are working hard on a daily basis to create the conditions with which to enhance Europe's role in human spaceflight and exploration also in view of the ISS lifetime extension."

ESA Director General Jean-Jacques Dordain said, "Not only do we have two astronauts in space, working as part of an international crew, but we also have hundreds of scientists and engineers on the ground using the operational science facilities on board. The benefits of the investments made by ESA Member States in the ISS are now a daily reality. We are gathering real science data from experiments conducted inside and outside the Station. Our contribution is an asset for Europe, upon which we will build up future exploration endeavours with our international partners."

Flight Day 2 – Saturday, 29 August 2009

The Discovery crew were woken for their first full day in space by *Back in the Saddle Again* by Gene Autry, which was played for Commander Rick Sturckow.

Ford, Forrester and Hernandez surveyed the orbiter's heat shield while Olivas, Fuglesang and Stott checked out the EVA spacesuits. In addition, the centreline camera was installed and the outer docking ring was extended. It was very routine stuff, which was just the way everyone liked it.

Flight Day 3 – Sunday, 30 August 2009

After it had performed the backflip to allow the ISS residents to photograph its heat shield, Discovery docked at 7:54 p.m. The hatches were opened at 9:33 p.m., and immediately after the brief welcoming ceremony Stott and Kopra exchanged their Soyuz seat liners.

Ford, Kopra, Forrester and Hernandez used Canadarm2 to remove the OBSS from its position on the payload bay sill and handed it off to the Shuttle's own arm to provide additional clearance for the transfer of the MPLM the following day. Meanwhile, Sturckow, Fuglesang and Stott transferred the spacesuits and tools to the Quest airlock in preparation for the first spacewalk.

Flight Day 4 – Monday, 31 August 2009

STS-128's Ford and Expedition 20's Barratt used Canadarm2 to lift the MPLM from the payload bay and attach it to the nadir port of the Harmony node. Then the two ESA astronauts, Fuglesang and De Winne, entered and prepared the pressurised cargo module for the transfer work that was to be carried out over the next six days.

Flight Day 5 – Tuesday, 1 September 2009

During the first of STS-128's spacewalk, Olivas and Stott removed an ammonia tank assembly from the station's truss and Canadarm2 stowed it in the payload bay for return to Earth. Then they went to the Columbus laboratory and retrieved the European Technology Exposure Facility (EuTEF) and the Materials International Space Station Experiment (MISSE).

Meanwhile, Hernandez, Kopra, Fuglesang and Barratt continued to unload the MPLM. One of the new items was a treadmill with the contrived name of Combined Operational Load Bearing External Resistance Treadmill (COLBERT). This was named after the comedian Stephen Colbert of Comedy Central's *The Colbert Report*. He had encouraged his viewers to vote in a NASA online poll to name Node 3. Although 'Colbert' garnered the most votes, NASA opted for the eighth most popular name, Tranquility. As a consolation, NASA astronaut Sunita Williams appeared on the television show and announced that the new treadmill was to be named after the host.

Flight Day 6 – Wednesday, 2 September 2009

This was a day of cargo unloading, as the Fluids Integrated Rack, Materials Science Research Rack-1 and the second Minus Eighty Laboratory Freezer (MELFI-2) were retrieved from the MPLM. Barratt installed the third of an eventual four NASA crew quarters. Placed in the Kibo laboratory, it would be used for the first time by Canadian Space Agency astronaut Thirsk. De Winne gave Stott Canadarm2 training for the Japanese H-II Transfer Vehicle. And Olivas and Hernandez were interviewed by CNN Espanol, Televisa Mexico, and the NBC-affiliated KCRA-TV in California.

Flight Day 7 – Thursday, 3 September 2009

Olivas and Fuglesang conducted the mission's second spacewalk to install the new ammonia tank on the P1 truss segment. They also installed a portable foot restraint on the truss and put covers on some of the lenses of the cameras of Canadarm2 in order to protect them from contamination when the arm was employed to berth the Japanese HTV-1 on 17 September.

While Padalka and Romanenko performed station maintenance in the Russian segment, the other crewmembers continued to unload the MPLM.

Frank De Winne and Christer Fuglesang prepare for the latter's EVA. (NASA)

Flight Day 8 – Friday, 4 September 2009

The Discovery crew took the morning off, while the Expedition crew continued station maintenance. Stott, Thirsk and De Winne studied procedures in preparation for the arrival of HTV-1. By the end of the day, half of the cargo had been unloaded from the MPLM.

Flight Day 9 – Saturday, 5 September 2009

During a busy spacewalk Olivas and Fuglesang began by deploying a payload attachment system on the S3 truss segment and replacing a failed rate gyro assembly. Then they split up. Olivas installed a GPS antenna while Fuglesang replaced a remote power control module, installed an insulation sleeve on a cable inside the truss, and installed a second GPS antenna. They reunited for the final task. This was to route two 60-foot-long cables for avionics systems along the exterior in readiness for berthing the Tranquility node in 2010. Unable to connect these cables in the absence of that module, they wrapped their ends in insulation. At the end of the EVA, the helmet-mounted video camera and headlight system became unlatched from Fuglesang's suit. Olivas assisted him to connect a tether to the apparatus and once they were back inside the airlock she inspected the latches.

Inside the station, Kopra and Thirsk replaced a bolt on the Common Berthing Mechanism in the Unity node while De Winne replaced a filter on the oxygen generating system.

Flight Day 10 – Sunday, 6 September 2009

Filling the MPLM with unwanted cargo for return to Earth was the main task for the Discovery crew while the station residents focused on maintenance.

In addition to Kopra and Olivas interviewing for KFOX-TV, KXAN-TV and the *Military Times*, toward the end of the crew day Fuglesang and De Winne met in the Columbus module for a 'special' on NASA TV that featured former ESA astronaut Jean-François Clervoy, Lotta Bouvain of Swedish television, Swedish Minister for Education Jan Björklund, the American-Finnish journalist and talk show host Mark Levengood, and Swedish opera singer Malena Ernman.

Flight Day 11 – Monday, 7 September 2009

With all the station's trash loaded into Leonardo, this was sealed and then Ford and Hernandez used Canadarm2 to transfer the MPLM back to the payload bay of the orbiter. Hernandez and Stott were interviewed by Telemundo, WTSP-TV in Florida, and Univision.

After the departure ceremony the hatches were closed at 9:30 p.m., with Stott on the ISS side and Kopra on the Shuttle side.

Flight Day 12 – Tuesday, 8 September 2009

After undocking, Ford flew Discovery around the ISS to enable the residents to provide video to the engineers on Earth to assist in checking for damage to the orbiter's exterior.

Flight Day 13 – Wednesday, 9 September 2009

The Discovery crew devoted the day to preparing their spacecraft for re-entry and ensuring its flight control systems and thrusters were operational. The final media interviews were conducted with CBS News, ABC News, and CNN. Once those were complete, the Ku-band antenna was stowed.

Flight Day 14 – Thursday, 10 September 2009

Unstable weather ruled out attempting a landing at the Kennedy Space Center in Florida, so Edwards AFB in California was activated.

Flight Day 15 – Friday, 11 September 2009

After mission managers called off the landing opportunities in Florida owing to thunderstorms and otherwise unstable weather conditions, STS-128 and its seven astronauts landed at Edwards at 7:53 p.m. CDT.

17.4 POSTSCRIPT

Subsequent Missions

At the time of writing, the Alissé mission was Christer Fuglesang's last spaceflight.

Christer Fuglesang Today

See the Celsius mission.

18

Magisstra

Mission

ESA Mission Name:	Magisstra
Astronaut:	Paolo Angelo Nespoli
Mission Duration:	159 days, 8 hours, 17 minutes
Mission Sponsors:	ESA
ISS Milestones:	ISS 25S, 59th crewed mission to the ISS

Launch

Launch Date/Time:	15 December 2010, 19:09 UTC
Launch Site:	Pad 1, Baikonur Cosmodrome, Kazakhstan
Launch Vehicle:	Soyuz TMA
Launch Mission:	Soyuz TMA-20
Launch Vehicle Crew:	Dimitri Yuriyevich Kondratiyev (RKA), CDR
	Paolo Angelo Nespoli (ESA), Flight Engineer
	Catherine Grace 'Cady' Coleman (NASA) Flight Engineer

Docking

Soyuz TMA-20

Docking Date/Time:	17 December 2010, 20:11 UTC
Undocking Date/Time:	23 May 2011, 22:45 UTC
Docking Port:	Rassvet

J. O'Sullivan, *In the Footsteps of Columbus*, Springer Praxis Books,
DOI 10.1007/978-3-319-27562-8_18

Landing

Landing Date/Time:	24 May 2011, 02:26 UTC
Landing Site:	near Dzhezkazgan, Kazakhstan
Landing Vehicle:	Soyuz TMA
Landing Mission:	Soyuz TMA-20
Landing Vehicle Crew:	Dimitri Yuriyevich Kondratiyev (RKA), CDR Paolo Angelo Nespoli (ESA), Flight Engineer Catherine Grace 'Cady' Coleman (NASA) Flight Engineer

ISS Expeditions

ISS Expedition:	Expedition 26
ISS Crew:	Scott Joseph Kelly (NASA), ISS-CDR Alexandr Yuriyevich Kaleri (RKA), ISS-Flight Engineer 1 Oleg Ivanovich Skripochka (RKA), ISS-Flight Engineer 2 Dimitri Yuriyevich Kondratiyev (RKA), ISS-Flight Engineer 3 Catherine Grace 'Cady' Coleman (NASA) ISS-Flight Engineer 4 Paolo Angelo Nespoli (ESA), ISS-Flight Engineer 5
ISS Expedition:	Expedition 27
ISS Crew:	Dimitri Yuriyevich Kondratiyev (RKA), ISS-CDR Catherine Grace 'Cady' Coleman (NASA) ISS-Flight Engineer 1 Paolo Angelo Nespoli (ESA), ISS-Flight Engineer 2 Andréi Ivanovich Borisenko (RKA), ISS-Flight Engineer 3 Aleksandr Mikhailovich Samokutyayev (RKA), ISS-Flight Engineer 4 Ronald John Garan Jr. (NASA), ISS-Flight Engineer 5

18.1 THE ISS STORY SO FAR

The volume and mass of the ISS had increased substantially since Christer Fuglesang's Allisé mission in August-September 2009.

In early October Soyuz TMA-16 arrived with the last 'space tourist' for the foreseeable future. Guy Laliberté, Canadian founder and CEO of Cirque de Soleil, travelled to the ISS with Expedition 21/22 crewmembers Maxim Surayev and Jeffrey Williams. With larger and overlapping station crews, along with contractual commitments to ESA, JAXA and CSA astronauts, there would no longer be room for space tourists on Soyuz missions.

November saw the arrival of a new Russian component. Launched by a Soyuz rocket and docked automatically to the zenith port of the Zvezda habitat module, the Poisk module, which was similar to Pirs, provided another airlock and docking capability.

STS-129 delivered two external storage units to the ISS in November. These ExPRESS Logistics Carriers (ELC) had an acronym within an acronym because ExPRESS meant Expedite the Processing of Experiments to the Space Station. ELC-1 was mounted on the

P3 truss and ELC-2 was mounted on the S3 truss. In addition to accommodating spare parts, these carriers had power and communications to run experiments.

Soyuz TMA-17 lifted off in December (the first Soyuz to be launched in that month for 19 years) to deliver a trio of flight engineers to the Expedition 22 crew. They went on to become the Expedition 23 crew when their colleagues returned to Earth in March 2010.

STS-130 delivered Node 3, named Tranquility, as part of Assembly Flight 20A. This module was built in Italy for NASA and was a development of the MPLM, two of which had been carrying cargo to and from the station since 2001. It arrived with the Cupola module installed at one end. Once the new node had been berthed at the port side of Unity, the Cupola was relocated to the nadir of Unity. The Cupola had seven windows that were protected by hinged shutters. At 80cm in diameter, the central circular pane was the largest yet flown in space. It would provide spectacular views of Earth passing below.

Soyuz TMA-18 launched in April 2010 to deliver three flight engineers to the Expedition 23 crew. They went on to become the Expedition 24 crew when their colleagues returned to Earth in June 2010.

Although listed as Assembly Flight 19A, STS-131 in April 2010 was a resupply mission, carrying MPLM Leonardo on its final round trip. When this particular module was next launched into space on STS-133, it would be as the modified and renamed the Permanent Multi-purpose Module (PMM). As the Space Shuttle Program wound down, the capacity to return large amounts of cargo to Earth would cease. Hence, in addition to delivering important supplies and spare parts to the ISS, this mission was to return bulky items of junk to Earth.

Whilst STS-132 in May 2010 might appear to have been merely a mission to deliver the Mini Research Module 1 (MRM-1) to expand the Russian segment of the ISS, this task masked layers of political, commercial and technical complications. Originally the Russians had planned to launch the large Docking and Storage Module (DSM) on a Proton rocket to reside at the nadir of the similarly-sized Zarya module, but this was cancelled due to financial difficulties. It was decided to produce a smaller module which NASA would deliver and Canadarm2 would install at the nadir port on Zarya. Rassvet would have a docking port on its end to preserve the station's capacity to accommodate the two Soyuz 'taxis' and two Progress freighters needed to sustain a crew of six. But there was more to this deal. The Russian Multipurpose Laboratory Module (MLM) is due to be launched on a Proton in 2017 (at the time of writing) and NASA were already committed to launch supporting equipment for this, so it was decided to launch this equipment with Rassvet, mounted internally and externally. As a result the MLM airlock, the MLM folded radiator and the elbow of the European Robotic Arm (ERA) were piggy-backed on Rassvet in preparation for the arrival of the MLM.

Soyuz TMA-19 launched in June 2010 to deliver three flight engineers to the Expedition 24 crew. They went on to become the Expedition 25 crew when their colleagues returned to Earth in September 2010.

Soyuz TMA-1M was the first of a new model of the spacecraft which was first introduced in 1967. The updates were internal and mainly to the avionics and docking systems. Enhancing the glass cockpit of the TMA variant, the TMA-M was 70 kg lighter, could be flown by one trained cosmonaut instead of two (thereby saving on training) and could

The ISS as seen from the departing STS-135 showing four Russian spacecraft docked: two Soyuz and two Progress. One Soyuz is docked at the Rassvet module. (NASA)

make a 6 hour 'fast rendezvous' (compared to the previous 48 hours). Soyuz TMA-1M delivered three flight engineers to Expedition 25. They went on to become the Expedition 26 crew when their colleagues returned to Earth in November 2010.

18.2 PAOLO NESPOLI

Early Career

See the Esperia mission.

Previous Missions

See the Esperia mission.

18.3 THE MAGISSTRA MISSION

Magisstra Mission Patches

The name Magisstra was chosen as part of an ESA competition that was won by Antonella Pezzani of Italy. Magister or Magistra means 'teacher' in Latin, so it represented the educational aspect of Paolo Nespoli's mission. In addition it means 'master' or 'expert', so it also indicated the very high level of astronaut competence demanded for each mission. Of course, as was becoming traditional for the ESA missions the spelling was altered to accommodate the name 'ISS'.

The logo for the Magisstra mission showed a human embracing a branch with leaves, the cogs of a machine, and a book, which together represented the elements of the mission: science, technology and education. The sunrise over the crescent Earth conveyed optimism about extending the operational life of the ISS to at least 2020.

The Soyuz TMA-20 crew with Paolo Nespoli on the right. (www.spacefacts.de)

The 'ISS/E3-3D/NESPOLI' banner was how the mission was identified by the Multilateral Crew Operations Panel, the senior forum for coordination and resolution of crew matters. 'E3' stood for the third long-duration European mission to the ISS and '3D' referred to the ESA-developed 3D camera with which Nespoli was to take unprecedented views of the station. The six stars represented both the number of crew members and the number of months that Nespoli was to be in space.

The Expedition 26 crew with Nespoli second from the right in the back row. (NASA)

The Expedition 27 crew with Nespoli on the left in the front row. (NASA)

Magisstra Mission Patch. (ESA)

The design of the Soyuz TMA-20 patch began with a drawing by Marina Korolenko of Murmansk which showed the constellation Ursa Major and the North Star. Dutch artist Luc van den Abeelen then incorporated this into a design that portrayed a Soyuz spacecraft flying between the North Star and the Southern Cross.

The Expedition 27 patch also included the Southern Cross. According to Dmitri Kondratyev, "The manned space programme of Russia doesn't stop in near Earth orbit. It is directed toward the future of humanity, which will master the distant boundaries of space up to the constellations of Ursa Major and Southern Cross, and the centuries-old lighthouse of humanity the Pole Star. The constellations of Ursa Major and Southern Cross occur in different hemispheres of the Earth. Despite the fact that Earth itself does not make it possible to simultaneously see constellations in both hemispheres … let all children everywhere see these constellations and feel the need to be discoverers of the new expanses of space. We believe this is the future of humanity as the united population of the planet Earth after mastery of the expanses of boundless space."

Soyuz TMA-20 Mission Patch. (www.spacefacts.de)

The ISS was prominently displayed in the foreground of the Expedition 26 patch to acknowledge the efforts of the entire team of astronauts who built and operated it, and the scientists, engineers and support personnel on the ground who provided the foundation for each successful mission. All their accomplishments were demonstrating the station's capabilities as a technology test bed and a science laboratory – and as a path to the exploration of our solar system and ultimately beyond. The ISS was depicted with the ESA Automated Transfer Vehicle named 'Johannes Kepler' docked to resupply it. The patch acknowledged the teamwork among the international partners USA, Russia, Japan, Canada, and ESA, plus the ongoing commitment to build, improve and use the ISS. Prominently displayed in the background was Earth as the focus of much of the research aboard the outpost in space. The two stars symbolised the Soyuz spacecraft, each delivering three people to work and live together as ISS Expedition 26. Their names were included and the patch was framed with the flags of USA, Russia, and Italy.

The Expedition 27 patch prominently showed the ISS orbiting Earth as an exercise in cooperation between the USA, Russia, Japan, Canada and the European Space Agency on

Expedition 26 Mission Patch. (NASA)

its mission for science, technology and education. The station was depicted in its completed state, with the latest addition of the Alpha Magnetic Spectrometer on the integrated truss structure and a number of docked vehicles. The Southern Cross constellation was also shown in the foreground, and its five stars, along with the Sun, symbolised the six international crewmembers on the station. Although it is one of the smallest modern constellations, the Southern Cross is one of the most distinctive and has cultural significance around the world. Its inclusion on the patch represented the manner in which it could inspire teams to push the boundaries both in space and on the ground.

Magisstra Mission Objectives

On 15 December 2010 ESA astronaut Paolo Nespoli was launched on Soyuz TMA-20 to undertake a long-duration mission aboard the ISS, serving as a flight engineer for Expeditions 26 and 27 with an intensive programme of experiments ranging from radiation monitoring to measurements that could improve oil recovery in petroleum reservoirs.

Expedition 27 Mission Patch. (NASA)

This would make him the fourth European astronaut to join a station crew and the third to serve a six month tour.

The mission objectives were as follows:

- Nespoli's duties aboard the ISS would include participating in the docking of two cargo ships, namely the second Automated Transfer Vehicle (ATV) and the second Japanese H-II Transfer Vehicle (HTV). He would be the prime operator for berthing the HTV.
- In May, Nespoli was to play host to an Italian crewmate for five days, because ESA astronaut Roberto Vittori was to fly on board STS-134 under a bilateral contract between NASA and the Italian Space Agency (ASI).
- Nespoli had been assigned more than 30 experiments. European space science was driven by a desire to improve life on our home planet. The scientific programme for this mission spanned human adaptation to the space environment, fluid physics, radiation, biology, and a variety of technology demonstrations. In addition, he was to carry out experiments for the US, Japanese and Canadian space agencies.

- As part of Nespoli's educational programme, children would be given the chance to follow an international initiative built around health, well-being and nutrition (Mission X: Train Like an Astronaut), as well as a special greenhouse activity in space. In addition, he was to use ESA's novel 3D camera to provide unprecedented images of the ISS.

Science

As part of the European scientific programme, Nespoli worked on five different research fields:

- **Human Research** – He was to serve as a test subject for human physiology experiments that addressed neuroscience, cardiovascular, metabolism, and fitness evaluation. In particular, he would undergo experiments designed to test how astronauts interpret visual information in a state of weightlessness and how this affects their perception. Measuring different parameters, ground-based European scientists were to study space data for how weightlessness alters the cardiovascular system, bone metabolism, and even brain functions such as memorisation or decision-making.
- **Fluid Physics** – The physical science experiments addressed diffusion phenomena in model fluid mixtures in order to understand the properties of oil fields, with possible applications in improving the recovery of oil from petroleum reservoirs. A second experiment was aimed at simulating geophysical fluid flows in microgravity in the hope of gaining clues about Earth's magma convection and the global-scale flows in the liquid cores of other planets.
- **Radiation** – The interactions between ionising radiation in space and human brain functions were a major concern in planning long missions in space. One example of this phenomenon is the 'light flashes' first reported by Buzz Aldrin during the Apollo 11 lunar mission in 1969. A multi-disciplinary research project had been devised to study this and other effects. Paolo was also to measure the nature and distribution of the radiation field inside the ISS, and the Sun's radiation was already being studied in detail by the SOLAR experiment on the exterior of the European Columbus laboratory.
- **Biology** – Nespoli would also be the main operator for several complex biology experiments which addressed specific paradigms, such as the immune response in plants during growth in microgravity.
- **Technology Demonstrations** – Nespoli was to help to assemble a new mechanism in order to demonstrate the space-based capability to identify maritime vessels. The Vessel Identification System was to test a way of tracking global maritime traffic from space by picking up signals from large international ships and all types of passenger carriers.

Nespoli was also to perform experiments on behalf of the American, Japanese and Canadian space agencies. Samples of his hair were collected to study the gene expression in a human body which had been exposed to a long period of weightlessness. Also state-of the-art instruments were to monitor his sleep patterns to gain insight into how to treat insomnia on Earth. Flames behave differently in space, forming tiny and almost invisible balls. Flame behaviour and combustion experiments were among his tasks, as was the study of a very resilient and lightweight material being stretched to very thin fibres in microgravity.

3D Experience Erasmus Recording Binocular 2

Nespoli was to use a novel ESA-developed 3D camera to show unprecedented stereoscopic views of the ISS. This Erasmus Recording Binocular 2 (ERB-2) was conceived by the Directorate of Human Spaceflight and developed by Cosine Research of Leiden in the Netherlands, and by Techno System Development in Naples, Italy. It exploited high-definition optics and advanced electronics to yield a vastly improved 3D video effect. In addition to enhancing our view of the ISS these almost-like-real images would support science operations.

Education

Nespoli was to share his experiences of being in space with people all across the planet. In particular, he was to promote two exciting educational activities:

- *Mission-X: Train Like an Astronaut* – Astronauts need to stay fit on the ground and in space, and Paolo undertook years of preparation and physical training prior to going to the ISS. ESA and several space organisations were using the example of space explorers to promote regular exercise and healthy nutrition in young people worldwide.
- *Greenhouse in Space* – Schoolchildren of 12–14 years were invited to join the Greenhouse in Space activity. This would involve attempting to grow both a flowering plant and a lettuce in space and monitoring their germination and growth cycles over a period of 2 months. Students on the ground would run the same experiment in parallel and compare their results with those obtained in space.

Preparation

On Thursday, 25 November Nespoli and his two Soyuz TMA-20 crewmates Dmitri Kondratyev and Catherine Coleman passed the Soyuz exams with flying colours, achieving 4.8 out of 5 for the entire crew when dealing with five simulated emergencies. It was always a worrying time for a prime crew because if they were to fail the test, the opportunity to fly the mission would pass to the backup crew. They spent the days leading up to the tests reading manuals and rehearsing the procedures, to ensure that they would know exactly what to do in every situation and which buttons to press when needed. Nespoli received perfect scores for rendezvous manoeuvring, station flyby and docking, and manual descent.

On Friday, the prime crew were examined in the ISS simulator while their backups received the Soyuz tests.

Timeline

Expedition 26 Week ending 19 December 2010

Kondratyev, Coleman and Nespoli launched on Soyuz TMA-20 at 20:09:25 CET on 15 December to rendezvous with the ISS and assume their roles as flight engineers of Expedition 26.

Their descent module was originally intended for Soyuz TMA-21, but when the module built for Soyuz TMA-20 was damaged in transit it had been replaced with just a two day delay in the delivery date.

ESA's Director of Human Spaceflight Simonetta Di Pippo said of the launch, "Ten years from the first expedition to the ISS, Paolo's mission marks the beginning of a new decade for this permanently inhabited space laboratory. With the great accomplishments in the last two years, its assembly is now virtually over. Paolo and his crewmates will inaugurate a 'Research Decade' on the ISS. I wish Paolo and his fellow explorers a safe trip aboard the Soyuz to reach the ISS and I am looking forward to their docking."

After a brief loss of communications due to a cable break between Mission Control in Moscow and other ground sites which didn't endanger the mission in any way, Soyuz TMA-20 docked at the port at the end of the Rassvet module at 21:12 CET on 17 December.

After the hatches were opened, Simonetta Di Pippo said, "It is always a strong feeling to see them entering the ISS and greet the rest of the crew. A very dense mission awaits Paolo and his crewmates with resupply missions, in particular the second ATV, named 'Johannes Kepler', a major scientific experiment deployment – AMS – and numerous scientific experiments. I am proud of the level of integration and importance Europe has reached in the ISS partnership, and the regular presence of European astronauts on the Station. I wish Paolo and the whole crew a pleasant stay and a fruitful mission."

Expedition 26 Week ending 26 December 2010

On 21 December, Nespoli and Coleman started the Neurospat experiment which was to record the response of the brain to weightlessness by using a laptop, a visor to focus on the computer without external visual distractions, and an electroencephalograph to record brain activity.

Expedition 26 Week ending 9 January 2011

On 6 January 2011 Nespoli had an interview with the *I fatti vostri* programme on Italian TV channel RAI2. He spoke with the presenter Giancarlo Magalli and Massimo Sabbatini, ESA Head of Erasmus and Human Space Flight Communication. After saying how he had been too busy to photograph the solar eclipse of 4 January, he explained how he had no fear of putting on weight, "Up here we don't usually get fat, also because we have more than 2 hours daily of physical exercises."

Expedition 26 Week ending 16 January 2011

On 13 January Paolo had a discussion with Tim Willcox, the presenter of the BBC World Service's show *Impact*, Simonetta Di Pippo and Tim Peake, Britain's first ESA astronaut and a member of the new 2009 class of astronauts, about the forthcoming arrivals of Japan's HTV-2 and ESA's ATV-2.

Expedition 26 Week ending 30 January 2011

Dmitri Kondratiyev and Oleg Skripochka performed a spacewalk on 21 January to install a high-speed data transmission system, remove an old plasma pulse experiment, install a camera for the new Rassvet docking module, and retrieve a materials exposure package.

Meanwhile, after trash had been loaded into Progress M-08M it undocked from Pirs on 24 January and de-orbited itself.

When HTV-2 came on 27 January, it was gripped by Canadarm2 and berthed at the Harmony nadir port. The HTV had two cargo sections, the Pressurised Logistics Carrier (PLC) and the Unpressurised Logistics Carrier (UPLC). Like the other vehicles, the PLC was accessible through a hatch. The cargo in the UPLC was to be unloaded by the Japanese robot arm and transferred to the external platform of the Kibo laboratory.

To conclude a busy week, Progress M-09M docked at the Pirs on 30 January with the usual stores for the station.

Expedition 26 Week ending 20 February 2011

The Greenhouse in Space project was initiated on 17 February by 750 children in the age range 12 to 14 years who gathered at the European Astronaut Centre in Cologne, Germany; at ESRIN in Frascati, Italy; at Cité de l'Espace in Toulouse, France; and at Ciência Viva, Agência Nacional para a Cultura Científica e Tecnológica in Lisbon, Portugal. From the ISS, Paolo explained the greenhouse in which he would grow Thale Cress in space while the students grew the same plants at their schools.

At 21:50 GMT on 16 February the second Automated Transfer Vehicle lifted off from Kourou in French Guiana. Some 64 minutes later the unmanned cargo ship was coasting in orbit ready to deploy the solar wings that gave it the distinctive 'X-wing' shape.

ESA Director General Jean-Jacques Dordain reported, "This launch takes place in a crowded and changing manifest for the ISS access – with HTV, Progress, ATV, and Shuttles coming and going. In October last year we fixed the ATV launch schedule with our international partners. We were able to keep to that schedule thanks to the expertise and dedication of European industry, Arianespace, ESA, and CNES and of our international partners. ATV-2 is the first of a production of four vehicles and is the result of technical expertise and political support from Member States of ESA and to international cooperation. We are now looking for the docking to ISS to declare success."

Simonetta Di Pippo added, "ATV 'Johannes Kepler' is inaugurating our regular service line to the ISS. Integration for the next vehicle, named 'Edoardo Amaldi', will be finished in Europe in August, and production is under way for ATV-4 and ATV-5."

The plan was for a total of five such spacecraft to fly to the station between 2008 and 2015, each carrying as much as 7,667 kg of cargo, which was more than three times the 2,230 kg payload of the Progress freighter.

Meanwhile, Kondratiyev and Skripochka performed a second EVA on 16 February to install two experiments relating to predicting earthquakes and gamma detectors for use during lightning storms. After retrieving a long-duration exposure experiment they released a pair of nanosats.

On 18 February, Canadarm2 moved HTV-2 from the nadir port of Harmony to the zenith port in preparation for the arrival of Shuttle Discovery. And on 20 February Progress M-07M departed from the aft port of Zvezda to clear the way for ATV-2.

Expedition 26 Week ending 27 February 2011

The automated docking of ATV-2 at the aft port of the Zvezda module at 16:59 CET (15:59 GMT) on 24 February 2011 was monitored by the ATV Control Centre in Toulouse, France.

Simonetta di Pippo announced, "With this smooth docking, 'Johannes Kepler' proves to be a great example of the wave of innovation 'Made in Europe'. We are more ready than ever to head into an era of autonomy in space exploration. Thanks to its flexibility, we can think of a wide range of new space vehicles. For example, the ATV could evolve into a future re-entry spacecraft to support future orbital infrastructures and exploration missions, transporting people and supplies into lunar orbit. This is very important for us and for all our partners in the ISS programme since, after the withdrawal of the Space Shuttle, the ATV will be the largest servicing vehicle left to support the ISS and it is our responsibility to deliver a proper service."

As well as delivering propellant, oxygen, and dry cargo in its pressurised bay, the ATV would be able to boost the orbit of the station.

On 26 February STS-133 docked with PMA-2 on the front of the Harmony node. It was the eighth and final trip for Leonardo, because the renamed Permanent Multi-purpose Module (PMM) was to be installed at the nadir port of the Unity node. With the Shuttle Program now winding down, the PMM carried vital spare parts to keep the station operating for many years. It had 14 racks, a spare common air assembly, a spare coolant fluid circulating pump, and a spare water storage tank. It also delivered a robot with a head, torso and arms that mimicked the human shape and was known as Robonaut-2.

STS-133 saw Steve Bowen become the first (and only) astronaut ever to fly back-to-back Shuttle missions by virtue of replacing Tim Kopra who had suffered a bicycle injury a month earlier. Another milestone was that during the mission, all types of ISS visiting spacecraft were docked to the station: an American Shuttle, two Russian Soyuz 'taxis', a Russian Progress, a European ATV and a Japanese HTV. As the maiden flight of a new spacecraft, a flyaround by Soyuz TMA-1M to take a photograph of the ISS with the Shuttle in place had been deemed too risky. Discovery undocked on 7 March, and when it landed at the Kennedy Space Center two days later it was retired from service.

On 27 February, Paolo undertook a major repair job when he installed a new water valve manifold in Columbus; this having been delivered by STS-133. The thermal system employed multiple valves to control the cold water flow from Columbus to Harmony, where heat exchangers cooled the water with radiators. The cooled water would then flow back into Columbus. In September 2010 the No. 8 Water ON-OFF Valve (WOOV) failed after months of suspect behaviour.

"The main repair included a procedure of some 80 steps," said Cesare Capararo of ALTEC, Lead Increment Flight Director for Columbus and the person responsible for overseeing the repair. "Each step had to be carefully checked before, during and after execution by Paolo and the ground team due to the complexity of performing maintenance on a running system."

Expedition 27 Week ending 20 March 2011

When Kelly, Kaleri and Skripochka departed for Earth in Soyuz TMA-1M on 16 March, Kondratyev became commander of the ISS. The station would be inhabited by three people until the arrival of the next Soyuz in April.

When ground scientists found fungus growing in the Arabidopsis cress greenhouse. It was decided to terminate the experiment and dispose of the greenhouse in order to prevent fungal contamination of the station. Nespoli explained, "Part of the experiment was indeed a success. We were indeed able to grow the plants and observe them. This is a lesson we leave to future astronauts, who I'm sure will be even better at it."

On 18 March ATV-2 at the rear of the Zvezda module boosted the station's orbit. In the previous week the Japanese HTV-2 was transferred back to the Harmony nadir port in preparation for its final unberthing on 20 March and de-orbiting on 30 March.

Expedition 27 Week ending 3 April 2011

On 30 March, Nespoli had a video link with 85 German students who had gathered at the European Astronaut Centre in Cologne to conclude the 8 week Mission X: Train Like an Astronaut challenge which promoted healthy nutrition and regular exercise.

Expedition 27 Week ending 10 April 2011

On 6 April, which was Nespoli's 54th birthday, Soyuz TMA-21 docked at the Poisk module with Aleksandr Samokutyayev, Andréi Borisenko and Ronald Garan, turning the existing three-person Expedition 26 into the six-person Expedition 27.

There was a double celebration on 12 April to mark both the 50th anniversary of Yuri Gagarin's historic first human spaceflight and the 30th anniversary of the first Space Shuttle mission.

Expedition 27 Week ending 1 May 2011

On 29 April, Progress M-10M docked at the Pirs module which had hosted its predecessor until that had departed one week earlier.

Expedition 27 Week ending 8 May 2011

Paolo Nespoli's 78-year-old mother, Maria Motta, died on 3 May. In line with the preferences that he had expressed prior to the mission for such an event, he was notified immediately in a private call. In addition to expressing their personal condolences, Jean-Jacques

Dordain, ESA Director General, and Thomas Reiter, former astronaut and ESA's Director for Human Spaceflight and Operations, assured Nespoli of the agency's support during this difficult time.

"We all are in our minds very close to Paolo and his family," said Michel Tognini, former ESA astronaut and Head of the European Astronaut Centre, "and we try to give him sufficient strength in space to cope with this difficult situation and overcome a severe personal loss which is already very difficult in normal conditions, but for Paolo, being still on the International Space Station, it is even harder. There are probably no words to relieve this sadness but our thoughts are with him, his family and his friends."

The funeral was held on 4 May in Verano Brianza, near Milan, and half an hour later, as the ISS flew over Italy, the crew stopped work and observed a minute's radio silence.

Thomas Reiter represented ESA at the funeral and Italian ESA astronaut Samantha Cristoforetti delivered the condolences of the European astronaut corps.

Expedition 27 Week ending 22 May 2011

On 18 May, Endeavour docked at PMA-2 to deliver the Alpha Magnetic Spectrometer (AMS-2) and Italian ESA astronaut Robert Vittori. See the next chapter for an account of Vittori's DAMA mission on board the ISS.

On 21 May the joint ISS/Shuttle crew of 12 people conducted a video link to the Foconi Room of the Vatican Library to speak with Pope Benedict XVI.

Thomas Reiter, ESA's Director for Human Spaceflight and Operations, handed the microphone to Enrico Saggese, President of the Italian Space Agency, who in turn introduced Dmitri Kondratyev the current ISS commander. The Pope talked of the importance of science, exploration and protecting our environment. Vittori had been given a coin by the Pope that he carried to space. He gave it to Nespoli to return to Earth. Finally, the Pope expressed his condolences to Nespoli in Italian for the recent loss of his mother.

Expedition 27 Week ending 29 May 2011

Italian President Giorgio Napolitano spoke with Nespoli and Vittori on 23 May, prior to the former's return to Earth on Soyuz TMA-20 on 24 May.

The Story behind the famous photos

The departure of Soyuz TMA-20 did not follow standard procedure. It had been decided to exploit an opportunity to photograph the station with the Space Shuttle Endeavour, the ATV 'Johannes Kepler', Soyuz TMA-21, and Progress M10-M docked, a situation that would not occur again. Unfortunately, the HTV had already departed.[19]

---

[19] In a similar situation, a Soyuz crew had taken a picture of Atlantis docked at the Mir space station in 1995.

After undocking, Kondratyev flew the Soyuz to a safe distance in order to enable Nespoli to take a number of pictures of the ISS, which had been tilted at 130 degrees in order to get the optimal lighting conditions.

"This was a complex and delicate manoeuvre that could have caused serious problems if it wasn't executed properly, but I felt it was worth the risk," explained Nespoli. "Taking these pictures was not as straightforward as aiming the camera and shooting. When we undocked we were already strapped in our seats wearing spacesuits. Our suits and the hatches had been leak-checked, and normally the seals aren't broken after undocking because retesting them costs oxygen – and there is not so much of it on board."

The ISS with Endeavour and the ATV-2 'Johannes Kepler' docked at opposite ends of the station. This photograph was taken by Paolo Nespoli aboard Soyuz TMA-20. (ESA)

The window where the pictures needed to be taken was in the orbital module, already then partly depressurised. So Nespoli removed his gloves, unstrapped from his seat, floated to the internal hatch, repressurised the other module and went in.

"I had to slide over Dmitry paying attention not to hit the manual controls, and go into the orbital module where I had prepared the cameras before hatch closure." Paying attention to the composition, checking the position of the Earth and for any internal reflections on the window, he alternated stills and video. "I really prayed that these would be good, since I was conscious of their value. But what was done was done – and I promptly forgot them when I had to concentrate on redoing correctly the hatch and suit leak checks and pick up the re-entry and landing procedures."

It was more than just an impromptu task on the flight plan. "Since I was a kid, photography had been a hobby dear to me and all through my life photography had brought me to unusual places and made me live unexpected experiences. Just like a photographer who has a gorgeous model in front of him, I was more concentrated on getting a good technical and artistic product than admiring it. I saw the view [through the tiny window] when changing from still to video images but I purposely limited looking because I knew that I would have been mesmerised by the beauty of it. Dreams are possible. We all should keep dreaming, as even the most impossible dream sometimes can become a reality."

18.4 POSTSCRIPT

Subsequent Missions

Paolo Nespoli is scheduled to fly to the ISS in May 2017 as a member of Expeditions 52/53.

Paolo Nespoli Today

See the Esperia mission.

19

DAMA

Mission

ESA Mission Name:	DAMA
Astronaut:	Roberto Vittori
Mission Duration:	15 days, 17 hours, 38 minutes
Mission Sponsors:	ESA/ASI
ISS Milestones:	ISS ULF6, 62nd crewed mission to the ISS

Launch

Launch Date/Time:	16 May 2011, 12:56 UTC
Launch Site:	Pad 39-A, Kennedy Space Center
Launch Vehicle:	Space Shuttle Endeavour (OV-105)
Launch Mission:	STS-134
Launch Vehicle Crew:	Mark Edward Kelly (NASA), CDR
	Gregory Harold 'Box' Johnson (NASA), PLT
	Edward Michael 'Mike' Fincke (NASA), MSP1
	Gregory Errol Chamitoff (NASA), MSP2
	Andréw Jay 'Drew' Feustel (NASA), MSP3
	Roberto Vittori (ESA), MSP4

Docking

STS-134

Docking Date/Time:	18 May 2011, 10:14 UTC
Undocking Date/Time:	30 May 2011, 03:55 UTC
Docking Port:	PMA-2, Harmony Forward

J. O'Sullivan, *In the Footsteps of Columbus*, Springer Praxis Books,
DOI 10.1007/978-3-319-27562-8_19

Landing

Landing Date/Time:	1 June 2011, 06:35 UTC
Landing Site:	Runway 15, Shuttle Landing Facility, Kennedy Space Center
Landing Vehicle:	Space Shuttle Discovery
Landing Mission:	STS-134
Landing Vehicle Crew:	Mark Edward Kelly (NASA), CDR Gregory Harold 'Box' Johnson (NASA), PLT Edward Michael 'Mike' Fincke (NASA), MSP1 Gregory Errol Chamitoff (NASA), MSP2 Andréw Jay 'Drew' Feustel (NASA), MSP3 Roberto Vittori (ESA), MSP4

ISS Expeditions

ISS Expedition:	Expedition 27
ISS Crew:	Dimitri Yuriyevich Kondratiyev (RKA), ISS-CDR Catherine Grace 'Cady' Coleman (NASA) ISS-Flight Engineer 1 Paolo Angelo Nespoli (ESA), ISS-Flight Engineer 2 Andréi Ivanovich Borisenko (RKA), ISS-Flight Engineer 3 Aleksandr Mikhailovich Samokutyayev (RKA), ISS-Flight Engineer 4 Ronald John Garan Jr.(NASA), ISS-Flight Engineer 5
ISS Expedition:	Expedition 28
ISS Crew:	Andréi Ivanovich Borisenko (RKA), ISS-CDR Aleksandr Mikhailovich Samokutyayev (RKA), ISS-Flight Engineer 1 Ronald John Garan Jr.(NASA), ISS-Flight Engineer 2

19.1 THE ISS STORY SO FAR

Since Paolo Nespoli began his Magisstra mission in December 2010, STS-133 and Soyuz TMA-21 had visited the ISS (see the Magisstra mission).

19.2 ROBERTO VITTORI

Early Career

See the Marco Polo mission.

Previous Mission

See the Marco Polo and Eneide missions.

19.3 THE DAMA MISSION

DAMA Mission Patches

Italian elementary and middle school students were invited to suggest a name and a logo for Roberto Vittori's visit to the ISS aboard Endeavour on the STS-134 mission, with more than 2,000 responses from around 200 schools being received.

In addition to enabling Vittori to achieve his third trip to the ISS, this mission was to deliver the Alpha Magnetic Spectrometer. This fundamental science instrument gave the hint to a young pupil to use the initial letters of 'dark matter' to form the winning acronym DAMA. Following the same idea, an 11-year-old girl from Modena supplied the logo that Roberto would wear on his spacesuit. AMS-2 had inspired her to sketch an antimatter particle emerging from the tail of the Space Shuttle, with the Italian flag in the background.

The STS-134 crew with Roberto Vittori on the right. (NASA)

DAMA Mission Patch. (www.spacefacts.de)

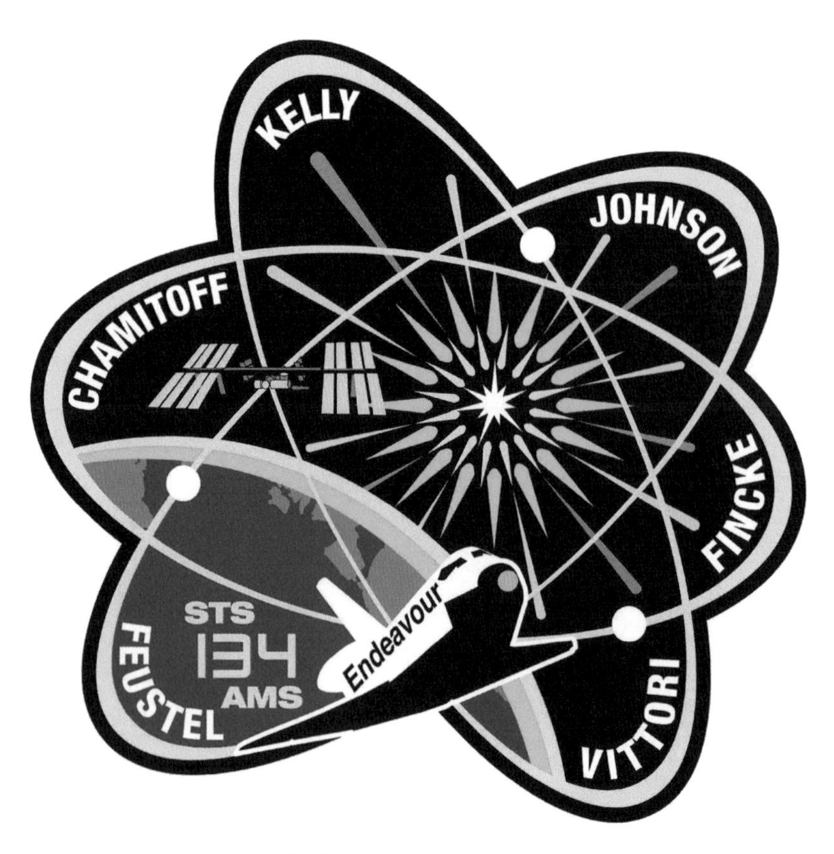

STS-134 Mission Patch. (NASA)

The STS-134 crew patch highlighted research on the ISS regarding the fundamental physics of the universe. The shape of the patch was inspired by the international atomic symbol, and represented the atom with electrons orbiting around the nucleus. The burst near the centre referred to the origin of the universe in the Big Bang. The Space Shuttle and ISS, flying together into the sunrise over the limb of Earth, represented the dawn of a new age of our understanding of the universe.

DAMA Mission Objectives

As Assembly Flight ULF6, the payload bay of Endeavour on the STS-134 mission carried the Alpha Magnetic Spectrometer and an Express Logistics Carrier loaded with ISS spare parts. This would be the final voyage for Endeavour, which was to be retired upon its return to Earth.

The mission was also to set a European record, because Roberto Vittori, one of the six astronauts aboard the Shuttle, would be the first ESA astronaut to fly to the station for the third time. He was a Mission Specialist on Endeavour's crew and at the ISS he would be visiting Paolo Nespoli, who was several months into his long-duration ESA Magisstra mission.

The STS-134 objectives were:

- The Alpha Magnetic Spectrometer (AMS-2) was a state-of-the-art cosmic-ray detector which was to investigate fundamental issues concerning matter and the origin of the universe. It was not only the largest and most complex science instrument ever to be sent up to the ISS, the $2 billion project was also the largest international collaboration on a single scientific experiment in space.
- An Express Logistics Carrier (ELC-3) was to deliver two S-band communication antennas, a high-pressure oxygen tank, an extra ammonia coolant reservoir, a number of circuit breakers, and a new component for the two-armed Dextre robot. The four spacewalks to undertake the installation work would be the final sorties of the Space Shuttle era.

Science

Even though STS-134 was primarily an ISS logistics flight, Vittori had been assigned a programme of experiments. He was to serve as a test subject for two ESA-led human physiology experiments which were to help scientists to investigate the impact of weightlessness on the human body. These involved him providing pre-mission and post-mission data in order to highlight possible changes resulting from flying in space.

ZAG

ZAG (Z-axis Aligned Gravito-inertial force) was to study how weightlessness affected an astronaut's perception of motion and tilt, as well as their level of performance both before and immediately after spaceflight. The tests included analysis of motion perception and eye movements while using a track-and-tilt chair.

OTOLITH

OTOLITH looked at the adaptation of the vestibular balance system and the eyes, which are strongly interconnected, for maintaining an astronaut's capacity to carry out tasks in the weightless state. This experiment assessed the sensitivity of the inner-ear organs to gravity before and after short-duration spaceflight.

ASI Experiments

The following experiments were funded by the Italian Space Agency:

- IAPE (Italian Astronaut Personal Eye) was to demonstrate a small autonomous, programmable micro-vehicle for supporting crew intra-vehicular and extra-vehicular activities. The device was powered by lithium ion batteries and its micro-controller received inputs from a unit which was based on gyroscopes.
- BIOKIS (Biokon In Space) was a multidisciplinary scientific experiment in which different biological samples such as algae, yeast, and water bears (tardigrades) would be evaluated to identify any genetic alteration resulting from a short period of weightlessness.
- IENOS (Italian Electronic Nose for Space exploration) was an air quality monitor. With three sensor units, it was to map the composition and cycles of the atmosphere on board the ISS for correlation with human activity.
- IFOAM (Italian Foam Shape Memory) was to gather data on the shape-memory properties of foam samples which were being considered for future use as energy absorbers and lightweight space actuators. Three foaming-phase samples had been prepared on the ground with different shapes, and the task was to heat them to assess their shape-recovery capabilities in space.
- NIGHT VISION was to study how natural antioxidants can protect light-sensitive algae cells similar to the human retina against space radiation. The experiment would use wild-type and mutant C. Reinharditii algae. A comparative evaluation would be made after the flight. The results were expected to assist in improving the night vision of astronauts as well as aircraft pilots.
- VIABLE ISS would evaluate bacterial and fungal development on the internal surfaces of the ISS as a means of controlling biological contamination and the quality of drinkable water. The results would assist research on long-duration missions in closed environments.

AMS-2

The primary payload was the Alpha Magnetic Spectrometer, a state-of-the-art cosmic-ray detector to examine fundamental properties of matter and the origin of the universe. It complemented the Large Hadron Collider at CERN by providing scientists with a new opportunity to study antimatter and so-called dark matter, the nature of which was a mystery. Dubbed the Hubble Space Telescope of cosmic rays, AMS-2 was to detect primary cosmic rays which had been accelerated by strong magnetic fields while travelling long distances through space.

A predecessor had been tested in 1998 aboard the last Shuttle to visit the Mir space station. Even with a short two-week mission, that instrument had provided enough information to make physicists reconsider some of their theories. The instrument to be installed on the ISS would have a sensitivity three orders of magnitude better than its predecessor. It was expected to operate until 2020 (longer if the ISS remained in service beyond this date) collecting cosmic particles a rate of 25,000/second and downlinking 6 megabytes of data per second.

Jean-Jacques Dordain, Director General of ESA, said, "The international science community has high expectations of the data to be collected by AMS-2 to understand key questions such as the nature of the universe's invisible mass. In this, it beautifully complements the observations of ESA's Planck space observatory, which is measuring the fraction of the invisible mass to a high degree of precision, and the Herschel satellite that is observing the effects of this invisible mass on young galaxies. AMS is a perfect example of the uniqueness of the ISS to promote fundamental scientific research in various disciplines, such as life sciences, Earth observation, materials sciences and physics. With the recently approved extension of the ISS to 2020, we now have the capability to offer an international laboratory to scientific communities around the world to push the frontier of knowledge."

Timeline

Flight Day 1 – Monday, 16 May 2011

Endeavour was launched from the Kennedy Space Center at 7:56 a.m. CDT for its final flight and the penultimate mission of the Space Shuttle Program. On board was the last non-NASA astronaut to fly on a Shuttle, Roberto Vittori who, having previously flown on Soyuz spacecraft, was making his third visit to the ISS.

The identity of the Shuttle commander had been in doubt because Richard Sturckow was prepared to replace Mark Kelly after Kelly's wife, congresswoman Gabrielle Giffords, was seriously injured in an attempted assassination on 8 January at a public meeting in Tuscon, Arizona. However, the mission was delayed due to technical issues and Giffords' remarkable recovery meant that she as able to attend the launch of her husband in May.[20]

Flight Day 2 – Tuesday, 17 May 2011

The first full day in space saw the crew perform the now standard inspection of the orbiter's thermal protection system, prepare the docking system, and check out the spacesuits in preparation for EVA work at the ISS.

[20] In the Shuttle Program's first launch pad fatality, engineer James Vanover died in a fall from Level 215 of the launch pad on 14 March 2011. As Vanover was suffering from depression, it was determined by Florida's Brevard County Medical Examiner to have been a case of suicide.

In addition, Vittori and Johnson latched the robotic arm onto the ELC-3 in readiness to place the carrier on the station's port truss structure. Feustel examined the Sensor Test for Orion Rel-nav Risk Mitigation (STORRM) equipment. This was to gather data during the initial rendezvous and docking for use in developing procedures for the new Orion spacecraft.

Flight Day 3 – Wednesday, 18 May 2011

After the obligatory backflip during which Nespoli, Coleman and Kondratyev on board the ISS took high resolution pictures of the Shuttle's thermal protection system, Endeavour finally docked at 5:14 a.m. CDT. Once the welcomes were over, Nespoli gave the visitors a safety briefing. He and Fincke then handed the ELC-3 to Canadarm2, operated by Johnson and Chamitoff, who in turn installed the carrier on the station's P3 truss segment. Later, Fincke and Feustel transferred their spacesuits to the Quest airlock while their colleagues set about the task of moving cargo into the station.

Flight Day 4 – Thursday, 19 May 2011

The Shuttle crew were awakened by *We All Do What We Can Do* by Shuttle thermal protection system engineer Dan Keenan and launch pad engineer Kenny McLaughlin. It was created to honour those who worked on the Shuttle Program.

After Johnson and Chamitoff used Canadarm2 to install AMS-2 on the S3 segment, the Principal Investigator, Professor Samuel Ting of the Massachusetts Institute of Technology, congratulated the astronauts by a radio link from the ISS flight control room in Houston.

At one point Kelly, Johnson, Feustel, Fincke and Garan participated in an interview with *The PBS NewsHour*. Kelly and Coleman also talked with National Public Radio, Associated Press, Reuters and Fox News to maintain public awareness of the mission. In addition, Fincke spent an hour with Feustel and Chamitoff preparing tools for their spacewalk.

Expedition 27's Kondratyev, Nespoli and Coleman completed a Soyuz drill in preparation for their departure planned for 23 May.

When the pictures of Endeavour's thermal protection system were examined they showed several damaged areas, and Mission Control began to consider requesting further detailed inspections.

Flight Day 5 – Friday, 20 May 2011

During the first spacewalk, Feustel and Chamitoff retrieved the Materials International Space Station Experiments (MISSE) 7A and 7B from ELC-2 on the starboard truss, where they had been exposed to space since their installation in November 2009. Feustel installed and connected the new MISSE 8 and Chamitoff installed a light on the Crew Equipment Translation Aid (CETA) cart. They also installed a cover on the starboard Solar Alpha Rotary Joint (SARJ) and vented nitrogen in preparation for adding ammonia to the photovoltaic cooling loop of the P6 segment during the next spacewalk. Finally, they installed antennas for the External Wireless Communication System on the Destiny module. At that point the spacewalk was curtailed because of a problem with a carbon dioxide sensor in Chamitoff's spacesuit.

Meanwhile, Johnson and Vittori transferred apparatus and supplies from Endeavour's mid-deck to the station.

Mission Control asked for a detailed inspection of the Shuttle's heat shield between the right main landing gear door and the ET disconnect door.

Flight Day 6 – Saturday, 21 May 2011

The highlight of the day was a gathering by the joint crew in the Kibo module to participate in a video link with Pope Benedict XVI.

Johnson, Fincke and Vittori used Canadarm2 and the OBSS to carry out the detailed inspection of Endeavour's thermal protection system requested by Mission Control, and transmitted the images for analysis.

In preparation for their next spacewalk, Feustel and Fincke spent the night 'camped' in the Quest airlock.

Flight Day 7 – Sunday, 22 May 2011

Kelly and Fincke began the day with a link to the Mesa Verde Elementary School in Tucson, Arizona, where 400 kindergarten through fifth grade students and STS-110 astronaut Lee Morin were gathered to talk to the astronauts. Later, Vittori and Nespoli took a call from Giorgio Napolitano, the President of the Italian Republic.

On their second spacewalk Feustel and Fincke rerouted an ammonia jumper cable between cooling loops of the P3 and P4 truss segments. In addition, they added ammonia to the leaking P6 cooling loop. Fincke lubricated the port SARJ and Feustel placed a camera cover on the Dextre manipulator. Fincke installed two radiator grapple bar stowage beams on the S1 truss segment.

Inside, Johnson and Vittori continued the transfer of supplies between Endeavour and the station.

In anticipation of Kondratyev, Nespoli and Coleman returning to Earth the next day Kondratyev passed command of the ISS to Andréy Borisenko, with Expedition 27 now becoming Expedition 28.

Flight Day 8 – Monday, 23 May 2011

Kondratyev, Nespoli and Coleman bade farewell to their ISS colleagues and Shuttle visitors, sealed themselves into the Soyuz TMA-20 spacecraft and undocked. This was the first and only time that a Soyuz departed the ISS while a Shuttle was in attendance. See the previous chapter for how Nespoli took the famous pictures of the orbital complex in this configuration.

Meanwhile the Shuttle crew installed a new filter in the station's Oxygen Generation System to provide continuous carbon dioxide scrubbing whenever the oxygen system's recirculation loop was running, and Fincke and Feustel prepared for the third spacewalk of the mission. In particular, they reviewed the new In-Suit Light Exercise which was designed to eliminate the need to 'camp' in the airlock overnight. The suited protocol involved spending a total of 50 minutes simply resting before and after a similar period of exercise in which the legs and arms were raised. In addition to purging nitrogen from the blood this new procedure would also save oxygen. It was to be used by future ISS spacewalkers.

Paolo Nespoli and Roberto Vittori floating in ATV-2 'Johannes Kepler'. (NASA)

Flight Day 9 – Tuesday, 24 May 2011

While Kelly replaced a desiccant bed in an ISS carbon dioxide removal assembly, Fincke and Garan replaced a remote power controller module for experiment data transmission. Fincke and Chamitoff also replaced parts in the oxygen generation system. After a few hours of off-duty time, Johnson and Chamitoff pursued NASA's public relations by talking to KPIX-TV and KGO-TV in San Francisco, and Sacramento's KFBK Radio. Later on Kelly, Fincke and Chamitoff gave interviews to *The Daily* iPad news app, NEWSRADIO 1020 KDHA in Pittsburgh, Houston's KTRK-TV, and the *Pittsburgh Tribune-Review*.

Flight Day 10 – Wednesday, 25 May 2011

The morning's wakeup call was *Countdown* by Rush, which had apparently been composed after the band watched the STS-1 launch. Mission Control played it as a tribute to the Space Shuttle Program, which had only one more mission scheduled.

For the new In-Suit Light Exercise (ISLE) procedure, Feustel and Fincke breathed pure oxygen for an hour as the Quest airlock was lowered to 10.2 pounds per square inch, then donned their spacesuits and performed light exercise in order to increase their metabolic rate and thereby purge nitrogen from their blood. Once outside, they installed a power and data grapple fixture on the exterior of the Zarya module to permit Canadarm2 to walk onto the Russian module for the first time, extending its range and utility. They also installed a video signal converter on Zarya, and ran power cables from the US segment to that module to provide a backup means of routing power from the US solar arrays to the Russian segment. To wrap up, they finished the installation of an external wireless communications system antenna that had been cancelled during an earlier EVA when Chamitoff's suit had suffered a sensor glitch.

In other activities, Johnson and Vittori spent much of their day stowing equipment and supplies.

Kelly took part in interviews with four television stations in Tucson, Arizona and, later, the two crews held an international news conference.

Flight Day 11 –Thursday, 26 May 2011

The joint crew was awakened by a rendition of the song *Fun, Fun, Fun* modified with Shuttle-themed lyrics by the all-astronaut band Max Q. Once the day got underway, the 50-foot-long OBSS was used to inspect Endeavour's wings and nose. This inspection was done with the Shuttle docked because the OBSS was to be left behind on the station to serve as an extension to Canadarm2 whenever occasion required. As this was Endeavour's final mission, the OBSS wouldn't be required again to inspect the orbiter.

Fincke and Chamitoff prepared for the fourth and final spacewalk of the mission. However, it had been decided to 'camp out' overnight in the Quest airlock rather than use the In Suit Light Exercise routine because it consumed more of the carbon dioxide cleansing lithium hydroxide canisters.

Flight Day 12 – Friday, 27 May 2011

During the final spacewalk, Fincke and Chamitoff relocated the OBSS to the S1 truss segment and it now became the ISS boom assembly. They retrieved a power and data grapple fixture from the truss, and took it to the boom. They removed an electrical flight grapple fixture, which the Shuttle arm had used, and replaced it with the fixture they had retrieved, to enable Canadarm2 to grasp the end of the boom and to employ it as an extension. To wrap up, they released launch restraints on the ELC-3 for the spare part that had been brought for Dextre. On repressurising the Quest airlock at 6:39 a.m. they concluded the final Shuttle EVA at the ISS. Earlier, at 4:02 a.m., they had passed the 1,000th hour of spacewalk activity for station assembly and maintenance.

At 7 p.m., Fincke also broke Peggy Whitson's record for the longest time spent in space by a US astronaut when he reached 377 days.

Flight Day 13 – Saturday, 28 May 2011

The wakeup call of *Will You Carry Me?* was performed by Michael FitzPatrick who had worked as an Electrical, Environmental, Consumables and Mechanical flight controller in Houston for more than 20 years and supported 80 Shuttle missions.

Kelly, Johnson and Garan took part in a video link to middle school students, teachers and others gathered at the University of Arizona in Tucson. Johnson spoke with television stations in Michigan and Ohio. Kelly and Feustel prepared two of three EVA spacesuits for return to Earth in Endeavour. The third one was resized to fit ISS Flight Engineer Ron Garan, for an EVA that was planned for the July visit by Atlantis. In routine maintenance, Fincke and Chamitoff replaced a desiccant bed on the carbon dioxide removal assembly. Feustel and Vittori continued to transfer items. Kelly and Johnson studied spinal elongation during spaceflight. Later on, Johnson spoke with KPRC-TV in Houston and the Voice of America.

Flight Day 14 – Sunday, 29 May 2011

On the final full day of joint operations, Fincke finished the maintenance work on the carbon dioxide removal assembly while Johnson and Feustel talked with WJRT-TV in Flint, Michigan, WJBK-TV in Detroit, WKYC-TV in Cleveland and WXMI-TV in Grand Rapids, Michigan. As a final contribution, Endeavour raised the station's orbit by 960 metres.

After the hatches were closed, Vittori and Chamitoff checked out the rendezvous tools and Feustel once again set up the Sensor Test for Orion Relative Navigation Risk Mitigation (STORRM) apparatus that had been used during the approach to the station.

As Endeavour undocked at 10:55 p.m. CDT, Expedition 28's Ron Garan continued the tradition of ringing the ship's bell and radioed, "Endeavour departing. Fair winds and following seas."

After the fly around, Kelly manoeuvred Endeavour on a rendezvous-like course to a point about 290 metres directly below the station, then performed a separation burn to initiate the departure. The STORRM sensors tracked the ISS until contact was lost, obtaining a good data set for the engineers.

On the ISS the Expedition 28 crew of Borisenko, Samokutyaev and Garan began a period of rest prior to resuming their regular schedule.

Flight Day 15 – Monday, 30 May 2011

The wakeup call was *Dreams You Give* by Brain Plunkett, which was the second-place winner in the Space Shuttle Program's Original Song Contest that had attracted 1,350 entrants.

The crew started their day by giving interviews with ABC News, CBS News, CNN, NBC News, and Fox News Radio. Then Kelly, Johnson and Vittori verified the craft's thrusters and flight control systems. Another task was packing up for return to Earth. The entire crew took part in a tribute to the orbiter Endeavour, which was about to be retired from service.

Flight Day 16 – Tuesday, 31 May 2011

The STS-134 crew awoke to *Sunrise Number 1*, which was performed by the band Stormy Mondays, the winner of the Space Shuttle Program's Original Song Contest.

Endeavour closed its payload bay doors at 9:49 p.m. The de-orbit burn was performed just after midnight CDT, so the landing on Runway 15 at the Kennedy Space Center occurred at 1:35 a.m. on Wednesday, 1 June.

When the wheels had stopped, CapCom Barry Wilmore radioed, "Your landing adds a vibrant legacy to this vehicle that will long be remembered. Welcome home Endeavour."

"It's sad to see her land for the last time," Kelly replied, "but she leaves a great legacy."

19.4 POSTSCRIPT

Subsequent Missions

At the time of writing, the DAMA mission was Roberto Vittori's last spaceflight.

Roberto Vittori Today

See the Marco Polo mission.

20

Promisse

Mission

ESA Mission Name:	Promisse
Astronaut:	André Kuipers
Mission Duration:	192 days, 18 hours, 58 minutes
Mission Sponsors:	ESA
ISS Milestones:	ISS 29S, 66th crewed mission to the ISS

Launch

Launch Date/Time	21 December 2011, 13:16 UTC
Launch Site:	Pad 1, Baikonur Cosmodrome, Kazakhstan
Launch Vehicle:	Soyuz TMA
Launch Mission:	Soyuz TMA-3M
Launch Vehicle Crew:	Oleg Dimitriyevich Kononenko (RKA), CDR
	André Kuipers (ESA), Flight Engineer
	Donald Roy Pettit (NASA) Flight Engineer

Docking

Soyuz TMA-3M

Docking Date/Time:	23 December 2011, 15:19 UTC
Undocking Date/Time:	1 July 2012, 04:48 UTC
Docking Port:	Rassvet

J. O'Sullivan, *In the Footsteps of Columbus*, Springer Praxis Books,
DOI 10.1007/978-3-319-27562-8_20

Landing

Landing Date/Time:	1 July 2012, 08:14 UTC
Landing Site:	near Dzhezkazgan, Kazakhstan
Landing Vehicle:	Soyuz TMA
Landing Mission:	Soyuz TMA-3M
Landing Vehicle Crew:	Oleg Dimitriyevich Kononenko (RKA), CDR
	André Kuipers (ESA), Flight Engineer
	Donald Roy Pettit (NASA) Flight Engineer

ISS Expeditions

ISS Expedition:	Expedition 30
ISS Crew:	Daniel Christopher Burbank (NASA), ISS-CDR
	Anatoli Alekseyevich Ivanishin (RKA), ISS-Flight Engineer 1
	Anton Nikolayevich Shkaplerov (RKA), ISS-Flight Engineer 2
	Oleg Dimitriyevich Kononenko (RKA), ISS-Flight Engineer 3
	Donald Roy Pettit (NASA), ISS-Flight Engineer 4
	André Kuipers (ESA), ISS-Flight Engineer 5
ISS Expedition:	Expedition 31
ISS Crew:	Oleg Dimitriyevich Kononenko (RKA), ISS-CDR
	André Kuipers (ESA), ISS-Flight Engineer 1
	Donald Roy Pettit (NASA), ISS-Flight Engineer 2
	Joseph Michael Acaba (NASA), ISS-Flight Engineer 3
	Gennadi Ivanovich Padalka (RKA), ISS-Flight Engineer 4
	Sergei Nikolayevich Revin (RKA), ISS-Flight Engineer 5

20.1 THE ISS STORY SO FAR

Soyuz TMA-2M was launched in June 2011 to deliver three flight engineers to the Expedition 28 crew and they became the Expedition 29 crew when their colleagues returned to Earth in September.

STS-135 marked the final flight of Atlantis and of the Space Shuttle Program. It had been added to the manifest after savings were made elsewhere and exploited the fact that the potential 'rescue flight' for STS-134 could be launched as ULF7 carrying the Raffaello MPLM to the ISS with extra supplies. To commemorate the end of an era, Shuttle commander Christopher Ferguson presented the ISS crew with the American flag which had been carried on STS-1 in April 1981. It was intended that the next crew to be launched from American soil would retrieve the flag.

Soyuz TMA-22 (the last of the TMA model to fly because future flights would use TMA-M) was launched in November 2011 to deliver three flight engineers to the Expedition 29 crew. Six days later they became the Expedition 30 crew when their colleagues returned to Earth.

20.2 ANDRÉ KUIPERS

Early Career

See the DELTA mission.

Previous Mission

See the DELTA mission.

20.3 THE PROMISSE MISSION

Promisse Mission Patches

The mission name Promisse was selected from over 200 proposals to an ESA competition, with the winner being 61-year-old Wim Holwerda of the Netherlands. Its appeal was its representation of the great expectations placed on the future of human spaceflight and exploration.

Promisse stood for Programme for Research in Orbit Maximising the Inspiration from the Space Station for Europe. The logo showed a stylised ISS orbiting Earth. The Promisse name included the orange letters 'ISS' to denote the Dutch aspect of the mission. The icons on the left represented the mission's three themes of science, technology and education. The globe represented the knowledge-based society focused on our planet. The electronic circuit denoted technology. The laboratory flask illustrated scientific research. The six stars represented both the ISS crew and the number of months that André was to spend in space.

The Soyuz TMA-3M crew with André Kuipers on the right. (www.spacefacts.de)

The Expedition 30 crew with Kuipers second from the right in the back row. (NASA)

The Expedition 31 crew with Kuipers second from the right in the back row. (NASA)

Promisse Mission Patch. (ESA)

The Soyuz TMA-3M patch depicted a trio of cosmonaut figures with their hands joined for the traditional celebration at the moment of successfully reaching orbit. The name of the spaceship was depicted in large Cyrillic characters and contained some play with letters, some doubling as both a letter and a number. Furthermore, the silhouette of the spacecraft formed part of one of the letters. Snaking through the design was the constellation of Scorpio, emphasising its primary star Antares, which was the call sign of the spacecraft commander Oleg Kononenko, who wanted a call sign that included an 'A' as a reference to his twin children Andréy and Alise. The cosmonaut figures and the constellation Scorpio were taken from a design by 11-year-old Alena Gerasimova of Petrozavodsk, Russia. The patch was designed by Dutch artist Luc van den Abeelen, who also designed Kuipers' Soyuz TMA-4 patch and the Expedition 31 patch amongst many others.

The Expedition 30 patch featured a fully assembled ISS above a sunlit crescent Earth. On the dark hemisphere, the portrayal of city lights was a reminder that mankind's presence on the planet is most readily apparent from space at night. It also commemorated how humans have transcended their early bonds throughout 50 years of space exploration.

Soyuz TMA-3M Mission Patch. (www.spacefacts.de)

The shape of the Expedition 31 patch was intended to represent a view of our spiral galaxy. The black background symbolised research into dark matter, this being one of the scientific objectives of the mission. At the heart of the patch were Earth and its Moon as thin crescents, Mars, and asteroids representing the focus of current and future exploration. The ISS was shown in an orbit around Earth, with a collection of stars for the Expedition 30 and Expedition 31 crews. The small stars symbolised the visiting vehicles that were to dock with the ISS during the mission.

Promisse Mission Objectives

ESA astronaut André Kuipers was to be delivered in December 2012 for the fourth European long-duration mission to what was now a fully operational ISS. His term as a member of an international six-person crew was scheduled to last almost 6 months.

The mission objectives were as follows:

Expedition 30 Mission Patch. (NASA)

Microgravity

Kuipers was himself a medical doctor who had been actively involved in microgravity research for at least a decade. As he posted on his ESA blog, "The data I will collect from myself can bring valuable information about the effects of weightlessness on the human body. This research may help to prepare for a future mission to Mars."

Scientific Experiments

He was to undertake around 30 ESA experiments covering human research, fluid physics, materials science, radiation, solar research, biology, and technology demonstrations. Most of the experiments were to be carried out in the Columbus laboratory, which would mark its fourth anniversary in orbit during his mission. Countermeasures against bone loss in weightlessness, the study of headaches in space, and mapping the radiation environment inside the ISS were among the experiments related to human exploration.

Expedition 31 Mission Patch. (NASA)

In addition, André was to carry out at least 20 experiments on behalf of NASA and the Japanese and Canadian space agencies involving almost 30 research facilities in the various laboratories of the ISS.

Flight Engineer

As a flight engineer, André had assignments ranging from station systems to payload operations. He had to be on hand to deal with visiting spacecraft. He was the prime crewmember for docking ESA's ATV-3 named 'Edoardo Amaldi'. He would also be involved in berthing the new Dragon spacecraft that had been developed by Elon Musk's company SpaceX as part of NASA's commercial resupply programme.

Education

André Kuipers would share some of the magnificent views of Earth from the ISS's Cupola and invite children to become involved in a wide range of educational activities. Spaceflight is uniquely able to inspire primary and secondary pupils to learn about biodiversity and climate change on Earth. André was to transmit to classrooms across Europe, demonstrating experiments on convection and wet foam formation. Being an advocate for health and well-being, he was also to encourage new generations of space explorers to stay fit by following the international education initiative Mission X: Train Like an Astronaut.

Science

Around 30 experiments were to be carried out during the Promisse mission, covering a wide range of disciplines. André had an extensive science, technology and education programme focused on life on Earth and looking ahead to future global human exploration missions.

Human Physiology

CARD

Experience had shown that one week in space is enough to alter the dilation of blood vessels, increase cardiac output, and lower blood pressure. A headward fluid shift gives astronauts a distinctive 'puffy face' and 'chicken legs' appearance. The CARD experiment tried to understand how weightlessness affects the regulation of blood pressure. André's cardiac output was to be measured repeatedly, along with sampling his blood, in order to investigate aspects of clinical conditions such as congestive heart failure.

SOLO

Astronauts lose bone density while in space. European scientists were researching salt retention and related human physiology effects by analysing blood and urine samples for the markers that indicate changes in bone metabolism. Samples were to be taken during two special diets followed by André; one a low-salt diet and the second a normal salt level diet. This metabolically controlled study was to provide insight into bone physiology, both in space and on Earth. This could be especially useful for determining the optimal sodium intake for a long mission without any negative effect on astronauts' health.

THERMOLAB

On entering a weightless state some of the body fluids, such as blood and lymph, flow very rapidly from the lower part body to the upper body. Changes in the heat balance are also linked to this fluid shift physiological effect. During André's daily exercise and rest period the ThermoLab experiment was a non-invasive method of accurately measuring how his body's core temperature adapted to the state of weightlessness.

EKE
The preservation of an astronaut's aerobic capacity is a major goal of exercise countermeasures on a space mission. A widely used measurement for endurance capacity is the maximal volume of oxygen used during exhaustive exercise. André was to be a subject of the EKE experiment designed to assess an alternative, more optimal method of measuring endurance capacity that would reduce the time that was spent in making this measurement.

IMMUNO
The aim of this experiment was to determine changes in hormone production and immune response during and after a stay on the ISS. Samples of saliva and blood would be taken from André's Russian crewmates in order to check for hormones associated with stress response and for carrying out white blood cell analysis. An increased understanding of the coupling between stress and the functioning of the immune system would have relevance for people on Earth.

VESSEL IMAGING
This experiment used ultrasound scans to evaluate changes in the properties of central and peripheral blood cell walls in weightlessness. By studying these changes on André's body during and after his 6 months of weightlessness, the experiment would help to optimise the countermeasures developed for long-duration space missions. It was also an opportunity to validate telemedicine concepts for routine medical checks.

SPACE HEADACHES
André was to regularly fill in questionnaires for a study of the incidence and prevalence of headaches during his time on board the ISS. Headache characteristics in humans on space missions are analysed and classified according to the International Classification of Headache Disorders.

ENERGY
Negative energy balance during spaceflight might affect many physiological functions. Changes in André's energy balance and expenditure were to be measured to assist in deriving an equation for energy requirements in weightlessness. This would in turn contribute to planning adequate, but not excessive cargo supplies for food.

NEUROSPAT
Reorganised perception in weightlessness was a unique demand that André's nervous system had to deal with in space. By recording his brain activity during virtual reality stimulations, this experiment aimed to detect the mechanisms involved in the altered behaviours in microgravity and to localise the crucial parts of the cerebral cortex involved. It could provide a new tool for testing changes in spatial cognition in normal ageing and pathological conditions.

SARCOLAB

Exposure to microgravity is known to lead to loss of muscle mass, function, and motor control. This experiment would determine the contractile characteristics of muscles that were particularly affected, namely the plantar flexor muscles in the lower leg, during static and dynamic contractions. André was to contribute by participating in muscle biopsy studies.

EDOS

Early Detection of Osteoporosis in Space was to study the mechanisms that underlie the reduction in bone mass which occurs in weightlessness. The results would help to evaluate the structure of weight and non-weight bearing bones by using computed tomography (CT) together with an analysis of bone biochemical markers in blood samples.

SPIN

This experiment was to compare pre-flight and post-flight testing of cosmonauts using a centrifuge and a standardised tilt test. Their ability to maintain an upright posture without fainting would be correlated with measures of otolith-ocular function; i.e. the mechanism which links the inner ear otoliths with the eyes in order to maintain a sense of balance.

Biology

KUBIK-ROALD2

This experiment was to study gene expression of the proteins involved in the metabolic control of the neurotransmitter Anandamide. Scientists wanted to determine the role of this lipid in the regulation of immune processes and in the cell cycle under microgravity conditions.

Fluid Science

GEOFLOW-2

This investigated the flow of an incompressible viscous fluid which was maintained between a pair of concentric spheres rotating around a common axis as a representation of a planet. It was of importance for astrophysical and geophysical problems such as global scale flow in the atmosphere, in the oceans, and in the liquid nuclei of planets. It followed up a previous experiment of the same name, but with a different experiment configuration and new scientific objectives.

MSG-SODI/COLLOID 2

This expanded upon the earlier Colloid experiment with the specific goal of observing nucleation and the early stages of aggregation in colloidal solutions. The focus was on materials of special interest in photonics, with emphasis on photonic crystals whose properties make them promising candidates for new types of optical components.

Materials Science

CETSOL-MICAST-SETA
These experiments were to examine different growth patterns and evolution of microstructures during crystallisation of metallic alloys in microgravity. The experimental results, plus numerical simulations, were to be used to optimise industrial casting processes.

Radiation Research

TRITEL
One of the many risks of spending long periods in the space environment is the exposure to cosmic radiation. This has great importance particularly during solar flares and higher solar activity. TriTel was to provide a 3D characterisation of the radiation environment in the Columbus module and thus help to estimate the absorbed dose and equivalent dose burden imposed upon ISS crewmembers.

DOSIS-3D
This experiment carried out 3D monitoring of the radiation environment in all segments of the ISS using a variety of active and passive detectors to determine the actual nature and distribution of the radiation field inside the station.

ALTEA-SHIELD
How the radiation in space affects brain functions is one of the major concerns when programming long missions in space. One example is the 'light flash' that occurs when an ionising particle passes through the eye or the optic nerve. This part of the long-running experiment was to test the ability of various materials to shield against radiation.

Solar Physics

SOLAR
The SOLAR experiment installed on the exterior of the Columbus laboratory was recording the Sun's electromagnetic radiation across a broad portion of its spectral range with unprecedented accuracy. Its data provided a valuable contribution to long-term analysis of the total solar irradiance. It also helped scientists to improve climate models and sharpen weather forecasts. Such data could also feed into the design of future satellites in order improve their tolerance to radiation effects and increase their useful life in orbit.

Technology

VESSEL ID SYSTEM
Using Columbus as a test platform, this ESA satellite receiver brought global sea traffic tracking within reach. Since its installation on the ISS, this experimental ship detector had

been pin-pointing more than 300,000 vessels every day to demonstrate the space-based capability for identifying ships on open seas within the field of view of the ISS.

METERON

The ISS is currently the most realistic environment that resembles a future human exploration mission. This experiment was to validate future human-robotic operations and carry out simulations of real-time mission operations including speed-of-light communications delays. Although Kuipers trained for this experiment it was not performed on the ISS until the Iriss mission of Andréas Mogensen in September 2015.

ERB-2

The Erasmus Recording Binocular was a high-definition 3D video camera sent to the ISS for recording and live streaming, to provide a new way to run interactive videoconferences from the orbital outpost. The ERB-2 experiment would exploit the high-definition optics and advanced electronics to provide a vastly improved 3D video effect for mapping the station.

NIGHTPOD

A two-direction ‘nodding’ mechanism was positioned in the Cupola to support a camera that took high-definition pictures of a particular point on Earth. This unit was also to be used in an educational role by teaching children and students about Earth observation.

Experiments with ISS partners

Kuipers was also to undertake more than 20 experiments on behalf of other space agencies in the ISS partnership. These studies would examine his bones, the micro-organisms living in his skin, and even his feelings in an effort not only to understand the effects and risks of human spaceflight, but also to help to benefit life on Earth.

This non-ESA programme included technology demonstrations (e.g. testing a robotic crewmate), materials science (growing crystals in weightlessness) and biology (growing cucumber seeds):

- Human Research:
  - *Pro-K*. Dietary intakes to predict and protect against changes in bone metabolism during and after spaceflight.
  - *Reaction self-test*. Psychomotor vigilance test.
  - *Integrated Cardiovascular*. Cardiac atrophy and diastolic dysfunction during and after long-duration spaceflight.
  - *Integrated Immune*. Validation of procedures for monitoring a crewmember’s immune function.
  - *Repository*. Biological specimens provide a means for investigating the physiological responses to spaceflight.

- ○ *Kinematics-T2*. Biomechanical analysis of treadmill exercise on the ISS.
- ○ *HAIR*. Biomedical analysis of human hair exposed to long-term spaceflight.
- ○ *MYCO*. Mycological evaluation of crewmember exposure to ISS ambient air.
- ○ *VO2 Max*. Evaluation of maximal oxygen uptake and estimates before, during and after long-duration missions.

- Fluid Physics, Materials and Combustion Science:
  - ○ *Protein Crystal Growth* (PCG). Growth of crystals by the counter-diffusion technique.
  - ○ *Marangoni experiment* (MEIS). Surface-tension driven flow.
- Technology Demonstrations:
  - ○ *Robonaut*. An on-orbit robotics capability within the ISS.
  - ○ *Photosynth*. Three-dimensional modelling of the ISS.
  - ○ *SS-HDTV*. Super sensitive HDTV system checkout and video downlink.
- Radiation Dosimetry:
  - ○ *Altea Dosi*. Measurement of cosmic radiation in the Destiny module.
  - ○ *Area PADLES*. Area Passive Dosimeter for Lifescience experiments in Space. A survey of the radiation environment on board the Japanese Kibo laboratory.
- Biology:
  - ○ *Dynamism of auxin efflux facilitators* (CsPINs). Studying the gravity response of dry cucumber seeds.
  - ○ *Microbe*. Identify the types of microbes in the Japanese Kibo laboratory by long-term sample collecting.
- Education and Earth Observation:
  - ○ *Crew Earth Observations* (CEO). Photography sessions over designated areas of Earth.
  - ○ *ISS knowledge acquired by middle schools* (EarthKam). Remote imaging and terrestrial research for students.
  - ○ *ISS Ham radio*. Amateur radio on the ISS.
  - ○ *Education Payload Operations Demonstrations* (EPO Demos). Videos for educational use recorded by astronauts on the ISS.
  - ○ *The Space Voice of the Open Mind* (Chuon). An investigation of the effects of the space environment on feelings, opinions and minds.

Timeline

Training Wednesday, 14 December to Tuesday, 20 December 2011

Throughout his training and during his stay on the ISS, André Kuipers kept a blog on the ESA website and here is a summary of his entries.

After the ceremonial cosmonaut breakfast, the prime and backup crews flew in separate planes to the Baikonur Cosmodrome in Kazakhstan. Kuipers noted that the plane's tables and sofas resembled a cosy living room and that a lot of delicious food had been prepared.

Kuipers' medic was Dr. Savin, who was very strict about the people allowed to come into contact with the crew. Anyone who had not being checked was required to wear a face mask and not to touch the cosmonauts.

The last days were taken up with reviewing the flight procedures and adding personal notes, plus signing photographs for distribution amongst the ground team and support personnel. Kuipers wrote, "We also spend a lot of time thinking about the personal items that we're allowed to take with us: one and a half kilos. Strict rules apply. No electronic gadgets, nothing that could be sold to collectors, and nothing that could jeopardise safety no matter how small the risk. I had brought two fly fishing hooks for a friend. I was only allowed to take them up if the hooks were cut off."

On 17 December the crew visited the launch pad and the Soyuz assembly building. They were able to climb into the actual spacecraft and note differences with the simulators. As there was no side door in the real descent capsule, they had to enter the hatch in the side of the orbital module and then climb down into the capsule.

While the backup crew checked out the Soyuz, the prime crew tried on their actual spacesuits and conducted leak tests with the media watching through a glass partition. Kuipers took note of the fact that it was awkward to stand up wearing the real spacesuit, just as it was in the training suit, because they were designed for sitting on the Soyuz couch. He wrote, "After the leak test we crawled into the Soyuz again. This reminds me of spelunking; that is, descending into caves. I carefully descended to the commander's chair, shut the hatch behind me, and slowly shuffled to my own couch. Once I was seated the hatch opened and the next cosmonaut entered. All tubes, cables and belts were attached as planned and we assumed our launch positions. We put on our gloves and closed our helmets for yet another leak check."

The temperature at Baikonur in December was minus 20 degrees so jogging wasn't an option, but the crews had access to a well-equipped gymnasium.

Simulation training of launch, approach, docking, and landing was continuous even after all of the exams had been passed in Star City.

There was another check of the capsule with it fully loaded with equipment and supplies. Kuipers wrote, "The orbital module is packed full. The three of us crawled inside, along with a technician. It was a tight fit, and warm. We had to be careful not to fall through the hatch into the descent module. Most importantly, we inspected all the items that had been loaded aboard. We checked every item on the list. I saw the ESA cargo such as the Vessel Imaging Experiment and the Nikon D3S camera that were fastened to the so-called divan. I need to be able to reach the Fuji 3D camera soon after launch, and also my medical questionnaire for an ESA experiment from the Netherlands. We're very pleased. In a short while this compartment will be our living room, bedroom, dining room, toilet, and attic for two whole days. Practically, it also functions as the hallway, with a door to our real home for half a year: the International Space Station."

Kuipers wrote about one of the many traditions of the cosmonauts at Baikonur, "Another tradition is placing our Soyuz TMA-3M and ISS Expedition 31 stickers on the windows of the cosmonaut bus. The stickers from my Soyuz TMA-4 and DELTA mission back in 2004 are still on the same window. Every other day, we go to the sauna in the new building that has been built next to the hotel. It is very relaxing to be with fellow cosmonauts and the doctors. As usual I brought Old Amsterdam cheese and Dutch stroopwafels, and my Russian, American and Japanese colleagues brought their own snacks to the table. Some traditions have to be cancelled due to the cold. For example, we won't be planting any trees."

In these final days before launch, even their family members had to remain behind glass partitions to prevent infecting the cosmonauts. After meetings with managers and instructors and receiving the formal approval of the commission to fly, the crew observed another tradition by watching the movie *White Sun of the Desert*. "I know it by heart by now, but it is a good opportunity to relax before going to sleep," Kuipers wrote.

As the launch on 21 December 2011 was set for 7 p.m., the crew were awakened at 9 a.m. and ate a light breakfast. They washed and then cleaned themselves with alcohol wipes. After lunch at noon they had a brief meeting with managers, the backup crew, and their wives.

Kuipers wrote in detail of other traditions:

> Speeches and a toast are given, with our glasses filled with water. There is a moment of silence before the glasses are thrown to pieces in one corner of the room. Then we leave the room and write our signatures on the doors. In the hallway, we are given a blessing by a Russian Orthodox priest who uses a lot of water.
>
> In a quick tempo we continue. Down the stairs, through a crowd of guests, outside and to the buses. At the launch site we can eat and relax but shortly it is time to put on our diapers, new underwear, and the electrodes for the electrocardiography. Then we put on our spacesuits and it is time to do a leak test of the suits. We do this in a room with a large glass window so that our family can see us. A special moment.
>
> And then the time has arrived to walk outside, hunched over in our spacesuits, to greet the public and the members of the commission. After a brief wave, we climb into the bus and drive to the launch platform. Shortly before arriving at the platform the bus stops and we all get out for the traditional pee against a wheel of the bus.[21] Gagarin did this, so now all astronauts do it. For those who find it hard to pee in a diaper, it is their last chance for hours…
>
> We walk a short distance from the bus to the rocket, which is visible for the first time as a complete unit. On the stairs to the lift we turn to wave, and then we go up to the entrance of the Soyuz. Our boots are removed. I enter first. In my stiff spacesuit, I let myself be lowered onto the commander's couch in the descent module. I close the hatch above me so that I can shuffle over to my own place to the left. Then I open the hatch for the technician who will assist me. I attach the ventilation and oxygen tubes and then the radio and ECG cable. I turn on the ventilation of my suit, then check the positions of the buttons and controls. Then the belts are tightened so that I am firmly secured in my cramped space. In the meantime, NASA astronaut Don Pettit and my commander Oleg Kononenko have clambered in and taken their positions. Once they attach their hoses, I turn on their ventilation as well.
>
> We are now two and a half hours before launch, and we check all the systems and establish radio contact. We also do another leak test of our suits. Regularly, we have breaks while ground control prepares for launch. At one point music is played to us

[21] According to Helen Sharman, female cosmonauts are not expected to perform this ritual.

in order to make the waiting more agreeable. I chose some songs, four of which are from Dutch artists. Pastorale by Ramses Shaffy and Liesbeth List seemed appropriate for a launch in a powerful rocket while my children look on. We close our helmets, and the time has come…

Expedition 30 Week ending 25 December 2011

Soyuz TMA-3M launched on 21 December carrying Russia's Oleg Kononenko, NASA's Don Pettit and ESA's André Kuipers. When they docked at the Rassvet nadir port on 23 December they joined Expedition 30 as a prelude to later becoming the Expedition 31 crew.

This was ESA's fourth long-duration mission on the ISS (the predecessors being Astrolab, Oasiss and Magisstra) and the fifth time that an ESA astronaut joined an Expedition (Léopold Eyharts spent 44 days on the station as a member of Expedition 16).

Expedition 30 Week ending 1 January 2012

Kuipers spent the first week adjusting to being back in space. The ISS had vastly expanded since his visit in April 2004 as a result of the addition of the ESA Columbus and Japanese Kibo modules. He was allocated personal quarters in the Harmony node.

Kuipers wrote about orientation in space and some of his science experiments, "Up and down do not exist in space. When you enter a module it mightn't be in the same way as you did last time. As a result you don't recognise the module immediately. We gradually got into our routine and I started to work on the ESA experiments and equipment. I've already worked with the Dutch-built Microgravity Science Glovebox, a box that allows you to conduct experiments in a controlled atmosphere. I took samples from the ROALD-2 experiment and put them in the ISS freezer that runs at minus 80 degrees (MELFI). The experiment investigates the role of gravity in certain proteins that are important for our immune system. The European window Cupola is fantastic. This observation post offers a 360 degree view of the ISS. When I am there I can see the Earth in its full glory pass below me, or above me, or next to me… it all depends on the angle with which I float into the Cupola. There is a beautiful view during the day cycle, as well as at night when the large cities show up as a glow of lights. During the night you can see fires, stars, sometimes a meteorite and an aurora. Aurorae are weak at the moment, but I hope it will get stronger during my stay. We were even welcomed by the unexpected trail of the comet Lovejoy that our commander Dan Burbank observed on the day of our launch. It is a beautiful sight to behold, just before the sunrise over Australia."

Expedition 30 Week ending 15 January 2012

On 10 January 2012 Kuipers sent a birthday card to the daughter of his mentoring professor at the University of Amsterdam. In conjunction with the Dutch postal service PostNL, he was able to use the new KaartWereld service to send it via the internet. He took a picture of himself in the Cupola and added the text:

> Hello Emma, Happy Birthday. It is very special to celebrate your birthday from space and for me to send you the first ever postcard from space. I took a picture of

> the ISS for you. I am enjoying the most magnificent views of our fragile planet from up here. When I'm back on Earth, shall we meet at ESA in Noordwijk? Greetings from the ISS, André.

The girl has a rare disease, MPS VI, or Maroteaux-Lamy syndrome, so the postcard was to raise awareness of metabolic diseases in children and also his ambassadorship for the WE Foundation.

André Kuipers' postcard from space. (ESA)

The Prime Minister of the Netherlands Mark Rutte and students at the Technical University of Delft spoke with Kuipers during a live link up on 12 January. The event was hosted by ex-astronaut Thomas Reiter, now ESA's Director of Human Spaceflight and Operations, together with ESA astronaut Frank De Winne.

When Mr. Rutte asked Kuipers how he had settled in, he replied, "It is wonderful. I have adjusted well to life in space, and it is really starting to feel like home. One of the most important things is to motivate schoolchildren. At that age, children become inspired and interested. If only a small number of students are influenced to continue their careers in science or engineering, then that is a good result of this mission."

Rutte remarked afterward, "That was one of the nicest events that I have done in my job as Prime Minister."

Later in the week, Kuipers posted on his blog about some of the experiments he was conducting, such as Vessel Imaging which used ultrasound to observe changes in his blood vessels. Or Neurospat which recorded his brain activity. He was eating a strict diet for 5 days with no snacks, to help create menus for future missions. He also undertook reaction tests every 4 days to examine the influence of weightlessness on the body and brain.

In relation to his clothing, he wrote, "Every day on board the ISS, I change into several different clothes. On Earth most people have a logical schedule. You go to work, go shopping at the weekend, visit friends or go to a sport club. In space things happen differently. Our schedule tells us what to do down to the last minute. Thankfully, there is some breathing space, but some events must be done on time to exploit the satellite connections with ground control. Our activities can be quite varied. One moment I am on the exercise bike, half an hour later I could be doing a live event with the American media. I might grab a quick lunch after the live event, another hour of exercise, and then return to the experiments. Sometimes it isn't worth changing clothes in between our different activities, so we end up floating around on television in our exercise clothes."

The crew celebrated Russian Christmas with their favourite foods. In the case of Kuipers this was Dutch cheese that had been delivered by Progress on a previous flight.

Kuipers commented upon the different rhythms and personalities of his crewmates, "I get on very well with my five colleagues. Each of us has his own rhythm. Don Pettit goes to bed early and gets up early. The Russians work late and even do their exercise afterwards. I try to sleep as long as possible in the morning. My schedule is fully booked, and it is important for me to get enough sleep. So far I have not had much luck. There is simply so much to do. A few nights a week we arrange to eat together. If we didn't, we'd hardly see each other due to our busy schedules. When we eat together I am reminded of camping on Earth. Although on a camping ground you can enjoy a warm shower, unfortunately we do not have a warm shower here…"

Expedition 30 Week ending 22 January 2012

The life of a spacefarer is not all glamour. On 19 January Kuipers posted about the not-so-glamorous housekeeping tasks, "Every Saturday morning there is no escaping it. All of us scour the whole of the ISS looking for rubbish and dirt. I vacuumed ESA's Columbus laboratory and the Japanese JEM [i.e. the Kibo module]. I cleaned the racks and panels with disinfectant wet-wipes. I discovered all sorts of things. Liquid stains, ear plugs, pills, Christmas decorations, and more. If you lose something on the ISS – and this happens a lot – the object is moved by the air that circulates in the Station. Near the air filters where the circulating air is sucked in, I often find objects which I had lost during the previous week. I found a big drop of water that had escaped in my own sleeping cabin. Not an actual drop, of course, because water forms perfect round spheres in microgravity."

Expedition 30 Week ending 29 January 2012

This week the ISS played host to the Spheres ZeroRobotics tournament. Student teams from all over America and Europe wrote software to control spherical robots on the station. This year the venture was called Asterospheres and involved extracting minerals from virtual asteroids. The winner was an alliance consisting of teams CyberAvo, Ultima and Lazy, a collaboration of three schools from Turin, Italy and Berlin, Germany. Pettit and Kuipers oversaw the matches in the Kibo module.

Also this week, Ivanishin and Kuipers measured the sound level in the Russian Zvezda module, which was always rather noisy. Kuipers also measured the amount of iodine present in the station's water supply and the bacteria growth in different places in the complex.

On 23 January, Progress M-13M undocked with a load of waste. Its departure freed up the nadir port of Pirs for the arrival of Progress M-14M on 29 January delivering fresh supplies.

Expedition 30 Week ending 5 February 2012

On 1 February, Kuipers wrote in his blog that whereas on Earth he might enjoy his snacks and skip some exercise, this wasn't allowed in space because his diet was strictly controlled and his exercise was essential to maintaining bone and muscle mass in the weightless environment.

He explained the facilities available, "Everything in the Station floats, so we cannot go jogging or lift weights. Special exercise machines have been developed. We have our own gymnasium on board. It is situated in Node-3 (Tranquility) which also houses the European Cupola module. A fine place, as we can enjoy the view from the Cupola window while we exercise. We exercise on a special treadmill that has straps for our shoulders and hips to keep us attached. We have an exercise bike with no saddle. The movements we make on the bike are similar to the exercises on an Elliptical trainer on Earth. We can lift weights using special cylinders that create a vacuum to simulate dumbbells. Devices like these are necessary, as a 100 kg weight on Earth weighs absolutely nothing in space – just like everything else around us. The worst part of exercising in space is that you cannot take a shower afterwards to wash off all of the sweat. There is no shower on board so we have to wash with wet towels. For six whole months! I'm getting used to it, but I know the first thing I'll do when back on Earth will be to take a long warm shower!"

Expedition 30 Week ending 19 February 2012

Kuipers was advised that a postponement of the next Soyuz launch would result in his ISS tour being extended by 6 weeks. This meant that he wouldn't be present at his mother's and children's birthdays, the graduation of his eldest daughter, and the planned family holiday.

In this week's blog, he deliberated on being a national hero. He had received a customised T-shirt from his wife, the band Marillion had dedicated a song to him, and he received video messages from a member of Pink Floyd as well as from Harry Sacksioni, the Dutch guitar virtuoso. In his absence, his wife Helen even accepted the ship's horn from the decommissioned Dutch navy ship *Southern Cross*. He wrote, "Some people put me on a pedestal as a hero and that makes me feel slightly uncomfortable. Especially when young colleagues in the space industry look up to me. I am only doing my job. I am part of a large team with many people working back on Earth. I simply try to perform as best as I can, taking care not to make mistakes. Just like everyone else does. But I do understand, because I used to feel the same. It is magical to float around and to live and work in space and to observe Earth and the wonders of the universe."

When asked via the internet how he sleeps, Kuipers posted a picture of his personal cabin in which he was able to email and phone his family, listen to music, or a science fiction radio play before sleeping.

André Kuipers' sleeping quarters in the Tranquility node. (ESA)

Expedition 30 Week ending 4 March 2012

On 27 February Kuipers wrote about the various scientific experiments he was conducting, "The last few weeks I've been on a diet. Not that I am too fat, but for some experiments I need to follow a strict diet, such as for the European SOLO experiment. Twice Dan Burbank and I ate a special diet for five days. The experiment is about salt and is called SOLO (SOdium LOad in microgravity). Dan and I eat a diet with normal salt content for a week (11.5 grams a day), and a diet with low salt the other week (2.9 grams a day). Scientists can work out the influence of salt on our bones. In space, you lose bone mass very quickly. You could say that we 'age' very quickly. This aspect of space benefits research on Earth. Many old people suffer from bone decalcification (osteoporosis). This experiment might help people on Earth with this disease. We now know that eating more salt increases the acid content in the human body and this might increase bone loss. The SOLO experiment will investigate whether eating less salt decreases the effects of bone loss. That is the theoretical part. In practice things on the ISS are a different story. Think about having a very strict diet as well as having to store all your urine in a bag. Here, going to the toilet is an expedition in itself; even more so if you have to store each pee. Getting blood is also a challenge in microgravity. I draw Don's blood and my own blood in ESA's Columbus module. A tricky process involving needles, tubes, and plasters. The samples that we take need to be frozen in the minus 80 degree freezers, called MELFI, that are in the Japanese laboratory Kibo. We also need to 'weigh' ourselves using a special spring-loaded device that moves us up and down. By measuring the delay in the spring mechanism we can calculate our body mass. The blood and urine samples will be returned on the next Soyuz spacecraft to America and Europe via Russia. The ISS is truly an international undertaking." Kuipers was the tenth astronaut to follow the special SOLO diet, with each test subject adding to the database of how bone density changes in space.

He also completed the DSC experiment in the European-built Microgravity Science Glovebox, looking at temperature changes in mixtures of different fluids. The results would be used to help to improve computer models used in oil drilling.

For the Neurospat experiment, Kuipers wore a cap which contained a network of electrodes for measuring his brain waves to determine whether the brain processes some tasks differently in space. This involved Don Pettit precisely positioning a total of 64 electrodes on Kuipers' skull, so it took a while to set up. Kuipers also finished the Roald-2 biology experiment on human immune cells. The cells had been collected from volunteers on Earth and they were to be chemically frozen at specific intervals in space. By looking closely at the cells after their return to Earth, scientists hoped to gain insight into the workings of the immune system.

The highlight of the week was a spacewalk by Shkaplerov and Kononenko to prepare the Pirs and Poisk docking ports for future operations. Rules during spacewalks meant that all astronauts must be able to reach their Soyuz (the one containing their seat liner) in the event of an emergency calling for evacuation. Soyuz TMA-3M, which was the 'lifeboat' for Kononenko, Kuipers and Pettit was docked at Rassvet and Soyuz TMA-22, at Poisk, was assigned to Shkaplerov, Ivanishin and Burbank. In this case, Burbank and Ivanishin returned to their craft because in an emergency the coupling mechanism above the airlock would be used as a temporary airlock, denying them access if they weren't already aboard. As an additional precaution, Kuipers retreated to his Soyuz for the duration of the spacewalk.

Expedition 30 Week ending 11 March 2012

On 5 March, Kuipers wrote a tutorial on space photography, "We have professional cameras here on the ISS. We have three Nikon D3S cameras equipped with 24, 28, 50, 85 and 180 mm lenses. It takes some practice to be able to use them well because the shutter speed, aperture, and ISO values need to be chosen very precisely. The D3S is very sensitive to light, you can set it to ISO 12,800, but then the image becomes grainy. The window is also a factor to reckon with. The 180 mm lens is of little use in the Cupola because it is impossible to focus fully on the Earth due to the protective plates in front of the windows. So I go to the Russian service module with the camera. I have to time it to when we are flying over the right area. It must not be too cloudy and I must be able to leave my work. Step 1 is to recognise the area quickly (not easy, we look out of the Cupola upside down and it is dark, so I can't check where we are in an atlas). Step 2 is to stabilise myself (we are floating up here, after all). And step 3 is to aim (follow the Earth with a steady hand, we are moving fast) and shoot."

When asked on his blog about cleaning his clothes, Kuipers replied, "I only have four polo shirts for Expedition 30, the permanent crew that I belong to. I retrieved two more Russian shirts from the Progress supply ship. They aren't my favourite colour, but at least they're clean. On average, I wear a clean set for a few weeks, sometimes even up to a month. I can wear new exercise clothes every other week. Remember, I have to exercise for two hours every day. Of course, we don't roll around in mud up here so things don't get very dirty. But after a month they do start to smell. Dirty clothes cannot be washed in space. They return as waste, where they are burnt up in the atmosphere. Just as waste is on Earth."

Kuipers spoke with students from schools in Spijkenisse and Schiedam in the Netherlands. As part of a Ruimteschip Aarde competition they had turned their classrooms into spaceships and he chose the winners based upon video entries. The radio contact was part of the ARISS programme in which ESA astronauts had been participating for many years.

Expedition 30 Week ending 25 March 2012

Kuipers received a delightful surprise during a routine video call with the Columbus Control Centre in Oberpfaffenhofen when his wife Helen, his youngest kids, and his in-laws put in an appearance having driven to Munich from their ski holiday in Austria in order to give him a treat.

After being awakened at 4 a.m. on Saturday 24 March, the crew gathered in the Russian segment of the ISS and sealed themselves in their Soyuz vehicles ready to undock and return to Earth if necessary. The previous day they had been advised that a piece of debris from an old Russian satellite would pass within 10 km of the ISS, but the data was not sufficient to calculate an evasive action. When it became clear that the object had passed safely by, they opened all the hatches again. The threat posed by space debris was becoming increasingly significant for the station.

Expedition 30 Week ending 1 April 2012

The third ESA Automated Transfer Vehicle, called 'Edoardo Amaldi', was launched from Kourou in French Guiana on 23 March. Kononenko and Kuipers carefully monitored its approach on 28 March, prepared to intervene, but it docked automatically with the aft port of the Zvezda module. A problem occurred afterward because the ATV was not receiving electricity from the ISS. The crew started to quickly unload the cargo. Fortunately this was fixed when half the unloading was completed, and the remainder of the cargo was able to be unloaded according to schedule. The craft would not have been able to operate for very long on its batteries. In that case it would have been necessary to unload the cargo rapidly and then undock the ship within a few days to enable it to de-orbit itself. As the assigned ATV loadmaster, Kuipers was in charge of unpacking and stowing all cargo.

Expedition 30 Week ending 8 April 2012

On 2 April Kuipers answered some questions received on his blog and twitter about food on the ISS. "Usually we are very busy, and everyone eats on their own. I eat cornflakes or oatmeal with brown sugar for breakfast. Sometimes I have scrambled eggs or dried fruit. I prefer strawberries. Usually we eat dried food and add water. I sometimes enjoy lunch in Node-3 with Dan and Don, if we have time. We have a table and a suitcase where we can warm up many different tins of meals. The American food doesn't include bread but does come with tortillas. I add apple syrup from my bonus container. We often eat all together in the evening. Especially on Friday, when we watch a film. The meals are very diverse and if I want to try something different I help myself to a Russian tin. At NASA's Space Foods Systems Laboratory we had tasting sessions where we told researchers what our favourite food was. My favourite? The Russians supply the best cheesecake with nuts. The Americans have the best swordfish."

Expedition 30 Week ending 15 April 2012

By 15 April, ATV-3 had boosted the station's orbit twice. Kuipers explained that even though the ISS was in orbit, atmospheric drag caused its orbit to fall by several tens of metres each day. "I could not see the engines burn because they are on the back of ATV. But I did see how the engines lit up a halo of gas particles and also how Service Module Zvezda's thrusters adjusted the orientation of the ISS so that its 'belly' faces Earth. A beautiful sight."

Expedition 30 Week ending 29 April 2012

On Wednesday, 25 April, Burbank transferred command of the ISS to Kononenko in preparation for the departure of the Expedition 30 crew. At the last minute, Kuipers took blood and collected saliva samples from Pettit and himself and stored them aboard the Earthbound ship.

After Soyuz TMA-22 left on 27 April, Kuipers and his colleagues became Expedition 31, which would be a crew of three until the arrival of Soyuz TMA-4M.

Expedition 30 Week ending 6 May 2012

Kuipers wrote how ESA had sent up a joke press release, "We laughed loudly during our last meal with the full Expedition 30 crew. Bert Vis, a space historian, made a fake press release in official ESA style announcing that a crew change had taken place. Instead of sending Joe

Acaba to launch with the Soyuz capsule in May, my daughter Megan would visit us. Don Pettit would have to stay another half year in space so Megan could take his place and return to Earth with me. He didn't mind at all, of course. The 'press release' included official portraits of the new crewmembers. It is great to receive jokes like this from Earth!"

Expedition 30 Week ending 13 May 2012

Pettit and Kuipers continued ground-based training for the arrival of a Dragon spacecraft by running a series of simulations of grappling the new freighter and berthing it to the station, as well as rehearsing emergency situations.

About the health and safety procedures in the US segment of the ISS, Kuipers wrote, "Every now and again we send videos of the American part of the ISS to health and safety inspectors on Earth. All corners, nooks and racks are filmed. Flight control looks at the film to ensure that all safety procedures are respected. We often move things around on the ISS, as we need room to work. But we must avoid placing things in front of 'fireports' or ventilation fans. Cables that can't be detached quickly must not run through hatches in case there is a leak that requires a hatch to be closed. Ground control asked us a few questions about our 'setup' and we changed a few things. Safety comes first. I also collected water samples. We need to do this every so often. It is a complicated procedure that requires different bags, filters and injection needles. This is the third time I have taken water samples, but I still read through the procedure in order to avoid making mistakes. I do many such tasks. I take samples of air quality, formaldehyde, oxygen, carbon dioxide, and bacterial or fungal growth on many surfaces. Everything is recorded and sent to ground control. We live in a closed system and we cannot afford any infections or contaminated water and air. Opening a window to air things out or flushing the plumbing with water is not an option."

On 8 May, Kuipers explained the experiment to investigate the enzyme 5-LOX. Scientists at the University of Teramo, the European Centre for Brain Research and the Santa Lucia Foundation had realised that the enzyme named 5-LOX becomes more active in weightlessness. In part, this enzyme regulates the life expectancy of human cells. Most cells divide and regenerate but the number of times they replicate is limited. Blood samples from two healthy donors were sent to the ISS, where one set was exposed to weightlessness for two days while the other was held in a small centrifuge to simulate Earth-like gravity. The samples were then frozen and sent back to Earth for analysis. As predicted, the samples exposed to weightlessness showed more 5-LOX activity than the ones that were centrifuged; with the latter being the same as control samples on the ground. Professor Mauro Maccarrone from the University of Teramo explained the significance, "We now have a target enzyme that could play a real role in causing weakened immune systems. The 5-LOX enzyme can be blocked with existing drugs, so using these findings to improve human health could be a close reality."

Expedition 30 Week ending 20 May 2012

The week was spent transferring cargo from ATV-3 and the Progress freighter in preparation for the arrival of Soyuz TMA-4M with Joe Acaba, Sergi Revin and Gennadi Padalka to restore the ISS to an Expedition of six crewmembers. Those currently on board were to return to Earth within 2 months of that date, so they tested their Soyuz seat liners. The human body can grow several centimetres taller as the spine relaxes in weightlessness. Although Ivanishin's liner required adjustment, Kuipers reported that his still fitted him "like a glove".

There was also a depressurisation drill. As Kuipers wrote in his ESA blog, "We had to find the leak while talking to ground control. First we checked the Soyuz. If something is wrong with the Soyuz we might have to abandon ship and return to Earth immediately. If the leak isn't in the Soyuz, we proceed to close all the modules one-by-one while calculating how much time we have left before air pressure is too low and we have to leave. We continue until we find the problem. In this case the leak was in the FGB, better known as Zarya, the Russian storage module. Many things become 'normal' up here, in a daily routine that doesn't require much thought. You forget that behind the thin aluminium there is no air. Exercises such as these are a good way to stay aware of the risks of a long stay space and, in case of emergency, to react adequately."

On 17 May, Soyuz TMA-4M docked at the recently vacated Poisk port. Gennadi Padalka, Sergei Revin and Joe Acaba joined Expedition 31 and restored the ISS to a six-person crew.

Regarding being reunited with a former colleague and working with new ones, Kuipers wrote, "I flew with Gennadi Padalka in 2004. This is his fourth long spaceflight so he felt at home straightaway and is very cheerful. It is Sergei's first time in space, however. He looks around a lot and is absorbing the new experience while moving around calmly. Joe was eager to learn immediately. Don and I have to introduce him to the Space Station. For example, he watches when I exercise on the bicycle. He also watches us do maintenance work, such as moving sensors that measure radioactivity or working with scientific racks. I feel like a veteran sometimes. Almost five months ago, I had the same introduction from Dan Burbank. Now I'm the one that is passing on the knowledge and explaining how everything works. Then it will be Joe's turn to make the ISS his home. Each crewmember has their own way of making themselves comfortable. Don, Anatoly and I are a 'bald crew'. The new Expedition members have told us that they won't cut their hair during their mission. That makes them a 'hairy crew'. New arrivals always make mistakes using the toilet or when washing. Sometimes one forgets to turn on the suction, or to brace oneself. Luckily we all laugh when it happens. We have all done it once."

Expedition 30 Week ending 27 May 2012

Once again Kuipers went on a diet, this time for the ENERGY experiment that was to analyse how he expended energy and correlate this with the diet. As with all physiology experiments, measurements were taken before, during and after flight.

As he explained, "The space part of this experiment lasts 11 days. I eat space food from a special package and register everything with bar codes and on written forms so scientists know exactly what I have eaten. It is a lot of work. In addition, I have to drink water with deuterium isotopes and regularly collect water and urine samples. The isotopes allow scientists to examine how my energy levels vary from day to day. I wear a mask that measures the amount of oxygen I absorb for 20 to 50 minutes at a time. I wear the mask in the European Columbus module at the Human Research Facility (HRF). I am not allowed to move for certain periods of time during the measurements. Fortunately I am allowed to watch television, so I watched *Star Trek* and *Blade Runner*. All of my movements are recorded during the 11 day experiment using an 'activity monitor' placed on my upper arm. Scientists working on the experiment want to know how to feed people on missions such as to the planet Mars. If an explorer's energy levels were getting too low, that would have great consequences for the mission. On the other hand we don't want to send too much food, since space launches are very costly. By conducting this experiment I am making a small contribution to future voyages to Mars. That is a nice thought."

On 25 May the first Dragon commercial freighter arrived. Designed, manufactured, launched and controlled by the private company SpaceX, this was a ground-breaking vehicle in both technological and commercial terms. This mission was called Commercial Orbital Transportation Services, Demo Flight 1 (COTS Demo 1). At the time of writing this book, six more successful Dragon missions had flown.[22] It could deliver 3,310 kg of cargo to the ISS, which was over 1,000 kg more that the Progress. It was unique among the family of ISS unmanned ferries (Progress, ATV, HTV) in that it was able to return 2,500 kg of cargo to Earth, splashing down in the ocean.[23] The demonstration vehicle gradually approached over a period of 2.5 hours and then halted 10 metres out. Pettit controlled Canadarm2 for the capture operation and then Kuipers berthed the newcomer on the nadir port of the Harmony node.

Kuipers wrote of this with pride, "Last Friday was a special day on my mission. Don and I docked the SpaceX's cargo ship Dragon to the Space Station. Dragon brings new equipment for the crew. On the 31st of May it will return to Earth with supplies from the others and myself. The Dragon mission is the operational highlight of my mission. But it is also a milestone for international spaceflight. This is the first time that a commercial spacecraft has flown to the ISS and docked with the Station. You could say a new era of spaceflight has begun. Soon private companies will take people to and from space."[24]

Expedition 30 Week ending 3 June 2012

Kuipers wrote about the personal moments that he enjoyed while on the ISS, "I am trying to create as many pleasurable moments as I can during the last weeks of my mission. It is amazing how time flies up here. I have only managed to fully enjoy the view from the Cupola observatory a couple of times. I leave my camera and float in the Cupola so that Earth is below me and next to me instead of above me. It is very special. I was recently able to see the whole of Europe in one go, simply by looking around, from Ireland to Cairo and from Norway to Morocco. These moments are magical. They will probably be the most memorable moments of my mission."

---

[22] Dragon flight Commercial Resupply Services 7 (CRS7) failed when the Falcon 9 rocket disintegrated 139 seconds into flight.

[23] Elon Musk, founder and CEO of SpaceX, designed the Dragon capsule with human spaceflight in mind. Hence safe re-entry was an inherent requirement.

[24] In conversation with the author, Kuipers expressed regret at not having performed a spacewalk, having qualified for it. But he was particularly proud of being the person to berth the first ever commercial spacecraft.

Expedition 30 Week ending 10 June 2012

In the previous weeks, Kuipers had worked with many life science experiments, investigating human behaviour in weightlessness. He also worked with the Fluid Science Laboratory to examine liquids in space. And of course, just like a laboratory on Earth, maintenance and repair was an essential part of keeping everything running smoothly. As he explained, "For example, I measured sound levels, took samples of water and air, and checked whether bacteria were growing near to ventilation ducts, towel racks, and so on. The samples were returned to Earth with the Dragon. Teams on Earth will interpret the findings and change procedures. I also hung up 23 new radiation sensors. The previous ones will return in my Soyuz. These measure how much radioactivity from space penetrates the interior of the Station. We are not protected by Earth's atmosphere in the ISS. People should not be exposed to too much radiation, so it is important to know which areas let radiation through."

Expedition 30 Week ending 17 June 2012

As the end of his mission approached, Kuipers adjusted his exercise regime to prepare to feel gravity again. He ran every day on the treadmill T2 and trained hard using the Advanced Resistive Exercise Device, the latter being a fitness device designed to train almost all muscles but especially the ones in the legs and back. These exercises also increased bone mass. He no longer needed to work out on the exercise bike. He wrote, "My fitness instructor on Earth said, 'When you climb out of the Soyuz you need to walk away and not ride a bicycle.' I do not mind running on T2 or training with ARED. I can watch films and listen to music. I saw all three *Terminator* films again and *Starship Troopers*. I watch action films because it helps me run. During weight training, I listen to music: heavy rhythms such as rock and trance, dance or techno."

Expedition 30 Week ending 24 June 2012

In his last few weeks on board the ISS, Kuipers began an experiment that investigated how well the Altea Shield was able to protect astronauts in the spacecraft from the radiation environment outside. "The procedures were complex. I had to hang large radiation sensors in the Station and connect them with cables. The cables float every which way: it looked like an Anaconda snake swimming around me. Altea Shield has six detectors that measure the radiation. For 40–60 days they will continuously measure space radiation that passes through the ISS to reach our bodies. On top of the detectors there are walls made of polyethylene and Kevlar, the same material that is used to make bulletproof vests. After testing we will know which material is better at keeping radiation away."

Expedition 30 Week ending 1 July 2012

Even as Kuipers' departure approached, the scientific tasks were seemingly endless. In one case he wore a 'cardiopress' for 24 hours that measured his blood pressure continuously. He wore a 'holter monitor' for 48 hours to measure his heart beat. These measurements were taken for the Integrated Cardiovascular experiment (ICV), a NASA project which required monitoring an astronaut's blood pressure and heart rhythm during long stays in space. It was not easy to work whilst wearing these monitors. "I wore the cardiopress, which was quite difficult. The batteries have seen better days so I had to change them every 45 minutes. Sleeping was difficult because I could hear and feel the blood pressure bands pump and let out air, and because the batteries were bleeping for attention. Also, I'm unable to use my left hand normally, since if it hits something the measurements go haywire and the machine gets confused. Nevertheless, I believe it is important to take part in these experiments. As a doctor myself, I am interested in the results and it helps scientists on Earth to know how the human body adapts to space."

The Thermolab experiment investigated how the human body regulates temperature. On Earth we rely on convection to cool down: as liquids and gases heat up, they become less dense and rise, taking heat away from our skin. Although there is no convection in weightlessness, the body adapts and does not overheat in space. The test involved recording an astronaut's temperature during a long mission to gain insight into this adaptation. It was undertaken in conjunction with a NASA study of the maximum oxygen intake while astronauts used an exercise bicycle. Kuipers wore two sensors on his forehead and chest to measure his temperature continuously. The sensor was first used by ESA astronaut Frank De Winne in 2009.

With only two days left, Kuipers wrote how ongoing maintenance work continued, "I managed to complete three important tasks before I have to leave. I inspected cooling fluid valves for the Columbus laboratory, fixed an American scientific experiment rack, and worked on the Japanese Ryutai rack. We do many maintenance tasks in the ISS. Maintenance must be done in a timely fashion to assure that the Station is used optimally for scientific experiments." He also loaded new software into one of the faulty Express racks so that it could be used again.

Kuipers spent some time communicating with the crews of other programmes who shared his sense of isolation.

> The first contact I had was with the science base Concordia, at the South Pole. They probably have a tougher time than we do up here. At minus 80 degrees Celsius it is extremely cold, and when it is summer in The Netherlands they live in permanent darkness. They cannot leave, because the fuel in their vehicles is frozen and aircraft cannot reach them. Compared to their experience, I think a space mission is quite varied.
>
> A base on the South Pole is attractive to me. It must be similar to living on another planet. It might be like living on Mars. I have a lot of respect for the people that live there. They are doing research and are away from home for long periods. We do have one thing in common: our view of the beautiful stars.
>
> The people that took part in NEEMO don't have a view of the stars. NEEMO is an acronym for NASA Extreme Environment Mission Operations. It is an experiment in an underwater base off the coast of Key Largo, Florida. At around twenty metres under the sea four people lived in a metal tank. One of them was ESA astronaut Timothy Peake. It looked enticing on video. Blue-green water and many fish swimming by. The team used 'spacewalks' under water to train for a landing on Mars or on an asteroid. They also had time for less serious experiments. Timothy

told me that he had taken a toy helicopter to see how it flies at three times the pressure. Good fun.

I wouldn't mind taking part in one of these programmes. Maybe I'll get the chance in the future. But now I must prepare for my trip back. If everything goes as planned I will have both feet firmly on the ground by Sunday morning, Dutch time.

At 4:47 a.m. (GMT) on 1 July 2012 the Soyuz TMA-3M spacecraft bearing Kononenko, Kuipers and Pettit undocked from the Rassvet module. It landed near Zhezkazgan in Kazakhstan several hours later.

Expedition 30 Post-Flight

As Kuipers observed, "My return to Earth with the Soyuz was an intense experience. After undocking we conducted some tests and orbited Earth one more time. Above the South Atlantic Ocean the braking motor fired for four minutes. 140 km above Egypt the spacecraft separated into three parts. Above Iraq we entered the atmosphere and the air started to glow. The first part was beautiful, we were surrounded by an orange cocoon. After that the ride was simply unpleasant. We were pushed deeper into our chairs and suffered 4.7 times normal gravity. The parachutes opened 10 km above the ground and our capsule shook violently. 15 minutes later we landed with a jolt. It felt like a serious car crash. Luckily, search and rescue arrived quickly to pull us out of the cramped capsule. I could not wait to get out, as my leg was stuck and I couldn't feel my foot anymore. Once I was sitting in the chair in front of the television camera I was laughing again. That was the moment that I thought: 'Yes, it is all over, everything will be fine.'"

After Kuipers had spoken to his wife by phone, the helicopter took the three men to Karaganda and then a plane took Kuipers and Pettit to Houston, where his family were waiting.

Compared to his previous return from space after just a week, his 6 months in space had taken its toll. He described dealing with gravity again, "When I landed in Houston the world still felt different, as if a large magnet was pulling at me continuously. For the first time I could feel what gravity really feels like, since I was used to the weightlessness on the ISS. Sometimes it is scary. When I was lying still in the airplane, and later in bed, I felt as if I was continuously falling backwards. The feeling of being paralysed and the fact that I could not walk was also unpleasant."

The science continued in the ensuing weeks, with him hooked up to electrodes and blood pressure monitors, giving blood, urine and saliva so that scientists could compare the data taken before, during and after the flight. As he wrote, "During the following days and weeks this research will continue so my mission is not over. I have sessions planned that I also had done on the ISS such as Neurospat and ICV. I must give debriefings where I recount my experiences of working in space. I must do reaction tests and answer many questions. In addition, a large amount of time has been reserved for exercising and physiotherapy. My muscles are clearly weaker after my mission and my coordination is terrible. I must exercise to regain my strength."

Later he wrote, "Luckily my recovery is going quickly. The first days I was mainly tired and sick. But I could celebrate the Fourth of July, American Independence Day, with my family. We all enjoyed immensely a fantastic fireworks display above Clear Lake. I saw the almost-full Moon rise very slowly, much slower than in space of course. I smelled the trees, the flowers and the smell of fresh food on my plate. It is beautiful in space and I would return if I had the chance. But here on Earth things are not so bad either."

20.4 POSTSCRIPT

Subsequent Missions

At the time of writing, the Promisse mission was André Kuipers' last spaceflight.

André Kuipers Today

See the DELTA mission.

The next generation

In May 2009 six new astronauts were selected as ESA astronauts and they began their basic training in September 2009.

They are:

- Samantha Cristoforetti of Italy
- Alexander Gerst of Germany
- Andréas Mogensen of Denmark
- Luca Parmitano of Italy
- Timothy Peake of United Kingdom
- Thomas Pesquet of France.

Since André Kuipers returned from the ISS in July 2012, all ESA missions have been flown by the new class of astronauts, starting with Luca Parmitano spending 165 days on board as a flight engineer for Expeditions 36 and 37 during his Volare mission in 2013.

That was followed by Alexander Gerst's Blue Dot mission as a member of Expeditions 40 and 41 in 2014, and Samantha Cristoforetti's Futura mission on Expeditions 42 and 43 in 2014–2015 during which she broke the record for the longest single spaceflight by a woman of any nation or agency. In the wake of Andréas Mogensen's 10 day Iriss mission in September 2015, Tim Peake, Britain's first ESA astronaut, launched on 15 December of that year to start his 6 month Principia mission.

With newcomer Thomas Pesquet and the veteran Paolo Nespoli scheduled to visit the ISS in 2016 and 2017 respectively, the future is bright for European spacefarers.

J. O'Sullivan, *In the Footsteps of Columbus*, Springer Praxis Books,
DOI 10.1007/978-3-319-27562-8

The ESA Astronauts Class of 2009. (ESA)

Bibliography

13 Dygen I Rymden, Efter 14 år på Jorden, Christer Fuglesang, Fri Tanke, 2007
An Astronaut's Guide to Life on Earth, Chris Hadfield, MacMillan, 2013
Bold They Rise, The Space Shuttle Early Years 1972–1986, David Hitt and Heather R. Smith. University of Nebraska Press, 2014
Drum Vlucht, Het verhaal van astronaut André Kuipers, Sander Koenen. National Geographic, 2012
Elon Musk, Ashlee Vance, Virgin Books, 2015
Europe's Space Programme, to Ariane and Beyond, Brian Harvey. Springer-Praxis, 2003
Gabby: A Story of Courage and Hope, Gabrielle Giffords and Mark Kelly, Simon & Schuster, 2011
How Columbus Learnt to Fly, Glimpses of a unique space mission, Thomas Uhlig, Alexander Nitsch, Joachim Kehr. ESA, 2013
International Space Station, 1998–2011 (all stages), Owners' Workshop Manual, David Baker. Haynes, 2012
Manned Spaceflight Log II-2006–2012, David J. Shayler and Micheal D. Shayler. Springer-Praxis, 2013
My Countdown, The Story Behind My Husband's Spaceflight, Lena De Winne. Apogee Prime, 2010
NASA Space Shuttle, 1981 onwards (all models), Owners' Workshop Manual, David Baker. Haynes, 2011
Praxis Manned Spaceflight Log 1961–2006, Tim Furness and David J. Shayler. Springer-Praxis, 2007
Reference Guide to the International Space Station. NASA, 2010
Ruimteschip Aarde, Ontdek Je Wereld, Met Een, Reis Door De Ruimte, Sander Koenen. Moon, 2012
Seize the Moment: The Autobiography of Britain's First Astronaut, Helen Sharman and Christopher Priest, Victor Gollancz, 1993

J. O'Sullivan, *In the Footsteps of Columbus*, Springer Praxis Books,
DOI 10.1007/978-3-319-27562-8

Soyuz, 1967 onwards (all models), Owners' Workshop Manual, David Baker. Haynes, 2014

Soyuz, A Universal Spacecraft, Rex D. Hall and David J. Shayler. Springer-Praxis, 2003

SpaceX, Making Commercial Flight a Reality, Erik Seedhouse. Springer-Praxis, 2013

The International Space Station, from Imagination to Reality, Rex Hall. British Interplanetary Society, 2002

The International Space Station, from Imagination to Reality Vol.2, Rex Hall. British Interplanetary Society, 2005

The Universe in a Mirror: The Saga of the Hubble Space Telescope and the Visionaries who Built it, Robert Zimmerman, Princeton University Press, 2010

Thomas Reiter, Leben in der Schwerelosigkeit, Hildegard Werth. Herbig, 2011

Un Passo Fuori, Umberto Guidoni. Editori Laterza, 2006

Wheels Stop, The Tragedies and Triumphs of the Space Shuttle Program 1986–2011, Rick Houston. University of Nebraska Press, 2013

Appendix 1

Charta of the European Astronaut Corps

J. O'Sullivan, *In the Footsteps of Columbus*, Springer Praxis Books,
DOI 10.1007/978-3-319-27562-8

Charta of the European Astronaut Corps

Our Vision

Shaping and Sharing Human Space Exploration
Through
Unity in Diversity

Our Mission

We Shape Space by bringing our European values to the preparation, support, and operation of space flights that advance peaceful human exploration.

we Share Space with the people of Europe by communicating our vision, goals, experiences, and the results of our missions.

Our Values

Sapientia: We believe that Human Space Exploration is a wise choice by and for humankind. Sapientia reflects our commitment to pursue our goals for the advancement of humanity.

Populus: We put people first, in two ways: First, the purpose of our missions is to contribute to a better future for people on Earth. Second Populus serves as a reflection of our respect for the people with whom we work: that we value their opinions, praise their work and compliment them for their support.

Audacia: We acknowledge that Spaceflight is a dangerous endeavour. While accepting the risks inherently involved in space travel we work to minimize these risks whenever we can. Audacia reminds us that the rewards will be unparalleled if we succeed.

Cultura: We continue the exploration started by our ancestors. Conscious of our history and traditions, we expand exploration into space, passing on our cultural heritage to future generations.

Exploratio: We value exploration as an opportunity to discover, to learn and, ultimately, to grow. We are convinced that humankind must embrace the challenge of peaceful human space exploration. We, the European Astronauts, are willing to take the next step.

Cologne, this fifteenth day of August twothousandone anno domini

Appendix 2

European missions to the ISS

J. O'Sullivan, *In the Footsteps of Columbus*, Springer Praxis Books,
DOI 10.1007/978-3-319-27562-8

Mission	Astronaut	Launch Date	Launch Vehicle	Landing Date	Landing Vehicle
STS-100	Umberto Guidoni	19 April 2001	Space Shuttle Endeavour, STS-100	1 May 2001	Space Shuttle Endeavour, STS-100
Andromède	Claudie Haigneré	21 October 2001	Soyuz TM-33	31 October 2001	Soyuz TM-32
Marco Polo	Roberto Vittori	25 April 2002	Soyuz TM-34	5 May 2002	Soyuz TM-33
STS-111	Philippe Perrin	5 June 2002	Space Shuttle Endeavour, STS-111	19 June 2002	Space Shuttle Endeavour, STS-111
Odissea	Frank de Winne	30 October 2002	Soyuz TMA-1	10 November 2002	Soyuz TM-34
Cervantes	Pedro Duque	18 October 2003	Soyuz TMA-3	28 October 2003	Soyuz TMA-2
DELTA	André Kuipers	19 April 2004	Soyuz TMA-4	30 April 2004	Soyuz TMA-3
Eneide	Roberto Vittori	15 April 2005	Soyuz TMA-6	24 April 2005	Soyuz TMA-5
Astrolab	Thomas Reiter	4 July 2006	Space Shuttle Discovery, STS-121	22 December 2006	Space Shuttle Discovery, STS-116
Celsius	Christer Fuglesang	10 December 2006	Space Shuttle Discovery, STS-116	22 December 2006	Space Shuttle Discovery, STS-116
Esperia	Paolo Nespoli	23 October 2007	Space Shuttle Discovery, STS-120	7 November 2007	Space Shuttle Discovery, STS-120
Columbus I	Hans Schlegel	7 February 2008	Space Shuttle Atlantis, STS-122	20 February 2008	Space Shuttle Atlantis, STS-122
Columbus II	Léopold Eyharts	7 February 2008	Space Shuttle Atlantis, STS-122	27 March 2008	Space Shuttle Endeavour, STS-123
Oasiss	Frank de Winne	27 May 2009	Soyuz TMA-15	1 December 2009	Soyuz TMA-15
Alissé	Christer Fuglesang	29 August 2009	Space Shuttle Discovery, STS-128	12 September 2009	Space Shuttle Discovery, STS-128
Magisstra	Paolo Nespoli	15 December 2010	Soyuz TMA-20	24 May 2011	Soyuz TMA-20
DAMA	Roberto Vittori	16 May 2011	Space Shuttle Endeavour, STS-134	1 June 2011	Space Shuttle Endeavour, STS-134
Promisse	André Kuipers	21 December 2011	Soyuz TMA-3M	1 July 2012	Soyuz TMA-3M

Index

J. O'Sullivan, *In the Footsteps of Columbus*, Springer Praxis Books,
DOI 10.1007/978-3-319-27562-8

H

I

T

57595964R00232

Made in the USA
Charleston, SC
16 June 2016